PROCEEDINGS

OF THE

1995 INTERNATIONAL CONFERENCE

ON

PARALLEL PROCESSING

August 14 - 18, 1995

Vol. III Algorithms & Applications
K. Gallivan, Editor
University of Illinois at Urbana-Champaign

Sponsored by
THE PENNSYLVANIA STATE UNIVERSITY

CRC Press
Boca Raton New York Tokyo London

Catalog record is available from the Library of Congress
ISSN 0190-3918
ISBN 0-8493-2619-2 (set)
ISBN 0-8493-2615-X (vol. I)
ISBN 0-8493-2616-8 (vol. II)
ISBN 0-8493-2617-6 (vol. III)
ISBN 0-8493-2618-4 (ICPP Workshop)
IEEE Computer Society Order Number RS00027

Additional copies may be obtained from:

CRC Press, Inc.
2000 Corporate Blvd., N.W.
Boca Raton, Florida 33431

PREFACE

We are pleased to introduce you to the proceedings of the 24th International Conference on Parallel Processing to be held from August 14-18, 1995. The technical program consists of 27 sessions, organized as three technical tracks, one in Architecture, one in Software, and one in Algorithms and Architectures. We have also put together three panels, and three keynote speeches. A workshop on "Challenges for Parallel Processing" has been put together by Prof. Dharma Agrawal to be held before the conference. Two tutorials will be held after the conference and have been organized by Prof. Mike Liu.

There will be three keynote speeches at the conference to start each day of the meeting. The first keynote speech will be entitled, "Future Directions of Parallel Processing", by Dr. David Kuck of Kuck and Associates. Prof. Hidehiko Tanaka from the University of Tokyo will present the second keynote speech on "High Performance Computing in Japan". The third keynote speech will be given by Prof. John Rice from Purdue University on "Problem Solving Environments for Scientific Computing".

There are three panels with the conference to end each day. The first panel is entitled, "Heterogeneous Computing" and has been organized by Mary Eshaghian. The second panel "SPMD: On a Collision Course with Portability", has been organized by Tom Casavant and Balkrishna Ramkumar. The third panel on "Industrial Perspective of Parallel Processing", has been organized by A. L. Narasimha Reddy and Alok Choudhary.

The technical program was put together by a distinguished program committee. Each paper was assigned to two program committee members and two external reviewers. Reviews of each paper were handled using an electronic review process. The program committee met on March 17, 1995 in Urbana, IL, and decided on the final program. The decision on acceptance and rejection of each paper was made on the basis of originality, technical content, quality of presentation and the relevance to the theme of ICPP. A summary of the disposition of papers by area is presented in the following table:

Area	Submitted	Accepted	
		Regular	Concise
Architecture	160	14(8.75%)	27(16.8%)
Software	90	18(20.0%)	15(16.7%)
Algorithms	115	16(13.9%)	16(13.9%)
TOTAL	365	48(13.2%)	58(15.9%)

All papers submitted from the University of Illinois were processed by Prof. Chitta Ranjan Das of Penn State University.

We would like to thank all the people responsible for the success of the conference. First, we would like to thank the University of Illinois for providing the infrastructure necessary for preparing the program for the conference including the support staff and the mailing facilities. We would like to express our gratitude to Carolin Tschopp and Donna Guzy for managing the processing of all the papers, for preparing the proceedings, and for arranging the program committee meeting. We would like to thank the entire program committee for doing such a diligent job in reviewing so many papers in such a fine manner. Finally, we are grateful to Prof. Tse Feng for providing the guidance and wisdom for running this conference.

Prith Banerjee, Program Chair
Constantine Polychronopoulos, Program Co-Chair
Kyle Gallivan, Program Co-Chair
University of Illinois
Urbana, IL-61801

Program Committee Members

Keynote Speakers

Speaker: **David Kuck, Kuck and Associates**
Topic: Future Directions of Parallel Processing

Speaker: **Hidehiko Tanaka, Univ. of Tokyo**
Topic: High Performance Computing in Japan

Speaker: **John Rice, Purdue University**
Topic: Problem Solving Environments for Scientific Computing

Panel Sessions

Panel I: Heterogeneous Computing
Moderator: **Mary M. Eshaghian, New Jersey Institute of Technology**

Panelists:

Gul Agha	Univ. of Illinois at Urbana-Champaign
Ishfaq Ahmad	The Hong Kong University of Science and Technology
Song Chen	New Jersey Institute of Technology
Arif Ghafoor	Purdue University
Emile Haddad	Virginia Polytechnic Institute and State University
Salim Hariri	Syracuse University
Alice C. Parker	University of Southern California
Jerry L. Potter	Kent State University
Arnold L. Rosenberg	University of Massachusetts
Assaf Schuster	Technion-Israel Institute of Technology
Muhammad E. Shaaban	University of Southern California
Howard J. Siegel	Purdue University
Charles C. Weems	University of Massachusetts
Sudhakar Yalamanchili	Georgia Institute of Technology

Panel II: SPMD: on a Collision Course with Portability?
Moderator: **Tom Casavant, University of Iowa**

Panelists:

Andrew Chien	University of Illinois at Urbana-Champaign
Alex Nicolau	University of California at Irvine
Balkrishna Ramkumar	University of Iowa
Sanjay Ranka	Syracuse University
David Walker	Oakridge National Laboratories

Panel III: Industrial Perspective of Parallel Processing
Moderator: **A. L. N. Reddy, IBM Almaden**

Panelists:

Tilak Agerwala	IBM
Ken Jacobsen	SGI
Bruce Knobe	Siemens-Nixdorf
Stan Vestal	Thinking Machines Corp.

Conference Awards

Daniel L. Slotnick Best Paper Award

A. Nowatzyk, G. Aybay, M. Browne, E. Kelly, M. Parkin, B. Radke, and S. Vishin, "The S3.mp Scalable Memory Multiprocessor"

Outstanding Paper Awards

S. P. Midkiff,   "Local Iteration Set Computation for Block-Cyclic Distributions"
A. Heirich and S. Taylor,   "A Parabolic Load Balancing Method"

List of Referees- Full Proceedings

Abdelrahman, T.
Abraham, S.
Adve, S.
Adve, V.
Ahamad, M.
Alexander, W. E.
Almquist, K.
Al-yami, A. M.
Armstrong, J. B.
Arrouye, Y.
Babbar, D.
Bagherzadeh, N.
Baker, J. W.
Baruah, S.
Bayoumi, M.
Beckman, C.
Beguelin, A.
Bekerle, M.
Bhuyan, L.
Bianchini, R.
Blough, D.
Boppana, R.
Bose, B.
Bruck, J.
Buddihikot, M. M.
Burr, J.
Carver, D. L.
Casavant, T.
Celenk, M.
Chae, S-H.
Chalasani, S.
Chalmers, A.
Chamberlain, R. D.
Chandy, J.
Chao, L-F.
Chatterjee, A.
Chatterjee, S.
Chaudhary, A.
Chaudhary, V.
Chen, P.

Chen, Y-L.
Cheng, A.M.K.
Cheng, K-H.
Chern, M-Y.
Chien, A.
Choudhary, A.
Clarke, E.
Conley, W.
Conte, T.
Cormen, T.H.
Craig, D.
Culler, D.
Cuny, J.E.
Das, C.R.
Das, S. K.
Das Sharma, D.
Davis, J.A.
Davis, T.
DeRose, L.
Deshmuth, R. G.
Dewan, G.
Dhagat, M.
Dhall, S.K.
Dietz, H. G.
Dincer, K.
Dowd, P.
Duato, J.
Dubois, M.
Dutt, S.
Dykes, S. G.
Efe, K.
El-Amawy, A.
Enbody, R.
Felten, E.
Feng, M. D.
Foster, I.
Fraigniaud, P.
Franklin, M.
Fu, J.
Fujimura, K.

Gallagher, D.
Ganapathy, K.
Gao, G.
Ge, Y.
Ghose, K.
Ghosh, J.
Ghosh, K.
Ghozati, S.
Gibson, G.
Gorda, B.
Grunwald, D.
Gupta, A.
Gupta, R.
Gupta, S.
Gyllenhal, J.
Haddad, E.
Haghighat, M.
Hanawa, T.
Hank, R.
Harper, M.
Harper, III, D.
Hassen, S. B.
Heath, M.
Herbordt, M.
Hermenegildo, M.
Hill, M.
Hirano, S.
Ho, C-T.
Holm, J.
Hou, R.
Hsu, D. F.
Huang, S.
Huang, Y-M.I.
Hwang, K.
Ibel, M.
Iwashita, H.
Jacob, J. C.
Jayasimha, D. N.
Jha, N. K.
Joe, K.

Johnson, D.
Jordan, H.
Joshi, B. S.
Kale, L.
Kavianpour, A.
Killeen, T.
Kim, H.
Kim, J-Y.
Kothari, S. C.
Krantz, D.
Krishnaswamy, D.
Krishnamoorthy, M. S.
Lai, A. I-C.
Latifi, S.
Lau, F.
Leathrum, J.
LeBlanc, T. J.
Lee, C-H.
Lee, C.
Lee, G.
Lee, J. D.
Lee, S.
Levy, H.
Lew, A.
Li, Q.
Lilja, D.
Lin, R.
Lin, W.
Lin, W-M.
Livingston, M.
Lo, V.
Lombardi, F.
Long, J.
Lough, I.
Loui, M.
Lu, M.
Lu, Y-W.
Maeng, S.
Mahgoub, I.
Mahmood, A.
Makki, K.
Marsolf, B.
Martel, C. U.
Mazumder, P.

McKinley, P.
McMillan, K.
Menon, J.
Michallon, P
Midkiff, S. P.
Mohapatra, P.
Moore, L.
Moreira, J.
Mouftah, H. T.
Mowry, T.
Mudge, T.
Mukherjee, A.
Mukherjee, B.
Munson, D.
Mutka, M.
Netto, M. L
Ngo, V. N
Nowatzyk, A.
Olariu, S.
Oruc, Y.
Padmanabhan, K.
Pai, M. A.
Palermo, D.
Palis, M. A.
Pan, Y.
Panda, D.
Park, J. S.
Parkes, S.
Passos, N. L.
Patil, S.
Patt, Y.
Pic, M. M.
Pinkston, T. M.
Pradhan, D.
Prasad, S. K.
Prasanna, V. K.
Quinn, M.
Radiyam, A.
Raghavendra, C. S.
Ramaswamy, S.
Ramkumar, B.
Rau, B.
Reeves, A.
Robertazzi, T.

Rodriguez, B.
Rogers, A.
RoyChowdhury, V.
Saha, A.
Saletore, V.
Samet, H.
Sarkar, V.
Scherson, I. D.
Schimmel, D.
Schouten, D.
Schwabe, E. J.
Schwan, K.
Sen, A.
Sengupta, A.
Seo, S-W.
Seznec, A.
Sha, E. H-M.
Shang, W.
Shi, H.
Shi, W.
Shih, C. J.
Shin, K.
Shoari, S.
Shu, W. W.
Siegel, H. J.
Singh, A.
Singh, J.
Singhal, M.
Sinha, B.
Siu, K.-Y.
Sohi, G.
Somani, A.
Srimani, P.
Stasko, J. T.
Stavrakos, N.
Stunkel, C.
Sun, T.
Sunderam, V.
Sundaresan, N.
Surma, D. R.
Suzaki, K.
Szymanski, T.
Taylor, S.
Teng, S-H.

Thakur, R.
Thapar, M.
Theel, O.
Theobald, K.B.
Thirumalai, S.
Tripathi, A.
Trivedi, K.
Tseng, Y-C.
Tzeng, N-F.
Ulm, D. R.
Vaidya, N.
Varma, A.
Varvarigos, E.
Vetter, J.
Wah, B.
Wang, F.
Wang, P. Y.
Wang, Y-F.
Warren, D. H. D.
Watson, D. W.
Weems, C.
Wen, C-P.
Wilsey, P.
Wittie, L.
Wojciechowski, I.
Wu, J.
Wu, K-L.
Wu, M-Y.
Wyllie, J.
Yalamanchili, S.
Yan, J.
Yang, Q.
Yang, T.
Yang, Y.
Yap, T. K.
Yew, P.
Yoo, S-M.
Youn, H-Y.
Yousif, M.
Zhang, X.
Zheng, S. Q.
Zwaenepoel, W.

Author Index - Full Proceedings

I = Architecture
II = Software
III = Algorithms and Applications

TABLE OF CONTENTS
VOLUME III -ALGORITHMS AND APPLICATIONS

(R): Regular Papers
(C): Concise Papers

Session 1C: Imaging Applications
Chair: S. Das, University of North Texas

Session 2C: Irregular Algorithms
Chair: K. Gallivan, University of Illinois at Urbana-Champaign

Building Surfaces from Three-dimensional Image Data on the Intel Paragon

Hoi Man Yip, Sai Chung Yeung, Ishfaq Ahmad, and Ting Chuen Pong

Department of Computer Science
The Hong Kong University of Science and Technology
Clear Water Bay, Hong Kong
hmyip@cs.ust.hk, iahmad@cs.ust.hk, tcpong@cs.ust.hk

Abstract

In this paper, we describe a scalable parallel convolution algorithm for the construction of surfaces from three-dimensional image data volume. The proposed algorithm exploits the decomposable characteristics of the Laplacian-Gaussian filter to design an efficient surface construction scheme using task as well as data parallelism. The task parallelism is achieved by partitioning the operations across groups of processors while the data parallelism is achieved by distributing the data across the processor within each group. For the boundary detection, the proposed strategy uses the direction of the convolution to guide the data redistribution within each processor group to minimize the communication overhead. An extensive experimental study has revealed that the proposed parallel strategy yields substantial speedup on the Intel Paragon with more than 100 processors. In particular, it performs very well at convolution of large image data volume. The proposed strategy is useful for a wide range of current applications, including three-dimensional medical imaging and spatiotemporal-surface reconstruction.

1 Introduction

Edge detection on two-dimensional images is one of the most important components for many conventional computer vision and image processing algorithms. In many imaging applications, particularly medical imaging [5, 8] and spatiotemporal analysis [4, 10] in computer vision, the images are, however, three-dimensional in nature. This 3-D data is usually composed of a set of image slices - the boundaries of objects in this 3-D data are surfaces.

In the medical imaging context, we call this surface detection *reconstruction*. In magnetic resonance imaging (MRI), the imaging device acquires a number of cross-sectional planes of data through the tissue being studied. Since all of these planes must be stacked back together to obtain the complete picture of what the tissue was like, MRI entails considerable amount of reconstruction as well as substantial visualization. For spatiotemporal analysis, the image slices are captured by a camera moving about a scene. It is easy to see that the projection of scene features will move about in the image. The movement of the scene features will sweep out a surface

as we watch it in space-time. The three-dimensional surface scheme can also be used to build the representation of the spatial and temporal structure of these surfaces.

This surface-construction process, however, is a computationally-intensive problem for large image data volume with fine resolution. Currently, a typical surface building program executing on a serial machine processes the image volume at about 1K-Hz voxel rate [4]. Such a slow processing response is a major barrier in many image processing applications. This is particularly true for real-time processing where a sequence of image frames must be processed in a very short time.

Essentially, building surfaces from three-dimensional image data consists of two phases, namely convolution and boundary detection. Our objective is to propose parallel algorithms for both phases on mesh-connected parallel architectures such as the Intel Paragon. We exploit the decomposable characteristics of the Laplacian-Gaussian filter to design an efficient two-level parallelism scheme for the 3D convolution algorithm. In our two-level parallelism strategy, task parallelism is partitioned across group of processors, and then within each group data is partitioned across processor meshes to achieve a data-parallel paradigm. Then, for the boundary detection, the proposed strategy uses the direction of the convolution to guide the data redistribution within each processor mesh to minimize the communication overhead.

The proposed parallel surface-construction algorithm is tested by an extensive experimental study on the Intel Paragon. Using more than 100 processors, the results confirm that the proposed strategy is scalable and yields a substantial improvement over the current state-of-the-art techniques.

This paper is organized as follows. We will review some related works in Section 2 and then describe our algorithm in Section 3. The experimental results are presented in Section 4, and Section 5 concludes this paper.

2 Previous Work

Parallel algorithms for 2D surface detection have been studies extensively [1, 3, 17]. Zhang and Deng [17] presented a parallel implementation and the performance of a special 2D edge detection technique called edge focusing on the Intel iPSC/860 and CM-5. Recently, some studies on developing 3D boundary detection algorithm have also been reported [13, 15]. However, in these studies, a master-slave parallel computational model is employed where the master node initiates, controls and schedules the computing processes on slave nodes. This model has a limited scalability because of serious synchronization problems for large computation, and therefore, cannot be implemented on a large number of processors. For the consideration of scalability and even workload distribution, an alternative model is proposed in our design for both convolution and surface-construction algorithms.

An important issue to be considered for parallel implementation is load balancing. In edge detection problem, uniform partition does not guarantee equal distribution of computation among the processors because the edges do not distributed evenly inside the image. Zhang and Deng [17] experimented with three different scheduling methods

Uniform data partitioning: Equally divides the image data file into multiple subregions.

Static scheduling scheme: Partitions the task equally into subtasks at run time.

Dynamic scheduling scheme: Divides the computation task into multiple subtasks with their data subregions which are put in a task queue. A task scheduler is required at execution time to assign the tasks in the queue to available processors.

The experiments indicate that the straightforward uniform data partitioning method is the most effective because the additional overheads introduced from the static and dynamic scheduling methods cost more than the savings from balancing the computation.

The most demanding operation in the surface construction problem is the convolution of Laplacian of Gaussian filter. Considerable body of literature can be found related to the 2D convolution algorithms on various architecture - see [6, 12, 14, 11, 7]. However, our objective is to design a 3D convolution parallel algorithm for building surfaces from image volumes which has been rarely done before. Although Wells *et al.* [16] and Arambepola [2] considered higher dimensional convolution, their techniques are not suitable for building surfaces.

Dykes et al. [6] investigated the problem of large-scale image convolutions on parallel architectures. A major problem with sequential convolutions is the heavy demand for memory accesses. High-speed data caches on typical workstations are generally no larger than 16KB. When image data and window data exceed the cache size, execution time increases due to cache misses. Distributed-memory multicomputers can reduce the effect of this problem because each processor has its own data cache, so the total cache size and bandwidth increase as more processors are used.

3 The Proposed Methodology

We will first discuss the issues of the parallel convolution in Section 3.1 and then the method of zero-crossing detection in Section 3.2.

3.1 Convolution of the 3-D Image Volume

As mentioned above, the most computationally-intensive operation in surface reconstruction is the convolution of Laplacian of Gaussian (LoG) filter. The number of multiplications required for convolution is shown in Table 1 where M is the length of the filter, N is the dimension of the image volume, and P is the number of processors. Here, we assume equal dimensions of the image volume.

	serial	parallel
original	$M^3 N^3$	-
decomposed	$3 M N^3$	$3 M N^3 / P$

Table 1: Number of multiplications for convolution.

In three dimensions, the LoG is defined as

$$\nabla^2 G(x, y, z) = \frac{1}{2\sqrt{2}\pi\sigma^6}(2 - \frac{x^2 + y^2 + z^2}{\sigma^2})$$
$$\times \exp\left(-\frac{x^2 + y^2 + z^2}{2\sigma^2}\right) \quad (1)$$

where σ is the space constant of the Gaussian. We have made use of the separable property of the Gaussian filter (as developed by Huertas and Medioni [9]) such that each convolutions can then be realized as successive one-dimensional row, column and depth convolutions, thus reducing the total number of operations required at each voxel from M^3 to $3M$, where the size of the filters is $M \times M \times M$ voxels. For a typical operator ($M = 30$), this approach reduces the amount of computation by a factor of 300.

The decomposable functional form of the LoG operator can be written as:

$$L(G(I)) = \partial_x^2 G(I) + \partial_y^2 G(I) + \partial_z^2 G(I) \quad (2)$$
$$= G''_x \otimes G_y \otimes G_z \otimes I$$
$$+ G_x \otimes G''_y \otimes G_z \otimes I$$
$$+ G_x \otimes G_y \otimes G''_z \otimes I \quad (3)$$

where $\otimes$ denotes convolution, G_x, G_y and G_z are one-dimensional Gaussian filters in x, y and z direction respectively, the G''s are the second derivatives of the appropriate Gs and I is the image volume. In other words, we transform the 3-D LoG convolution to nine 1-D convolutions.

3.1.1 Operation Partitioning

Essentially, our algorithm performs the following three steps for the calculation of convolution.

1. Divide the processors into 3 groups with each group in the form of a 2-D mesh with $M \times N$ processors.

2. Perform convolutions by all three groups of processor meshes in parallel.

$$G"_x \otimes G_y \otimes G_z \otimes I \quad - \quad Group\ 1$$
$$G_x \otimes G"_y \otimes G_z \otimes I \quad - \quad Group\ 2$$
$$G_x \otimes G_y \otimes G"_z \otimes I \quad - \quad Group\ 3$$

3. Obtain convolution result by summing up the corresponding pixel values of the convoluted image volume obtained by the three groups of processors.

The processor mesh actually performs the convolution of I with $G"(x)G(y)G(z)$ in the following steps serially (the convolutions of I with the other two sets of filters are similar):

1. $I_1 = convolve(I, G"(x))$
2. $I_2 = convolve(I_1, G(y))$
3. $I_3 = convolve(I_2, G(z))$

However, each step itself can further be done in parallel by data partitioning. Therefore, the convolution operations can accomplish substantial speedup by the two level parallelism scheme.

3.1.2 Data Partitioning

The data partitioning is achieved by dividing the data among the processors in such a way that it maps onto the structure of the paragon processor mesh. For example, when doing the convolution of I with $G"(x)$, the volume of image data (3-D block) is partitioned into rows (Y-direction) of image data block. Each row of image data block is further partitioned in the Z (depth) direction (Figure 1a).

The advantage of this design is to minimize the communications between processors because the data used in each iteration of convolution is located in the same processor. Similar matching designs are used for (2) and (3).

However, the orientation of the partitioned block of data is different from step to step. The partitioned block of data is oriented horizontally, vertically and depthly in step 1, step 2 and step 3, respectively. Therefore, each processor requires data which are not currently held when the program switches from (1) to (2) and (2) to (3). This requires data redistribution.

3.1.3 Data Redistribution

The advantage introduced by data redistribution can be significant for a large data volume and a long filter length. It should be noted that if the data blocks are not redistributed, a large amount of communication overhead will be incurred. Communication between adjacent processors is required whenever the overlapping region of the image data and filter cut across the boundary between processors during the iterations of convolution. It is obvious that this overhead becomes

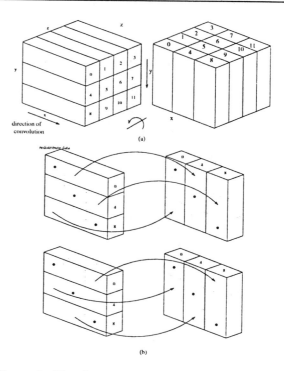

Figure 1: The data partition & the redistribution

more serious with an increased filter size. For example, consider a mesh with configuration **3x3** without data redistribution, processor 0 has to communicate with processor 4 and processor 8 in those boundary crossing iterations when it performs convolution in Y direction (Figure 1a).

After finishing convolution in the current direction, each processor sends and receive its data block to and from other processors which are required in the subsequent step. For example, as shown in Figure 1(b), after convolution is completed in the X direction, processor 0 has to receive image data block from processor 4 and processor 8, and send its image data blocks to processor 4 and processor 8 for convolution in the Y direction.

3.1.4 The Convolution Algorithm

Consider a processor mesh with $R \times C$ processors. Let P_{ij} denote processor (i, j) in the processor mesh, B^x_{ij}, B^y_{ij} and B^z_{ij} denote the data blocks held by P_{ij} at different stage of the data redistribution process, where $0 \leq i < R$, $0 \leq j < C$. One-dimensional convolutions in X-direction, Y-direction and Z-direction are performed on B^x_{ij}'s, B^y_{ij}'s and B^z_{ij}'s, respectively. During the redistribution process, each B^α_{ij} will be divided along the direction α into equal sized sub-blocks which is denoted by $B^\alpha_{ij}(k)$. The following is the three-dimensional LoG convolution algorithm.

Procedure Convolue3D(B)

for all processor groups G do [PARALLEL]

    for all P_{ij} in G do [PARALLEL]

      {

        Perform 1-D convolution on B_{ij}^x in X-direction.

        * data redistribution *\\

        for k ← 0 to R-1

          $B_{ij}^y(k) \leftarrow B_{kj}^x(i)$

        Perform 1-D convolution on B_{ij}^y in Y-direction.

        * data redistribution *\\

        for k ← 0 to C-1

          $B_{ij}^z(k) \leftarrow B_{ik}^y(j)$

        Perform 1-D convolution on B_{ij}^z in Z-direction.

      }

Sumup all B_{ij}^z's

End procedure

3.1.5 Parallel Summation of the Convolution Result

After finishing convolutions in all three directions, each group of processor mesh will hold the partial results. The result of the convolution can be obtained by summing up the corresponding pixel values of the convoluted image volume obtained by the three groups of processors.

In our algorithm, the Summation operation is also carried out in parallel as follows:

- Group 2 and Group 3 send the first slice of convoluted results to Group 1.
- Group 1 and Group 3 send the second slice of convoluted results to Group 2.
- Group 1 and Group 2 send the third slice of convoluted results to Group 3.

Therefore, group 1, group 2 and group 3 processors sum up the corresponding pixel values for the first, second and third slice of the convoluted image volume, respectively. The output from this 3-D convolution program will be used for surface reconstruction using zero-crossing detection algorithm which is described in the following section.

3.2 Surface Reconstruction

Our boundary detection algorithm is based on a surface building technique, named the *WeavingWall* [4], which is designed for the analysis of image sequences. This process operates over images as they arrive from a sensor, knitting together, along a parallel frontier, connected descriptions of images as they evolve over time.

The original Weaving Wall algorithm essentially consists of the following two phases:

Preprocessing: Blur the image volume using the 3-D Gaussian filter. The purpose of this step is to filter away noise and unnecessary detail. Compute the Laplacian of the blurred 3-D image. The Laplacian and the Gaussian operations can be combined by convolving the image volume with a Laplacian of Gaussian filter.

Reconstruction: The surface is defined at the zeros of Laplacian. We can construct the surface by searching the zeros of Laplacian in the 3-D volume. In the search process, a **2x2x2** window is used to scan through the 3-D signed volume. While the window is scanning, the existence and the orientation of the local surface facets will be determined by a well defined indexing system.

Based on the results of the convolution, each processor will perform the Weaving Wall reconstruction algorithm. Let us ignore the boundary case first. In the reconstruction process a small **2x2x2** window is employed to scan through the image signed volume. The direction of the scanning is from the first image to the last. For each image, the scanning is bottom to top, and within that, left to right.

At each step, the local surface structure inside the window will be determined. Notice that there are six neighbors for each voxel. If the sign of the Laplacian value is different for two neighbors, there is a facet between them. Hence, there are at most six facets for each voxel and twelve facets for each window.

Instead of determining all the facets of the window at the same time, we determine at most three new facets at each step. The position of these three facets are between the following voxel pairs (see Figure 2):

$$I_{ijk} \quad and \quad I_{i-1jk}$$
$$I_{ijk} \quad and \quad I_{ij-1k}$$
$$I_{ijk} \quad and \quad I_{ijk-1}$$

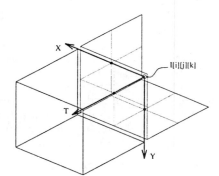

Figure 2: Three possible new facets for each movement of the window.

These new facets will be combined with the facets determined at previous step and linked together to form the local surface. There may be more than one local surface inside a window. Figure 3 shows an example of the local facets inside the window.

The black solid dot represents the voxel with positive sign while the white dot represents non-positive voxel. A saddle shape patch is reconstructed from the six facets on the left hand side.

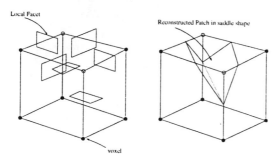

Figure 3: Local facets connectivity.

In order to handle the boundary cases, each processor will communicate with its neighbor processor. There are six surfaces for each local volume of a processor. One obvious and simple scheme is that each processor sends the voxel values of its six volume surface to the corresponding neighbors and receive from the same neighbors their boundary voxel values. The number of communication messages can, however, be further reduced by half if each processor sends only the voxels on three of the surfaces and receives the other three from its neighbors.

4   Results and Discussion

In this section, we present the results of the proposed parallel surface construction algorithm implemented on our recently acquired 140-node Intel Paragon. We first visualize the results produced by our program on different applications in Section 4.1, and then present the performance analysis in Section 4.2.

4.1   Results for Visualization

The data sets in our experiments can be classified into three categories.

Simple geometric primitives: This synthetic data set is generated by the analytical equation of the simple geometric primitives such as sphere and torus. These images are useful for testing the program in the development phase. Figure 4 shows anther reconstructed primitive which is a torus.

Real MRI slices of a human skull: This data set consists of twenty-seven real MRI slices of a human skull. Each slice contains a lot of details, e.g. the tissue of brain, inside the skull. The reconstructed skull is shown in Figure 5. It should be noticed that some of the details inside the skull are lost because the contrast of the tissue is too low.

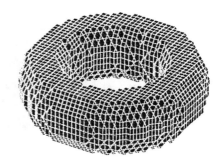

Figure 4: The reconstructed torus.

Figure 5: The top-front view of the skull.

Synthetic image sequences: This realistic data set is generated using a ray tracer called RayShade which is available in our system. Figure 6 is the resultant detected surface from a sequence of one hundred images taken by a camera undergoing forward motion. The surface indicates looming effect from the objects in the images.

4.2   Performance Analysis

The impact of various parameters on the performance has been studied that includes number of processors used, the size of the filter and the size of the image data volume. For these experiments, only the synthetic data was used.

4.2.1   Number of Processors

In this analysis, we study how the number of processors affect the execution time. We set the size of the image data volume to 120x120x60 and σ to 1 such that the filter size is 7. With increased number

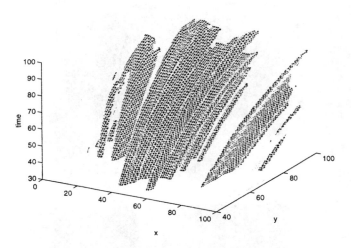

Figure 6: Spatiotemporal-surface detected from image evolution.

of processors used, the single convolution and surface construction problem is decomposed into more smaller subtasks which require less processing time for each processor. Therefore, we can use the execution time measured to compare the speedup associated with the increase in number of processors used.

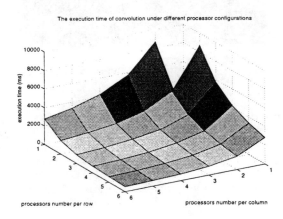

Figure 7: Execution time (ms) for convolution with filter length = 7.

The execution times (in *msec*) for convolution and size-crossing detection with different processor configuration (per group) are shown in Figure 7 and 8. There is a missing data point for the **1x1x3**[1] processor meshes. Without data partitioning, heavy demand for memory accesses is needed. Since the size of the data caches is limited, there were excessive

---
[1] **axbx3** means three groups of processors meshes with axb configuration

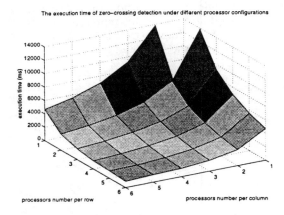

Figure 8: Execution time (ms) for zero crossing.

number of page faults as the image data and result data exceed the cache size. Hence the execution time increase tremendously. Therefore, we were not able to get the data point for **1x1x3** processor meshes because of the extremely large running time. Without loss of generality, we will use the **1x2x3** processor meshes as the base case for comparison.

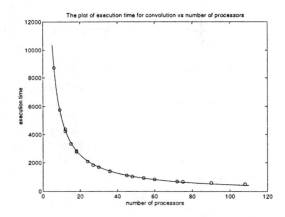

Figure 9: Execution time (ms) for convolution versus number of processors.

We can observe from Figure 9 and 10 that the execution time linearly decreases with an increase in the number of processors. This is because the data partitioning results in a almost linear decrease in the load on each processor with the increase in number of processors used. However, the execution time curves flatten off eventually because the communication overhead dominates the performance of the algorithm as the problem granularity becomes very small.

These experiments also reveal the success in the design of our algorithm in the minimization of com-

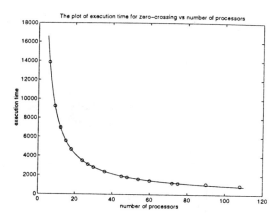

Figure 10: Execution time (ms) for zero crossing versus number of processors.

munication overheads as described in section 3.1. For example, the ratio for the **1x3x3** and **1x2x3** processor meshes is $1.51 = 8698msec/5747msec$ which is nearly equal to their Processor Ratio, $1 \times 3 \times 3/1 \times 2 \times 3 = 1.5$ and similar for other cases.

It is also interesting to note that the execution time is not much affected by the configuration of the processor meshes as long as the total number of processors used is the same. For example, the execution times for both **1x6x3** and **2x3x3** processor meshes are the nearly the same.

4.2.2 Size of the Filter
For the second experiment, an image data volume with size **120x120x60** was used as input data on **2x2x3** processor meshes, we altered σ^2 from 1 to 10 such that the filter size varied from 7 to 25. These results are presented in Figure 11.

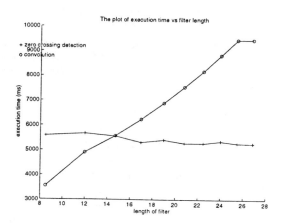

Figure 11: Execution time versus filter size.

Theoretically, with the same image data volume

and processor meshes configuration, the ratio of the filter sizes should be equal to the ratio of the execution times in convolution (see Table 1). However, the experimental results show that there is a significant difference between them. For example, with sigma equal to 1 and 10 such that the ratio of their filter sizes is $3.57 = 25/7$ which differs much from the ratio of their execution time, $9453msec/3560msec = 2.66$. The reason is that the experimental results not only include the timings for convolution, but also the timings for data redistribution and summation of convolution results (constant for the same data size and processor meshes configuration). As a result, the communication time becomes a more dominant portion of the execution time as the filter size is decreased. Therefore, it is reasonable that the ratio of the actual execution times for convolution is smaller than the theoretical ones. In addition, we also observe from Figure 11 that there is a linear relationship between the execution time and the filter size used.

The experimental results also indicate that the execution time decreases slightly with the increase in filter size for the zero crossing detection algorithm. The reason is that for larger filter, the image will be blurred in larger extent and the amount of details will be reduced. Hence, less processing time is required for surface reconstruction.

4.2.3 Size of Image Data Volume
Finally, we examine how the execution time of the algorithm is affected by different sizes of image data volume by setting the third dimension of the volume from 10 to 150 while the other two dimensions are both fixed at 150. The experimental results are illustrated in Figure 12.

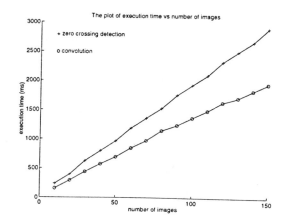

Figure 12: Execution time versus problem size.

Theoretically, with the same filter size and processor meshes configuration, the ratio of the data sizes should be equal to the ratio of the execution times in convolution (see table 1). However, the experimental results show that there is a significant difference between them. For example, with the third dimension of the data size equal to 10 and 150, the ratio of their execution time is 12.5=1945msec/156msec, differs from the ratio of their data size, 15=150/15. This is again due to the fact that the theoretical results do not take into account the timing used in data redistribution and summation of convolution results, but the experimental results do include the timing of these operations. This slow down factor becomes more prominent when transferring larger data blocks which incurs more overhead. Once again, we observe that there is an almost linear relationship between the execution time and the data size used. Similar observation can also be made on the zero-crossing detection algorithm.

5 Conclusions

In this paper, an efficient algorithm for the convolution of the three-dimensional image data volume has been proposed. The proposed strategy uses the direction of the convolution to guide the partitioning of image data volume to minimize of communication overhead. The two level parallelism strategy described illustrates how the operation partitioning of the processor meshes and the data partitioning scheme can achieve efficient speedup. The proposed algorithm also takes even workload distribution into consideration so that each processor does similar operations to minimize the idle times.

An extensive experimental study has revealed that that the proposed parallel convolution algorithm yields substantial speedup. In particular, it performs very well at convolution of large image data volume. The results achieved indicate the efficacy of the proposed convolution algorithm.

References

[1] A. Antola, N. Scarabottolo, M. Spertini, "A transputer-based self-tuning edge detection chain," *Proceedings of the Twenty-Fifth Hawaii International Conference on System Sciences,* Jan. 1992

[2] B. Arambepola, "Architecture for high-order multidimensional convolution using polynomial transforms," *Electronics Letters* vol.26, no.12 (7 June 1990) p801-2.

[3] J. H. Baek, K. A. Teague, "Parallel edge detection on the hypercube multiprocessor computer," *Proceedings of the Fourth Conference on Hypercubes, Concurrent Computers and Applications,* March 1989.

[4] H. H. Baker, "Building Surfaces of Evolution: The Weaving Wall," *International Journal of Computer Vision* vol.3, (May 1989) p51-71.

[5] M. Bomans, K. H. Hohne, et al., "3-D segmentation of MR-images of the head for 3-D display," *IEEE Trans. Medical Imaging,* Vol.9, No. 2, pp. 177-183, June 1990.

[6] S. G. Dykes, X. Zhang, Y. Zhou, H. Yang, "Communication and computation patterns of large scale image convolutions on parallel architectures," *Proceedings Eighth International Parallel Processing Symposium* (1994) p926-31.

[7] Zhixi Fang, Xiaobo Li, L. M. Ni, "On the communication complexity of generalized 2-D convolution on array processors," *IEEE Transactions on Computers* vol.38, no.2 (Feb. 1989) p184-94.

[8] K. H. Hohne and W. A. Hanson, "Interactive 3D segmentation of MRI and CT volumes using morphological operations," *J. Computer Assisted Tomography,* Vol. 16, No. 2, pp.285-294, 1992.

[9] A. Huertas, G. Medioni, "Detection of intensity changes with subpixel accuracy using Laplacian-Gaussian Masks," *IEEE Transactions on Pattern Analysis and Machine Intelligence* vol.8, no.9 (Sept.1986) pp.651-64.

[10] O. Monga, R. Deriche, J.-M. Rocchisani, "3D edge detection using recursive filtering: application to scanner images," *CVGIP: Image Understanding* vol.53, no.1 (Jan. 1991) p76-87.

[11] W. L. Nowinski, "Parallel implementation of the convolution method in image reconstruction," editor H.Burkhart, *CONPAR 90-VAPP IV. Joint International Conference on Vector and Parallel Processing. Proceedings.*

[12] S. Ranka, S. Sahni, "Convolution on mesh connected multicomputers," *IEEE Transactions on Pattern Analysis and Machine Intelligence* vol.12, no.3 (March 1990) p315-18.

[13] D. Raviv, "Parallel algorithm for 3D surface reconstruction," *Proceedings of the SPIE - The International Society for Optical Engineering* vol.1192, pt.1 (1990) p285-96.

[14] O. Schwarzkopf, "Computing convolutions on mesh-like structures," *Proceedings of Seventh International Parallel Processing Symposium* (1993) p695-9.

[15] S. Talele, T. Johnson, P.E. Livadas, "Surface reconstruction in parallel," *Proceedings of the Fourth IEEE Symposium on Parallel and Distributed Processing* (1992) p102-6.

[16] N. H. Wells, C. S. Burrus, G. E. Desobry, A. L. Boyer, "Three-dimensional Fourier convolution with an array processor," *Computers in Physics* vol.4, no.5 (Sept.-Oct. 1990) p507-13.

[17] X. Zhang, H. Deng, "Distributed Image Edge Detection Methods and Performance," *Proceedings. Sixth IEEE Symposium on Parallel and Distributed Processing* (1994) p136-43.

An Efficient Sort-Last Polygon Rendering Scheme on 2-D Mesh Parallel Computers *

Tong-Yee Lee, C.S. Raghavendra

School of EECS
Washington State University
Pullman, WA 99164

John B. Nicholas

Environmental Molecular Sciences Laboratory
Pacific Northwest Laboratory
Richland, WA 99352

Abstract

In this paper, we describe an efficient implementation of a sort-last parallel polygon rendering algorithm for 2D mesh parallel computers such as the Intel Delta and Paragon. Our goal is to provide a very fast rendering rate for extremely large sets of polygons, a requirement of scientific visualization, virtual reality, and many other applications. We implement and evaluate our scheme on the Intel Delta parallel computer at Caltech. Using 512 processors, our method renders close to one million triangles per second, comparable to the speed of current state-of-the-art graphics workstations.

1 Introduction

There have been many previous efforts to parallelize different rendering methods such as ray tracing, volume rendering and polygon rendering [1]. In this paper, we focus on developing a parallel polygon rendering (Z-buffer) method. A standard polygon rendering pipeline is described as follows. The vertices of polygons are illuminated by different light sources, transformed from 3D world space to 2D screen space, and truncated by a clipping pyramid. The polygons are then scan converted to pixel values and a Z-buffer hidden surface elimination is performed. Monlar *et al* [2] describe a framework for parallel rendering where the sort and redistribution of data occurs when transforming 3D objects (polygons) to 2D screen space (pixels). They delineate three types of parallel rendering algorithms: sort-first, sort-middle and sort-last.

In sort-first algorithms, each polygon is first preprocessed to determine which screen region it will be projected on. The primitive is then sent to the processor corresponding to this projected region, which performs all pipeline operations on this polygon. There is limited interest in sort-first algorithms, because they are very vulnerable to load imbalances. Most previous works have focused on sort-middle parallel algorithms. In this class of algorithm, each polygon's geometry transformation is first done locally, then the rasterization processor determines where the transformed

polygon will be sent. Sort-middle algorithms have high communication costs if the number of polygons or processors is high and are susceptible to load imbalances due to the non-uniform distribution of polygons on the screen [3, 5, 4, 6, 7]. Efficient load balancing and primitive communication scheduling are keys to achieving a high rendering rate [6, 7] in sort-middle algorithms.

Another popular class of parallel rendering is sort-last algorithms. This class of algorithm delays the data sort until the geometry processing and rasterization of all polygons is completed. The polygons that constitute a scene are then evenly partitioned and each partition is assigned to a processor. After each processor finishes rendering the allocated polygons, the sub-images created by the processors are merged into the final image. The advantage of sort-last parallel rendering algorithms is that they generally offer good load balancing and are easier to implement. However, they need more sophisticated composition hardware or other schemes to reduce the large communication overhead. In the previous work, image merging was achieved either in a direct composition [8, 9] a pipelined composition [10, 13], or a tree-like composition fashion [11, 12].

In this paper, we first outline our parallel polygon renderer built on the 2-D mesh Intel Delta parallel computer. We then describe our image composition scheme, and another well-known divide-and-conquer composition technique (binary-swap composition) proposed by Ma *et al* [14] and Karia [15]. We then analyze and experimentally evaluate both schemes, and show that our scheme is generally superior to binary-swap composition on a 2-D mesh parallel computer such as the Intel Delta. The Delta is a high-speed concurrent multi-computer, consisting of an ensemble of 512 computational nodes arranged as a 16 x 32 2-D mesh. In another performance study, we implement a recent method described by Crockett *et al* [3] on the Intel Delta. Our preliminary results show that our algorithm performs better for larger sets of polygons. For our experiments, we used five test scenes from the public domain dataset SPD [16], with sizes ranging from 150K to 500K triangles. Using

*This research is supported in part by the Boeing Centennial Chair Professor funds.

512 processors of the Delta, we achieved a rendering rate of about 1 million triangles per second.

2 Overview of Our Parallel Polygon Renderer

2.1 Dataset Partition and Distribution

The first step in a parallel rendering task is to partition the polygons in the scene and distribute them among the processors. Generally, there are two ways to partition the dataset: scattering and clustering [10]. Scattering simply assigns polygons to the processors in an interleaved fashion. Scattering gives good load balancing in many cases, and is well suited for sort-middle algorithms [3]. On the other hand, clustering assigns to each processor a group of polygons which are close to one another in storage. Since polygons stored in databases preserve geometric coherence to some extent, polygons in the same cluster are likely to fall into the same regions of the screen. Therefore, we can find a more compact bounding box to contain the polygons, and minimize communication if sort-last algorithms are used. However, clustering can lead to load imbalances during rendering.

We adopt a hybrid approach to partition the dataset. We first partition the whole dataset into a few larger groups. Each group is further divided into many smaller groups of polygons that are close to one another. In each round, these small groups are evenly assigned among the processors. In this way, we preserve some geometric coherence in each small group and achieve load balancing by scattering the larger groups.

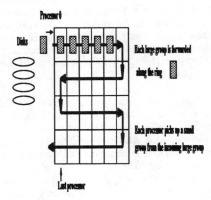

Figure 1: Data distribution is achieved in a pipelined fashion

We distribute polygons among the processors as follows: We logically organize the processors on the Delta in a snake-like ring topology, as shown in Figure 1. Processor 0 reads a large group of polygons from the disk. Processor 0 picks up one small group of polygons and then forwards this large group along the ring. The next processor picks up one small group and forwards

the large group along the ring. This method of sending the large group along the ring continues to the last processor. After reading a small group to its memory, processor 0 immediately reads the next large group from the disk and repeats the above process until all polygons have been imported from the disk. Thus, reading of data and distributing it to all processors is performed in a pipelined fashion.

2.2 Our Polygon Rendering Algorithm

In our parallel renderer each processor creates a full screen image on its local Z-buffer memory, then renders it as follows:

Rendering Loop

forall local polygons *do*
1. : compute normal vector for current polygon.
2. : do backface culling on current polygon.
3. : do lighting on each vertex of polygon.
4. : do perspective transformation.
5. : do clipping.
6. : do scan conversion.
7. : Z-buffer this polygon.
enddo

In our implementation, we optimize for speed. Therefore, we use simple Gouraud shading. After each processor finishes the entire loop, we do a global composition of all sub-images into a final image. Our resulting image has 512 x 512 resolution without anti-aliasing. All codes are written in C.

3 Parallel Composition

In most previous sort-last algorithms, more and more processors become idle as the composition proceeds — a waste of computational power. Recently, Ma *et al* [14] and Karia [15] presented divide-and-conquer image composition techniques (binary-swap) to improve parallel volume rendering. The divide-and-conquer method is well suited for parallel polygon rendering (Z-buffer). In this section, we describe the binary swap techniques, and our own divide-and-conquer scheme.

3.1 Composition by Binary-Swap

The aim of binary-swap (BS) composition is to exploit more parallelism in the composition stage and to keep every processor involved in all stages of the composition process. In BS, only half the image is swapped between a pair of processors and each processor pair composes the two opposite halves of sub-images at each composition stage. As the composition proceeds, the processors are responsible for smaller and smaller portions of image composition. In total, BS composition requires $\log N$ composition stages, and each processor keeps a fraction $(1/N)$ of the final image, where N is the number of processors in use.

Figure 2 shows an example of BS using four processors. Each processor's Z-buffer is divided into four disjoint areas $Z_{00}, Z_{01}, Z_{10}, Z_{11}$, where $Z_{**} = Z_{00} \bigcup Z_{01} \bigcup Z_{10} \bigcup Z_{11}$. In the first stage, processor

0 (P_{00}) sends its Z_{1*} (i.e., $Z_{10} \bigcup Z_{11}$) to its neighbor processor 1 (P_{01}) and receives a Z'_{0*} from processor 1. Conversely, processor 1 sends its Z_{0*} (i.e., $Z_{00} \bigcup Z_{01}$) to processor 0 and receives a Z'_{1*} from processor 0. Both processors complete this "send and receive" and then compose the received sub-image with the local sub-image. Meanwhile, processors 2 (P_{10}) and 3 (P_{11}) do a similar exchange of sub-images. In the second stage, processor 0 (P_{00}) sends its local Z_{01} to processor 2 (P_{10}) and receives Z_{00} from processor 2, where this incoming Z_{00} was composed earlier by processors 2 and 3 in stage 1. Processor 2 does the converse operation in the opposite direction. Similar sub-image exchanges and compositions occur between processors 1 and 3. At the end of the composition phase, each processor holds a composed sub-image equivalent to $1/N$ of the final image.

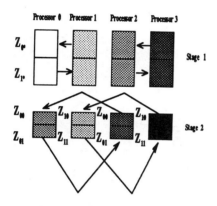

Figure 2: Binary-swap using four processors

3.2 Composition by Parallel Pipeline

In our parallel pipeline (PP) approach, the Z-buffer at each processor is divided into N ($Z_0, Z_1 \ldots, Z_{N-2}, Z_{N-1}$) portions for sub-images. Processors are organized on a circular ring and are denoted as $P_0, P_1 \ldots, P_{N-2}, P_{N-1}$. P_{next} and P_{prev} for a processor P_i will be $P_{(i+1)modN}$ and $P_{(i-1)modN}$. To circulate and compose the sub-images along the ring using the following algorithm, $N - 1$ stages are necessary.

Parallel Pipeline Composition
for all processors *do* in parallel
1. : set current composed area $B_{current}$ as Z_i in processor P_i
2. :*for* $j = 1$ *to* $N - 1$ *do*
3. : Each processor sends $B_{current}$ to its P_{next} processor
4. : Each processor receives a $B'_{current}$ from its P_{prev} processor
5. : set $k = (i - j) \bmod N$
6. : Each processor composes its incoming $B'_{current}$ with its local Z_k
7. : set newly composed Z_k as $B_{current}$
enddo

enddo

The sub-images are accumulated in a pipelined fashion along the ring, with each processor involved in each stage. At the end of composition phase, each processor P_i holds a fraction of the final image at partition $Z_{(i+1)modN}$. Figure 3 shows an example of our scheme using three processors. The Z-buffer in each processor is divided into three disjoint parts, Z_0, Z_1 and Z_2. In stage 1, processor P_0 sends its Z_0 to processor 1 and receives a Z'_2 from processor P_2. Processor P_0 composes Z'_2 with its local Z_2 to form a new Z_2. In stage 2, processor P_0 sends its new Z_2 to processor P_1 and receives a Z'_1 (was already composed by processors P_1 and P_2 in stage 1) from processor P_2 to compose with its local Z_1. The resulting new Z_1 in processor P_0 is a portion of the final image. Similar sub-image exchanges and compositions occur in a pipelined fashion between processors P_1 and P_2.

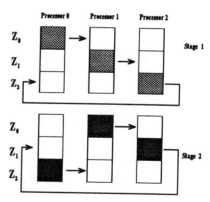

Figure 3: Parallel pipeline using three processors (1 × 3)

For 2-D mesh parallel computers like the Delta and Paragon, we logically group the 2-D mesh ($r \times c$) into many sub-rings. In the first phase of our algorithm, the PP composition is executed along one dimension, say within each column independently (as a sub-ring of r processors). On each processor, the local Z-buffer is divided into r equal sub-images. It takes $r - 1$ steps to circulate the sub-images along the ring and to accumulate the result in a pipeline fashion to produce temporary sub-images distributed among the processors. After the first phase, each processor holds a temporary sub-image that contains the accumulated result along the entire column. In the second phase, a similar composition process is repeated, but now along each row independently (as a sub-ring of c processors) using the sub-image that all of the processors in that row share in common as the entire image. At the end of the second phase, the image has been composed, with the final image being distributed among all N processors. In total, our algorithm takes $r + c - 2$ steps to form a final image on 2-D mesh parallel computers.

3.3 Analysis and Experimental Evaluation

Before we analyze both BS and PP parallel image composition techniques, we need to point out that similar techniques have been used for global vector combining [17]. Barnett *et al* named their techniques recursive-halving (similar to PP) and a bucket scheme (similar to BS). Here, we briefly evaluate both schemes on the 2-D mesh assuming that a full image data of Z_i is sent in each stage. We use notation as in [17] to analyze the methods.

On the Delta, we can simply model the communication cost to send and receive a message of L bytes between two processors at any distance by $\alpha + \beta L$, where α is the startup latency per message, β is $max(\beta_{free}, \beta_{sat}\chi)$. β_{free} is the transfer time per byte without link conflict, β_{sat} is the transfer time per byte with link conflict, and χ is number of messages contending for a saturated link. Therefore, the costs of BS and PP are:

$$Cost_{BS}(Z, N) = \sum_{i=0}^{\log N - 1} [\alpha$$
$$+ \frac{Z}{2^{i+1}} max(\beta_{free}, \chi(i)\beta_{sat}) + \frac{Z}{2^{i+1}}\gamma] \quad (1)$$

$$Cost_{PP}(Z, N) = \sum_{i=1}^{r-1} [\alpha + \frac{Z}{r}\beta_{free} + \frac{Z}{r}\gamma]$$
$$+ \sum_{i=1}^{c-1} [\alpha + \frac{Z}{N}\beta_{free} + \frac{Z}{N}\gamma] \quad (2)$$

Here, N (*i.e.*, $r \times c$) is the number of processors in use, Z is the resolution of a full image, γ is the combined cost per pixel, and $\chi(i)$ is the number of message contentions in stage i. In both schemes, the amount of data transferred per processor is approximately the same (less than Z). This differentiates between the BS and PP methods and balanced tree algorithms. The latter require processors to send an amount of data equal to a full or many times Z [10, 12].

From equations 1 and 2, we note that BS takes fewer stages in composition but suffers from link contention. With image resolution ranging from 64×64 (low resolution) to 1024×1024 (high resolution), a full image might take 32K to 8M bytes. When larger messages transfer on the links or more processors are used, the saturation on the links will degrade the communication performance severely.

Next, we experimentally evaluate this observation and show that PP is better than BS, using different resolutions and mesh sizes, even though PP needs more composition stages. In our implementation, each pixel value is eight bytes long and consists of a float

z depth value (4 bytes) and a color quadruple (red, green, blue, alpha; 14bytes). We perform our experiments for different mesh sizes and image resolutions ranging from 64×64 to 512×512. Figure 4 shows that the composition timings for different image resolutions using mesh sizes of 16×32 and 4×4. This figure clearly indicates that PP performs better than BS. For PP, performance ranges from 0.02 to 0.55 seconds per image. In contrast, BS requires 0.03 to 1.7 seconds for the same size of images.

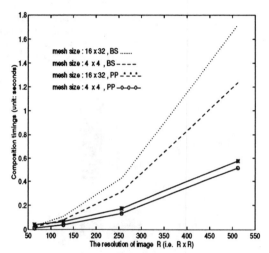

Figure 4: Composition timings for rendering different image sizes using 16×32 and 4×4 mesh sizes

3.4 Optimization using a Bounding Box

In BS, each processor is responsible for large image areas in the early stages; however, pixels are sparse in these areas. This sparsity in composition area decreases as the composition proceeds, since more processors contribute to each area. On the other hand, in our PP method, since each composed area is bounded by $O(Z/r)$ in the first $r - 1$ stages and by $O(Z/N)$ in the last $c - 1$ stages, the sparsity at each stage is relatively less than for BS.

We can avoid sending "inactive" pixels if we can look up an arbitrary active pixel very quickly, and determine the amount of active pixels on the fly. A pixel location is "active" at a given processor if at least one pixel has been rendered to it; otherwise it is "inactive". Special memory access hardware is usually necessary for this purpose [13]. Ma *et al* [14] suggested using a bounding box at each composition stage to include all active pixel areas. Each processor would binary-swap pixels only within this box. This technique works very well for volume rendering, since local sub-images are rendered from a block of continuous voxel data. Therefore, the number of "inactive" pixels within the bounding box is low. However, in polygon rendering, polygons are spread over the screen randomly, and it is hard to contain them in a small bounding box. Therefore, in early stages the sparsity in each bounding box can be very high, as illustrated in Figure 5. A large number of "inactive" pixels leads to redundancy in both communication and computation, which can severely degrade the rendering rate.

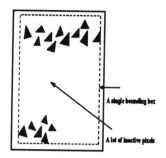

Figure 5: A sparse single bounding box in binary-swap

In PP, the local Z-buffer is divided into many fixed portions (*i.e.*, Z/r or Z/N). In our implementation, we used a single bounding box for each portion at each composition stage. As shown in Figure 6, in our implementation we used two smaller bounding boxes for these two disjoint parts (Z_0 and Z_3) instead of using a single larger box as in BS. Note that we used empty bounding boxes for Z_1 and Z_2 in this example. In comparison with Figure 5, the number of "inactive" pixels is significantly reduced in Figure 6.

In the following section, we show that our scheme, composed either in full or in the bounding box, outperforms a sort-middle scheme [3]. In this paper, our experimental results were obtained by using a single bounding box in each composed area, as shown in Figure 6.

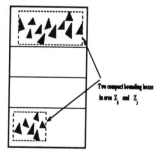

Figure 6: More compact bounding boxes for different areas in the parallel pipeline method

4 Performance Evaluation and Preliminary Results

To compare its performance with PP, we also implemented a recent method proposed by Crockett, *et al* [3]. For simplicity, we sent triangle data instead of trapezoid data. We used a scattering scheme similar to that given in [3] to evenly distribute the triangles

Scene	Number of Triangles	Size of Dataset
Mountain	524288 (512K)	24.0 Mbytes
Tree	425776 (416K)	19.5 Mbytes
Tetra	262144 (256K)	12.0 Mbytes
Lattice	235200 (230K)	10.7 Mbytes
Teapot	159600 (155K)	7.3 Mbytes

Table 1: Number of triangles and data size of the five test scenes.

Scene	32 processors	128 processors
Mountain	11.41	14.31
Tree	9.88	13.48
Tetra	6.39	9.17
Lattice	5.87	9.09
Teapot	4.55	8.08

Table 2: Time (in seconds) required to distribute the triangles for the five test scenes.

among processors. For PP in the data distribution phase, we used the hybrid scheme described previously.

To perform our experiments, we used five datasets from Eric Haines's SPD database. [16] The advantage of the SPD database is that we can tessellate the datasets in any resolution to create datasets of various sizes. The SPD database has been used in many previous studies and is believed to be a good representation of real data. Table 1 shows the sizes of the different datasets in our tests. These datasets represent different object distributions in the image screen, with sizes ranging from 150K triangles (medium-size model) to 500K triangles (large-size model). Table 2 shows our triangle distribution timings for each dataset, using 32 and 128 processors. Figures 7 - 11 show the rendering results for the five scenes. In our implementation, each large group consists of 2000 triangles and the data for each triangle is 48 bytes.

Figure 7: Teapot

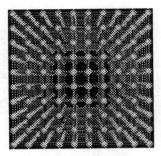

Figure 8: Lattice

Figure 11: Mountain

Figure 9: Tetra

These five scenes were run on a variety of processors, rendering the image screen at 512 × 512 resolution without anti-aliasing. Each scene was illuminated by a single light source and shaded by Gouraud shading. The reported timing was obtained by averaging rendering times for 20 frames of images. We controlled the viewpoint to allow "zoom-in" and "zoom-out" effects in these 20 frames. The purpose of this is two-fold. First, it represents a more general image distribution. Second, it gives a better evaluation of our (sort-last) scheme and the sort-middle scheme proposed in [3]. The sort-last techniques favor images whose objects concentrate on a small portion of screen (i.e., need less data communication in composition). On the other hand, sort-middle algorithms [3] favor images whose objects are evenly distributed over the whole screen (i.e., hot spot communication is removed). In our evaluation, we start rendering an

image whose objects are projected close to the center of the screen, and continuously zoom-in until objects show in most areas of the screen. The different image distributions give a fair evaluation of both schemes.

First, Figure 12 shows the rendering rate of PP with bounding box optimization. The rendering rate increases as more processors are used. Using 16 × 32 processors, our peak performance is about 1.3 million triangles/sec for the "Tree" scene while our lowest performance is about 500 K triangles/sec for the "Teapot" scene. The rendering rate for the other scenes is about 1 Million triangles/sec. Tables 3 and 4 show a time breakdown of our renderer for these two scenes. From these data, we see that the rendering time decreases almost linearly as the number of processors increases for both scenes. The composition timings in each scene are almost fixed regardless of the number of processors. Rendering both scenes using 512 processors, the composition time dominates the overall cost (87 % for the "Teapot" scene and 80.5 % for the "Tree" scene). Figure 12 shows performance does not drop off using up to 512 processors. However, we can expect performance will begin to decline beyond 512 processors. As the rendering time approaches zero the total time is dominated by the composition time, which will slightly increase and finally slow down the rendering rate. Furthermore, since the composition time remains constant, our rendering rate would be better for the same size image if we had a larger number of triangles: there would be more rendering computation, with the same composition cost. The rendering rates for the "Tree" and the "Mountain" scenes are better than those of the other three scenes. In the "Tree" scene, we achieve peak performance due to the large number of "inactive" pixels. Using bounding box optimization, many pixels are deleted. Thus, we get a better rendering rate.

Figures 13 and 14 show rendering rates for our PP without bounding box optimization and the sort-middle algorithm [3]. Compared with Figure 12, these performance graphs show our PP scheme with bounding box optimization outperforms the sort-middle scheme by a factor of four to eight. The peak performance of the sort-middle scheme is less than 300K triangles/sec, higher than Crockett's reported results for non-uniform scenes on the iPSC/860 [3]. Without

Figure 10: Tree

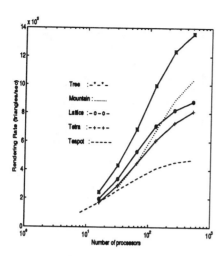

Figure 12: The rendering rate for five test scenes using different numbers of processors and our PP with bounding box optimization

P	16	32	64	128	256	512
R	0.705	0.3678	0.196	0.107	0.066	0.045
C	0.228	0.254	0.279	0.282	0.286	0.293

Table 3: The time (in seconds) for the "Teapot" scene (i.e., P is the number of processors, R and C are timings for both rendering and composition)

bounding box optimization, the rendering rate of our scheme is slowed down by 20 % - 50 %, due to the differing distributions of sparse pixels in different scenes. Therefore, the bounding box optimization is quite important to our scheme.

Figure 15 shows the effects of the "hybrid" data distribution on rendering performance. We experimentally compared rendering rates using "scatter" and "hybrid" data distribution and our results show "hybrid" can lead to better rendering performance. For larger numbers of processors, this improvement might be less, since each small group contains only a small number of triangles.

P	16	32	64	128	256	512
R	1.588	0.812	0.416	0.221	0.120	0.061
C	0.161	0.17385	0.205	0.208	0.224	0.252

Table 4: The time (in seconds) for the "Tree" scene (i.e., P is the number of processors, R and C are timings for both rendering and composition)

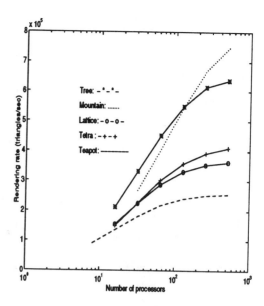

Figure 13: The rendering rates for five test scenes using different numbers of processors and our PP scheme

5 Conclusions and Future Work

This paper presents an efficient sort-last parallel polygon rendering scheme. After detailed evaluation, we decided to use the PP composition method instead of the better known BS scheme. On the 2-D mesh Intel Delta, our PP composition speed is 2-3 times faster than the BS technique, for different image sizes and mesh sizes. In addition, our PP composition scheme does a better job of removing inactive pixels from the rendering calculation than does the BS scheme.

We achieve a rendering rate of approximately 1 million triangles/sec for our five test scenes. Our scheme is useful for rendering the large number of polygons usually required in scientific visualization. The bounding box optimization is important to our scheme; rendering performance is improved by 25% to 50 %. Future work will include using more compact bounding boxes in each composed area or using a hierarchical composition scheme to further reduce the composition time. We are also considering ways to merge portions of images while processors are rendering their polygons.

Acknowledgments

This research was performed in part using the Intel Touchstone Delta System operated by California Institute of Technology on behalf of the Concurrent Supercomputing Consortium. Access to this facility was provided by Pacific Northwest Laboratory (PNL), a multi-program laboratory operated for the U.S. Department of Energy by Battelle Memorial Institute under Contract DE-AC06-76RLO 1830. We would also like to give our special thanks to Rik Littlefield at PNL and Michael Barnett at the University of Idaho for useful comments on our composition scheme.

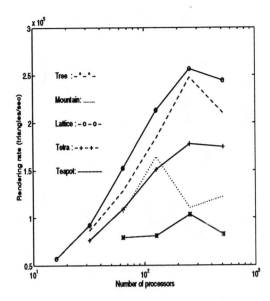

Figure 14: The rendering rates for five test scenes using different numbers of processors and a sort-middle scheme

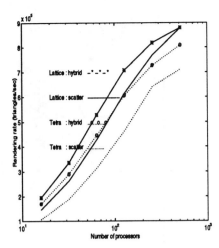

Figure 15: The rendering rate for the "Lattice" and the "Tetra" scenes using different data distribution methods

References

[1] *The Proceedings of 1993 Parallel Rendering Symposium*, IEEE press, San Jose, CA, October 1993.

[2] S. Monlar *et al*, "A Sorting Classification of Parallel Rendering," *IEEE Computer Graphics and Applications*, pp. 23-32, July 1994.

[3] T. W. Crockett *et al*, "Parallel Polygon Rendering for Message-Passing Architectures," *IEEE Parallel and Distributed Technology*, pp. 17-28, Summer 1994.

[4] S. Whitman, "A Task-Adaptive Parallel Graphics Renderer," *Proc. 1993 Parallel Rendering Symp.*, ACM Press, pp. 27-34, 1993.

[5] D. Roble, "A Load-Balanced Parallel Scanline Z-buffer Algorithm for the iPSC Hypercube," *Proc. First Int'l Conference PIXIM 88*, Editions Hermes, Paris, pp. 177-192, 1988.

[6] D. Ellsworth, "A New Algorithm for Interactive Graphics on Multicomputers," *IEEE Computer Graphics and Applications*, July 1994.

[7] F. Ortega *et al*, "Fast Data-Parallel Polygon Rendering," *Proc. Supercomputing'93*, IEEE Computer Society Press, Los Alamitos, Calif., pp. 709-78, 1993.

[8] Kubota Pacific Computer, Denali Technical Overview, version 1.0, March 1993.

[9] Evans and Sutherland Computer Corporation, Freedom Series Technical Report, October 1992.

[10] S. Monlar, "*Image-Composition Architectures for Real-time Image Generation*," Ph.D dissertation, Univ. of North Carolina at Chapel Hill, October 1991.

[11] J. Li and S. Miguet, "Z-buffer on a Transputer-Based Machine," *Proc. Sixth Distributed Memory Computing Conference*, IEEE Computer Society Press, pp. 315-322, 1991.

[12] R. Heiland, "Object-Oriented Parallel Polygon Rendering," *Proceedings of Gviz'94*, Richland WA, pp. 19-26, 1994.

[13] M. Cox *et al*, "A Distributed Snooping Algorithm for Pixel Merging," *IEEE Parallel and Distributed Technology*, Summer, pp. 30-36, 1994.

[14] K. Ma *et al*, "Parallel Volume Rendering Using Binary Swap Composition," *IEEE Computer Graphics and Application*, pp. 59-67, July 1994.

[15] R. J. Karia, "Load Balancing of Parallel Volume Rendering with Scattered Decomposition," *Proc. Scalable High Performance Computing Conference*, Knoxville, TN, May 1994.

[16] E. Haines, "A Proposal for Standard Graphics Environments," *IEEE Computer Graphics and Applications*, 7(11): 3-5, November 1987.

[17] M. Barnett, R. Littlefield *et al*, "Global Combine on Mesh Architectures with Wormhole Routing," *Proceedings of 7th International Parallel Processing Symposium*, April 1993.

Antipodality-Based Time-Optimal Diameter Computation on Meshes with Multiple Broadcasting*

V. Bokka, H. Gurla, S. Olariu, J. L. Schwing

Department of Computer Science, Old Dominion University, Norfolk, VA 23529-0162

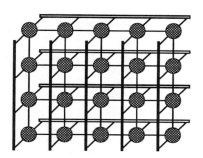

Figure 1: An MMB of size 4×5

Abstract

In this work we address the problem of computing the diameter of a set of points in the plane. Specifically, we show that once we fix a positive constant ϵ, an arbitrary instance of size m, ($n^{\frac{1}{2}+\epsilon} \leq m \leq n$) of the problem above, stored in the first $\lceil \frac{m}{\sqrt{n}} \rceil$ columns of a mesh with multiple broadcasting of size $\sqrt{n} \times \sqrt{n}$ can be solved time-optimally in $\Theta(\frac{m}{\sqrt{n}})$ time.

1   Introduction

The mesh with multiple broadcasting (MMB, for short) is obtained by enhancing the well known mesh with row and column buses as illustrated in Figure 1. For details see [8, 9]. The MMB underlies the DAP family of computers [13] and has received a great deal of attention in the literature [1, 2, 6, 8, 9, 11, 12].

The main contributions of this work is to present an antipodality-based time-optimal algorithm for computing the diameter of a set of m points in the plane. This problem is motivated by, and finds applications to, pattern recognition, computer graphics, computational morphology, image processing, computer vision, and VLSI design [7, 10].

2   Geometric Preliminaries

Specifying an n-vertex polygon P in the plane amounts to enumerating its vertices in *clockwise* order as $p_1, p_2, \ldots, p_n$ ($n \geq 3$), in such a way that $p_i p_{i+1}$ ($1 \leq i \leq n-1$) and $p_n p_1$ define the edges of P. A polygon is *simple* if no two of its non-consecutive edges intersect. A simple polygon is *convex* if its interior is a convex set [14].

The convex hull of a set of points in the plane is the smallest convex set that contains the original set [14]. Recently, the authors have demonstrated that

the task of computing the convex hull of m points in the plane can be solved time-optimally on an MMB of size $\sqrt{n} \times \sqrt{n}$. More precisely, the following result was proved in [5].

Proposition 2.1. For any positive constant ϵ, the task of computing the convex hull of a set of m ($n^{\frac{1}{2}+\epsilon} \leq m \leq n$) points in the plane can be solved time-optimally in $\Theta(\frac{m}{\sqrt{n}})$ time on an MMB of size $\sqrt{n} \times \sqrt{n}$. $\square$

A line λ is a *supporting line* for a polygon P if the interior of P lies completely in one halfplane determined by λ. Vertices p_i and p_j of a convex polygon P are termed *antipodal* if P admits parallel supporting lines through p_i and p_j.

The *diameter* of a convex polygon [14] is the largest Euclidian distance between any pair of its vertices. A classic theorem in combinatorial geometry [14] asserts that the diameter of a convex polygon is the largest distance between antipodal pairs. Given a convex polygon $P = p_1, p_2, \ldots, p_n$ and a point ω in the plane, there is a well-known one-to-one map between the vertices of P and some wedges about ω. This map is obtained by considering every edge $p_{i-1} p_i$ of P as a vector originating at p_{i-1} and by translating this vector parallel to itself until the origin becomes ω. Clearly, this transformation associates with every vertex p_i of P the unique wedge w_i centered at ω and delimited by the vectors $p_{i-1} p_i$ and $p_i p_{i+1}$. The above map affords a deep insight into a number of properties of antipodal pairs. The following geometric results will be instrumental in devising our time-optimal diameter algorithm.

Proposition 2.2. [14] Vertices p_i and p_j are antipodal if and only if some line through ω intersects both wedges w_i and w_j. $\square$

Proposition 2.3. [14] For every i ($1 \leq i \leq n$), the set of subscripts of vertices antipodal to p_i form a cyclic interval. $\square$

The *antipodality* matrix A associated with P is a 0/1 matrix of size $n \times n$ having its rows and columns labeled by the vertices of P and such that $A[i, j] = 1$ if and only if vertices p_i and p_j are antipodal. Since antipodality is symmetric, the matrix A itself is symmetric. By Proposition 2.3, the 1's in every row of A occur consecutively in cyclic order. We refer to a row of A as *dense* if the number of 1's it contains is

*Work supported, in part, by NSF grant CCR-9407180

larger than 3; otherwise the row is called *sparse*. The following result is key in our time-optimal algorithms.

Lemma 2.4. If the vertex p_i is antipodal with vertices $p_j, p_{j+1}, \ldots, p_k$ ($| k - j | > 3$), then all the rows $j + 1, \ldots, k - 1$ of the antipodality matrix are sparse. $\square$

The following crucial observation shows that if a vertex v_i is antipodal with "many" other vertices of the polygon, then information about most of these pairs will be recorded in sparse rows of the antipodality matrix. Specifically, we have the following result.

Lemma 2.5. If the vertex p_i is antipodal with vertices $p_j, p_{j+1}, \ldots, p_k$ ($| k - j | > 3$), then information about every antipodal pair (p_i, p_l) with ($j + 1 \leq l \leq k - 1$) is recorded in some sparse row of the antipodality matrix. $\square$

Lemmas 2.4 and 2.5 have the following consequence.

Corollary 2.6. For the task of identifying all the antipodal pairs, no information is lost by ignoring, in dense rows of the antipodality matrix, all the 1's except for the extreme ones. $\square$

Note that the vertices corresponding to extreme 1's in a dense row need not occur in sparse rows of the antipodality matrix. The *sparse* antipodality matrix is obtained from A by removing in every dense row all the 1's except for the leftmost and rightmost. For simplicity, we continue to refer to the sparse antipodality matrix as A.

Let p_i be an arbitrary vertex of P. A wedge w_j ($i \neq j$) is said to be *left* (resp. *right*) with respect to the wedge w_i if w_j stays in the left (resp. right) halfplane determined by a directed line through ω that sweeps w_i. A wedge which is neither left nor right is termed *mid*. Observe that by virtue of Proposition 2.2, mid wedges correspond to vertices antipodal to p_i. Propositions 2.2 and 2.3 have the following corollary.

Lemma 2.7. With respect to a wedge w_i, the sequence $w_{i+1}, w_{i+2}, \ldots w_{i-1}$ consists of right wedges, followed by mid wedges, followed by left wedges. $\square$

3 Computing the Diameter of a Set

Our platform is a mesh $\mathcal{R}$ with multiple broadcasting of size $\sqrt{n} \times \sqrt{n}$. Fix an arbitrary positive constant ϵ. The input to the algorithm is a set S of m points in the plane, with m in the range

$$n^{\frac{1}{2}+\epsilon} \leq m \leq n. \tag{1}$$

The input is distributed in some order, one point per processor, in the first $\frac{m}{\sqrt{n}}$ columns of $\mathcal{R}$. The goal is to compute the diameter of S and to return it in processor $P(1,1)$. We show that the task at hand can be performed in $O(\frac{m}{\sqrt{n}})$ time.

Before discussing the details, it is appropriate to give the reader a general idea of the various stages of the algorithm. Stage 1 is a preprocessing stage and involves computing the convex hull P of S. By using the algorithm of Bokka *et al.* [5], this task can be

performed in $O(\frac{m}{\sqrt{n}})$ time. It is well-known [14] that the diameter of S is exactly the diameter of P. Consequently, from now on, we concentrate on the task of computing the diameter of P.

To understand the motivation for the next stages of the algorithm, recall that by virtue of Corollary 2.6, in every dense row of the antipodality matrix we can safely ignore from further processing all vertices with the exception of those corresponding to the leftmost and rightmost 1. We shall abuse the terminology a bit referring to a vertex of P as *dense* or *sparse* depending on whether or not the corresponding row in the antipodality matrix is dense or sparse. The task of identifying and processing dense and sparse vertices will be carried out in the next two stages.

Specifically, let A_i be the set of vertices antipodal to vertex p_i corresponding to a 1 in the sparse antipodality matrix. We know that if p_i is sparse, then A_i contains at most three vertices. In case p_i is dense, Lemma 2.5 and Corollary 2.6 guarantee that A_i contains exactly two vertices corresponding, respectively, to the leftmost and rightmost 1 in the antipodality interval of p_i.

Since we do not have enough bandwidth to compute the sets A_i directly, the task specific to Stage 2 is to compute for every vertex p_i an *approximation* to A_i, denoted by C_i. C_i is guaranteed to contain all vertices in A_i along, possibly, with other vertices. The C_i's are referred to as *candidate sets*. In Stage 3, using the information computed in Stage 2, we refine the candidate sets C_i by removing vertices which are not antipodal to p_i. Finally, the goal of Stage 4 is to compute, for every vertex (dense or sparse), the largest distance to an antipodal vertex in A_i and to return the largest of these values as the diameter.

We now present the implementation details of the four stages outlined above. At this point, we view the original mesh $\mathcal{R}$ as consisting of submeshes $R_{i,j}$ of size $\frac{m}{\sqrt{n}} \times \frac{m}{\sqrt{n}}$ with $R_{i,j}$ involving processors $P(r, s)$ such that $(i - 1)\frac{m}{\sqrt{n}} < r \leq i\frac{m}{\sqrt{n}}$ and $(j - 1)\frac{m}{\sqrt{n}} < s \leq j\frac{m}{\sqrt{n}}$. We refer to the submeshes $R_{i,i}$ as *diagonal*. Occasionally, it will be useful to view the mesh $\mathcal{R}$ as consisting of vertical submeshes $R_1, R_2, \ldots, R_{\frac{n}{m}}$ with R_i involving the submeshes $R_{1,i}, R_{2,i}, \ldots, R_{\frac{n}{m},i}$. In this notation, the input is stored by the processors in the submesh R_1.

Stage 1. Using the algorithm of [5] the convex hull P of S is computed and returned in row-major order in R_1. We assume, without loss of generality, that all the points in S belong to the convex hull. With this assumption the vertices of P are $p_1, p_2, \ldots, p_m$. The processor storing vertex p_i computes in constant time the corresponding wedge w_i, in the form of an ordered pair consisting of the two vectors $p_{i-1}p_i$ and p_ip_{i+1}. For further reference, we note that $p_1, p_2, \ldots, p_{\frac{m}{\sqrt{n}}}$ are stored in the first row of R_1, $p_{\frac{m}{\sqrt{n}}+1}, p_{\frac{m}{\sqrt{n}}+2}, \ldots, p_{\frac{2m}{\sqrt{n}}}$

are stored in the second row, and so on.

Stage 2. We select a set of sample vertices of P by retaining every $\frac{m}{\sqrt{n}}$-th vertex of P. The assumed layout of P guarantees that the vertices in the sample occur in top-down order in the $\frac{m}{\sqrt{n}}$-th column of R_1. It is important to note that by virtue of Proposition 2.3 if s_t and s_u are sample vertices antipodal to vertex p_i, then all vertices of P between s_t and s_u are antipodal (in P) to p_i.

This observation motivates the task specific to Stage 2, namely to determine, for every vertex p_i of P, the sample vertices which are antipodal (in P) to p_i. For this purpose, we replicate the set of samples in all rows of $\mathcal{R}$, by first moving the samples to the processors on the main diagonal of $\mathcal{R}$, from where they are replicated in all rows using vertical buses.

We describe the processing that takes place in the first row of the mesh, all the other rows being handled, in parallel, in the same way. Begin by broadcasting p_1 horizontally to the entire first row. Upon receiving p_1, every processor $P(1,j)$ compares the wedges w_1 and $w_{\frac{jm}{\sqrt{n}}}$ and determines whether $w_{\frac{jm}{\sqrt{n}}}$ is right, mid, or left with respect to w_1. To simplify the exposition, we shall refer to the wedges $w_{\frac{jm}{\sqrt{n}}}$ $(1 \leq j \leq \sqrt{n})$ as *sample wedges*. Three cases may occur:

Case 1. Some samples wedges are mid wedges with respect to w_1.

As a consequence of Lemma 2.3 and of our previous data movement, these sample wedges are stored by consecutive processors $P(1,j), P(1,j+1), \ldots, P(1,k)$. It is now a easy to identify the first and last processor storing such a wedge. Taking turns, $P(1,j)$ and $P(1,k)$ broadcast (j,mid) and (k,mid) back to $P(1,1)$.

Case 2. Exactly one sample wedge is mid wedge with respect to w_1.

Let processor $P(1,j)$ detect that the sample wedge it holds is mid with respect to w_1. As in Case 1, $P(1,j)$ broadcasts the packet (j,mid) back to $P(1,1)$.

Case 3. No sample wedges are mid wedges with respect to w_1.

In this case, Lemma 2.5 guarantees the existence of a unique processor $P(1,j)$ with the property that the sample wedge $w_{\frac{jm}{\sqrt{n}}}$ is right and the wedge $w_{\frac{(j+1)m}{\sqrt{n}}}$ is left with respect to w_1. Processor $P(1,j)$ broadcasts the packet (j,right) back to $P(1,1)$.

Upon receiving the packet(s) of information, $P(1,1)$ will take the following action. In Case 1, Proposition 2.2 combined with our way of choosing the samples guarantee that p_1 is dense. Note that Lemma 2.5 along with our way of selecting the set of samples imply that the leftmost and rightmost vertices antipodal to p_1 can only occur in rows j and $k+1$, respectively. Accordingly, the candidate set C_1 is set to $[j, k+1]$. In Case 2, by Proposition 2.2, the only vertices that can be antipodal to p_1 lie in rows j and $j+1$. Now the candidate set C_1 is described implicitly by $[j, j+$

1]. In Case 3, by Lemma 2.5, the only vertices that may be antipodal to p_1 lie in row j. Accordingly, the candidate set C_1 is denoted by $[j,j]$. Clearly, the processing of p_1 takes O(1) time.

The remaining vertices of P in the first row of R_1 are handled similarly, one after the other. Since there are only $\frac{m}{\sqrt{n}}$ convex hull vertices in the first row, Stage 2 runs in O($\frac{m}{\sqrt{n}}$) time.

Stage 3. The goal of this stage is to refine the sets C_i computed in Stage 2 into A_i. For this purpose, we replicate in O($\frac{m}{\sqrt{n}}$) time the contents of R_1, column by column, to all the R_j's ($2 \leq j \leq \frac{n}{m}$). The main reason for this data movement is to use the processors in R_j to handle the convex hull vertices originally stored in $R_{j,1}$. The reader will not fail to observe that by virtue of the previous data movement these vertices have been replicated in the diagonal submesh $R_{j,j}$. Moreover, the submeshes $R_{j,j}$ can be perceived as independent submeshes with multiple broadcasting. For definiteness, we discuss the processing that takes place in R_j, all the other R_k's being processed similarly.

Recall that for a convex hull vertex p_i in $R_{j,j}$ the candidate set C_i involves vertices from *at most* two rows of R_j. We shall refer to the smallest index of such row as *low row-rank* of p_i. The second row index, if any, will be referred to as *high row-rank* of p_i. The processing begins by sorting the vertices in $R_{j,j}$ in row-major order of their low row-rank. Further, the rows of $R_{j,j}$ will be termed *pure* or *impure* depending on whether or not their vertices have the same low row-rank. Pure and impure rows are handled differently as we are about to describe. We begin by identifying every row of $R_{j,j}$ as pure or impure. For this purpose, every processor in the last column of $R_{j,j}$ broadcasts the low row-rank of the vertex it holds horizontally in its own row. Upon receiving this information, every processor in the first column of $R_{j,j}$ has enough information to determine whether the corresponding row is pure or impure.

Next, we process pure rows with low row-ranks at most $\sqrt{n}$, as follows. Let $r_1, r_2, \ldots, r_t$ be the low row-ranks of the pure rows in $R_{j,j}$. By broadcasting one by one the rows $r_1, r_2, \ldots, r_t$ of R_j vertically we ensure that row r_k will be replicated in every pure row with low row-rank r_k. Since there are at most $\frac{m}{\sqrt{n}}$ pure rows in $R_{j,j}$, this data movement takes O($\frac{m}{\sqrt{n}}$) time. Further, in every pure row, the vertices received are moved horizontally, in lock step. This data movement ensures that every vertex in a pure row determines whether it is dense or sparse. In case they are sparse, they also retain the identity of (at most three) vertices to which they are antipodal. In case they are dense, they retain the identity of the leftmost and rightmost vertex in their antipodality interval. It is easy to confirm that the overall processing time of pure rows is bounded by O($\frac{m}{\sqrt{n}}$).

Impure rows are processed differently. Consider an arbitrary impure row of $R_{j,j}$ and let $r_1, r_2, \ldots, r_t$ be the low row-ranks of the vertices in this row. Since the sequence of vertices in $R_{j,j}$ was sorted in row-major order of their low row-rank, the vertices having the same low row-rank occur consecutively. For further reference, such a set of queries is termed a *run*. It is important to note that for any p $(1 \leq p \leq \sqrt{n})$, at most two impure rows contain vertices whose low row-rank is p. The impure rows are processed sequentially. Each of them is broadcast vertically throughout the submesh R_j. This ensures that all vertices in an impure row are moved to the row of the submesh R_j that equals their low row-rank. Since there are at most $\frac{m}{\sqrt{n}}$ impure rows in $R_{j,j}$, this data movement takes $O(\frac{m}{\sqrt{n}})$ time. Our previous observation guarantees that no processor stores, as a result of this data movement, more than two vertices. Now using a simple data movement, whose details are omitted, in each row of R_j, the vertices determine whether they are dense or sparse. In case they are sparse, they retain the identity of (at most three) vertices to which they are antipodal. In case they are dense, they retain the identity of exactly two vertices: the leftmost vertex and rightmost vertex to which they are antipodal.

The same processing is repeated for the high row-rank of vertices in $R_{j,j}$. It is easy to see that, as a result, all the dense and sparse vertices have retained the identity of at most three vertices of P to which it is antipodal. Furthermore, the running time of Stage 3 is bounded by $O(\frac{m}{\sqrt{n}})$.

Stage 4. In every submesh $R_{j,j}$ each processor storing a vertex (dense or sparse) determines in $O(1)$ time the largest distance of the corresponding vertex to any of its antipodal vertices. Next, the maximum of the distances over all the vertices in $R_{j,j}$ is computed in $O(\frac{m}{\sqrt{n}})$ time in the obvious way. We assume that the processor in the north-west corner of $R_{j,j}$ stores the local maximum.

Further, using horizontal buses, every such processor broadcasts the local maximum to the corresponding processor in the first column of $\mathcal{R}$. As a result of this data movement, the first column contains $\frac{n}{m}$ partial results. Using a simple data movement, these values are compacted in the topmost $\frac{n}{m}$ positions of the first column and their maximum is computed using the semigroup algorithm of [11] in $O(\log \frac{n}{m})$ time. As shown in [5] the running time of this stage is bounded by $O(\frac{m}{\sqrt{n}})$. Thus, we have the following result.

Theorem 3.2. For every choice of a positive constant ϵ, the diameter of a set of m $(n^{\frac{1}{2}+\epsilon} \leq m \leq n)$ points in the plane stored in the first $\frac{m}{\sqrt{n}}$ columns of a mesh with multiple broadcasting of size $\sqrt{n} \times \sqrt{n}$ can be obtained in $O(\frac{m}{\sqrt{n}})$ time. Furthermore, this is time-optimal. $\square$

References

[1] D. Bhagavathi, P. J. Looges, S. Olariu, J. L. Schwing, and J. Zhang, A fast selection algorithm on meshes with multiple broadcasting, *IEEE Trans. Parallel and Distributed Systems*, 5, (1994), 772–778.

[2] D. Bhagavathi, S. Olariu, W. Shen, and L. Wilson, A time-optimal multiple search algorithm on enhanced meshes, with applications, *J. of Parallel and Distr. Comp.*, 22, (1994), 113–120.

[3] D. Bhagavathi, S. Olariu, J. L. Schwing, and J. Zhang, Convex polygon problems on meshes with multiple broadcasting, *Parallel Processing Letters*, 2 (1992) 249–256.

[4] D. Bhagavathi, H. Gurla, R. Lin, S. Olariu, J. L. Schwing, and J. Zhang, Square meshes are not optimal for convex hull computation, *Proc. of ICPP*, 1993, III, 307–311.

[5] V. Bokka, H. Gurla, S. Olariu, J. L. Schwing, A time- and VLSI-optimal convex hull algorithm for meshes with multiple broadcasting, *Proc. Frontiers'95*, .

[6] Y. C. Chen, W. T. Chen, G. H. Chen and J. P. Shen, Designing efficient parallel algorithms on mesh connected computers with multiple broadcasting, *IEEE Trans. Parallel and Distributed Systems*, 1, (1990) 566–570.

[7] R. O. Duda and P. E. Hart, *Pattern Classification and Scene Analysis*, Wiley and Sons, New York, 1973.

[8] V. Prasanna-Kumar and C. S. Raghavendra, Array processor with multiple broadcasting, *Journal of Parallel and Distributed Computing*, 2 (1987) 173–190.

[9] R. Lin, S. Olariu, and J. L. Schwing, An efficient VLSI architecture for digital geometry, *Proc. of International IEEE Conference on Application-Specific Array Processors*, San Francisco, August 1994, 392–403.

[10] T. Lozano-Perez, Spatial Planning: A Configurational Space Approach, *IEEE Transactions on Computers*, C-32 (1983) 108–119.

[11] S. Olariu, J. L. Schwing, and J. Zhang, Time-optimal convex hull algorithms on enhanced meshes, *BIT*, 33 (1993) 396–410.

[12] S. Olariu and I. Stojmenović, Time-optimal proximity problems on meshes with multiple broadcasting, *Proc. IPPS*, 1994, 94–101.

[13] D. Parkinson, D. J. Hunt, and K. S. MacQueen, The AMT DAP 500, 33rd *IEEE Comp. Soc. Internat. Conf.*, 1988, 196–199.

[14] I. M. Yaglom and V. G. Boltyanski, *Convex Figures*, Holt, Rinehart, and Winston, New York, 1961.

Parallel Implementations of Block-Based Motion Vector Estimation for Video Compression on the MasPar MP-1 and PASM

Min Tan, Janet M. Siegel, and *Howard Jay Siegel*
Parallel Processing Laboratory
School of Electrical Engineering
Purdue University
West Lafayette, IN 47907-1285, USA

Abstract — *Video compression is important for video transmission and storage. This paper presents two parallel implementations of a compression technique called block-based motion vector estimation. Results from a 16,384 processor MasPar MP-1 (a SIMD machine) and the 16 processor PASM prototype (a partitionable SIMD/MIMD mixed-mode machine) are discussed. The trade-offs of using different modes of parallelism and different data distribution schemes are examined.*

1. Introduction

Compression techniques for video sequences have been the focus of research for video transmission and storage [4]. One well known technique is *block-based motion vector estimation* [2, 3], in which the current frame of a video sequence is divided into disjoint rectangular blocks of pixels. For each block in the current frame, a block of pixels in the previous frame is found that represents the closest match for it, according to some criterion such as the mean absolute difference (*MAD*). The spatial distance between the positions of the two blocks in their corresponding frames defines a two-dimensional motion vector associated with the block in the current frame. Because of the high redundancy that may exist between successive frames, an accurate estimation of the motion vectors makes compression possible. If maximum displacements of W_r pixels and W_c pixels are allowed for the row and column directions respectively, there are $(2W_r+1)(2W_c+1)$ locations to search in the previous frame for the closest match for each block in the current frame[*].

Suppose the intensity of the pixel with coordinates (i,j) in video frame n is $I_n(i,j)$. Each video frame is divided into disjoint blocks with size $B_r \times B_c$, where B_r and B_c are the number of pixels in the row and column directions for each rectangular block, respectively. A block of $B_r \times B_c$ pixels is denoted by the coordinate (k, l) of its upper left corner pixel [5]. The size of the search subimage, which is referred to as the *displacement window* in this paper, is $(2W_r+B_r)(2W_c+B_c)$. The MAD between the block $(k, l)[n]$ of the current frame n and the block $(k+x, l+y)[n-1]$ of the previous frame $n-1$ is
$$\text{MAD}[(k,l)[n], (k+x,l+y)[n-1]] =$$
$$(\sum_i \sum_j |I_n(k+i, l+j) - I_{n-1}(k+i+x, l+j+y)|)/(B_rB_c),$$
where $0 \le i < B_r$ and $0 \le j < B_c$.
The motion vector $mv(k,l)$ of the block $(k,l)[n]$ is given by (mv_x, mv_y), such that
$$\text{MAD}[(k,l)[n], (k+mv_x, l+mv_y)[n-1]] =$$
$$\min \{\text{MAD}[(k,l)[n], (k+x,l+y)[n-1]]\},$$
where $-W_r \le x \le W_r$ and $-W_c \le y \le W_c$.

Blocks at the edges of each frame require fewer operations for finding the corresponding motion vectors than do interior blocks, because the displacement window regions are constrained by image boundaries. The computational savings made possible by this property are easily realized using serial algorithms, but are more difficult to exploit using parallel algorithms. The choice of block and displacement window sizes predicate what type of parallel algorithm should be used.

A number of serial search algorithms have been proposed for block-based motion vector estimation (e.g., [3, 5]). To achieve real-time compression, a parallel implementation is desirable. The goal of this paper is to evaluate two parallel implementations, the rectangular and the row stripe subimage methods, using two parallel machines. The two machines are the MasPar MP-1 [6], an SIMD machine with 16,384 *PEs* (processing elements, i.e., processor/memory pairs), and *PASM* (*P*Artitionable-*SI*MD/*MI*MD) reconfigurable parallel processing system small-scale prototype with 16 PEs in the computational engine [8].

2. Parallel Implementations

In the *rectangular subimage method*, it is assumed that N PEs are logically arranged as an $N_r \times N_c$ grid. The previous and current video frames are mapped onto the PEs by superimposing the two images onto the PE grid. Thus, if each frame is of size $M_r \times M_c$ (M_r and M_c are the number of pixels in the row and column directions, respectively), then each PE stores two $(M_r/N_r) \times (M_c/N_c)$ rectangular subimages. All PEs search for motion vectors simultaneously, each PE for one of its local blocks.

When searching the previous-frame subimage for the motion vectors of the blocks at the edge of the current-frame subimage, pixels from spatially adjacent previous-frame subimages need to be transferred. Assume that $M_r/N_r > W_r$ and $M_c/N_c > W_c$ (a reasonable as-

This research was supported by NRaD under subcontract number 20-950001-70. It used equipment supported by the National Science Foundation under grant number CDA-9015696.

[*]To be consistent with this terminology from the video compression field, X_r will be the number of items in a row (i.e., the number of columns) and X_c will be the number of items in a column (i.e., the number of rows).

sumption for this problem domain). PE i requires $M_c/N_c \times W_r$ pixels of the previous-frame subimage from the two PEs directly adjacent to it along its leftmost and rightmost columns and $M_r/N_r \times W_c$ pixels of the previous-frame subimage from the two PEs directly adjacent to it along its topmost and bottommost rows. PE i must also receive $W_r \times W_c$ pixels of the previous-frame subimage from each of the four PEs diagonally adjacent to it. Thus, a total of $2M_cW_r/N_c + 2M_rW_c/N_r + 4W_rW_c$ inter-PE data transfers are required per PE to find the motion vectors for all blocks of the current-frame subimage. All N PEs can transfer pixels of the previous-frame subimage simultaneously with appropriate network support (both the MasPar MP-1 and PASM have this type of network support).

If α is the time to perform an absolute difference calculation and an addition between two integers, γ is the time to perform a comparison between two integers, and τ is the inter-PE data transfer time for one pixel value, then the computation time using the rectangular subimage method is (detailed derivation is in [9])

$$M_rM_c(2W_r+1)(2W_c+1)(\alpha B_rB_c+\gamma)/(NB_rB_c). \quad (1)$$

The corresponding communication time is

$$(2M_cW_r/N_c+2M_rW_c/N_r+4W_rW_c)\tau. \quad (2)$$

It follows that the communication time depends on N_r and N_c. The best PE configuration depends on the video frame size (i.e., M_r and M_c) and the displacement window size (i.e., W_r and W_c) [9].

An alternate method for mapping the previous and current video frames among the N PEs is based on distributing consecutive rows of pixels and is referred to as the *row stripe subimage method*. Both previous and current $M_r \times M_c$ video frames are divided into N subimages of size $(M_r \times M_c/N)$. Each PE stores $M_r/B_r \times M_c/(NB_c)$ blocks of both the previous and current video frames.

The blocks close to the image border do not require a search of the whole displacement window space. In the row stripe method, no calculation is performed by any of the PEs on exterior column blocks for part of the displacement window search space, because these blocks are distributed evenly among all the PEs. By allowing all PEs to realize the savings on the exterior column blocks, the computation time is decreased. The rectangular subimage method cannot take advantage of this savings because the exterior blocks of the current video frame are distributed unevenly among the PEs. The PEs with no exterior blocks must search the whole displacement window space and will dominate the computation time required for the parallel implementation.

For the $2M_c/NB_c$ column border blocks of the current-frame subimage, the search space of the displacement window is $(W_r+1)(2W_c+1)$. Suppose W_r is a multiple of B_r and let $p = W_r/B_r$. For the $2M_c/NB_c$ blocks that are i ($0 \le i < p$) blocks away from the column border blocks, the search space is $(W_r+iB_r+1)(2W_c+1)$. For the rest of the blocks, the search space is $(2W_r+1)(2W_c+1)$. Thus, the computation time using the row stripe subimage method is

$$(2W_c+1)(\alpha B_rB_c+\gamma) M_c/(NB_c) \times$$
$$[(M_r/B_r)(2W_r+1)-W_r(p+1)]. \quad (3)$$

Compared with the computation time using the rectangular subimage method (shown in Eq.(1)), the time saved by taking advantage of the exterior blocks of the current video frame is

$$(2W_c+1)(\alpha B_rB_c+\gamma)M_cW_r(p+1)/(NB_c). \quad (4)$$

For the row stripe subimage method, PE i requires $M_r \times W_c$ pixels of the previous-frame subimage from PE $[(i-1) \bmod N]$ and $M_r \times W_c$ pixels of the previous-frame subimage from PE $[(i+1) \bmod N]$. It follows that the communication time is

$$2M_rW_c\tau. \quad (5)$$

This is in contrast to Eq. (2) for the rectangular subimage method.

Another method is based on distributing consecutive columns of pixels among the N PEs. This is referred to as the *column stripe subimage method*. Detailed discussion of the trade-offs between these two stripe methods is in [9].

Comparing the two different schemes for distributing the previous and current video frames, the rectangular subimage method usually requires fewer inter-PE data transfers compared with the row stripe subimage method. But the row stripe subimage method has the potential to decrease the computation time due to the reduced search space of the displacement window for the exterior column blocks. Because the computation time involved for each possible displacement position (i.e., $\alpha B_rB_c+\gamma$ for the calculation of the MAD) can be large when B_r and B_c are large, the computation time saved by using the row stripe subimage method can be significant. This trade-off between two data distribution schemes is analogous to that in [1].

3. Timing Results

The SIMD implementation on the MasPar MP-1 concentrates on the comparison between the estimated execution time (based on Eqs. (1, 2, 3, 5)). and the experimental values obtained by running the parallel programs on the machine. For both methods, the estimated values of both computation and communication times differ from the experimental data by at most 2.0% [9].

PASM is a partitionable SIMD/MIMD system concept being developed as a design for a large-scale dynamically reconfigurable parallel processing system [8]. A small-scale proof-of-concept prototype (16 PEs in the computational engine) has been built at Purdue University. PASM can be dynamically reconfigured to form submachines of various sizes. Each submachine can independently perform mixed-mode parallelism. PASM uses a flexible multistage interconnection network for inter-PE communication.

Fig. 1 shows the timings for SIMD, SPMD, and mixed-mode implementations of the rectangular subimage method ($M_r = 64$, $M_c = 128$, $W_r = 4$, $W_c = 6$, and $B_r = B_c = 4$) on PASM. *SPMD* (single program - multiple data stream) mode is a special form of MIMD mode

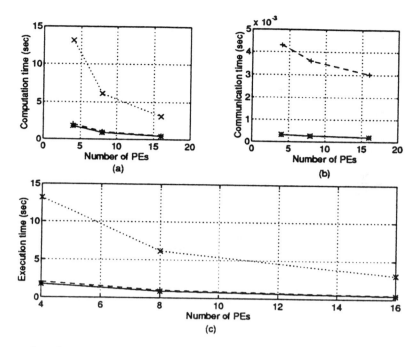

Fig. 1: The (a) computation time, (b) communication time, and (c) execution time for SIMD (dotted line), SPMD (dashed line), and mixed-mode (solid line) programs using the rectangular subimage method.

where all the PEs execute the same program in an asynchronous fashion, each on its own data. The three data points are for $N = 4$ ($N_r = N_c = 2$), 8 ($N_r = 4$ and $N_c = 2$), and 16 ($N_r = N_c = 4$).

As shown in Fig. 1(a), the SPMD mode implementation requires less computation time then the SIMD mode implementation. One of the advantages of SPMD mode is the ability to execute the "then" and "else" clauses of data conditional statements across different PEs simultaneously [7]. No PE is idle during the execution of a particular data conditional statement. In contrast, for SIMD mode, the "then" and "else" clauses of data conditional statements must be broadcast to PEs serially. Because finding the motion vector for each block of the current-frame subimage requires many comparisons (involved in α and γ), this part of the algorithm is implemented in SPMD mode in the mixed-mode implementation.

In the mixed-mode implementation, the loop control variables are stored and computed by the control unit (CU) in SIMD mode. Thus, the mixed-mode implementation can utilize CU/PE overlap, which occurs in this case when the CU performs the overhead associated with loop control in SIMD mode while the PEs execute the loop bodies in SPMD mode [7]. Because of this effect, the computation time for the mixed-mode implementation is somewhat less than the pure SPMD implementation.

One advantage of SIMD mode is the implicit synchronization that occurs after every instruction broadcast from the CU to the PEs, thus, communication protocols with less overhead than SPMD are used during inter-PE data transfers [7], as demonstrated by Fig. 1(b). Therefore, inter-PE communication is performed in SIMD

mode in the mixed-mode implementation.

For both the rectangular and the row stripe subimage methods, the mixed-mode implementation has the smallest total execution time compared with the pure SIMD and pure SPMD implementations. Given this example, the total SPMD execution time is close to the mixed-mode time because the computation time (versus the communication time) dominates the execution. Other values for the machine and algorithm parameters could be chosen to show a larger difference.

Because the mixed-mode implementation performs best, all timing results for the rest of this section are for mixed-mode. The comparison between the estimated values for the timings of both data distribution schemes and the experimental values obtained by running the parallel programs on PASM is included in [9]. The estimated values of both computation and communication times (based on Eqs. (1, 2, 3, 5)) differ from the experimental data by at most 1.7%.

From the analyses shown in Section 2, the computation time for the row stripe subimage method is always smaller than the rectangular subimage method. This is illustrated by the experimental results shown in Fig. 2(a), in which $M_r = M_c = 64$, $W_r = W_c = 2$, and $B_r = B_c = 2$. In terms of the communication time, when N is increased, the rectangular subimage method has fewer inter-PE data transfers while the number of inter-PE data transfers stays the same for the row stripe subimage method (as shown in Fig. 2(b)).

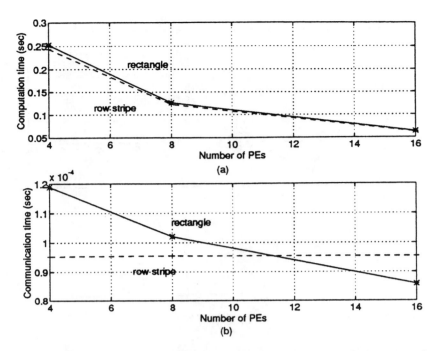

Fig. 2: The trade-offs between the rectangular subimage method (solid line) and the row stripe subimage method (dashed line); (a) computation time and (b) communication time.

4. Summary

This paper summarizes from [9] two parallel implementations of block-based motion vector estimation on two parallel machines. The timing results on the MasPar MP-1 and PASM demonstrate that this video compression technique can successfully exploit parallelism. The SIMD implementations on the MasPar MP-1 present an achievable speedup through the use of a commercially available massively parallel processing system.

Through the study of the mixed-mode implementation of this video compression technique on the PASM prototype, the trade-offs of using different modes of parallelism and different data distribution schemes are examined. The use of mixed-mode parallelism has some advantages over pure SIMD and pure SPMD implementations. With equivalent processing capability, a mixed-mode system can achieve higher speedup due to the flexibility of using more than one mode of parallelism.

References

[1] N. Giolmas, D. W. Watson, D. M. Chelberg, and H. J. Siegel, "A parallel approach to hybrid range image segmentation," *6th Int'l Parallel Processing Symp.*, Mar. 1992, pp. 334-342.

[2] "Coding of moving pictures and associated audio," *Committee Draft of Standard ISO11172*, Int'l Standards Organization/Motion Pictures Expert Group 90/176, Dec. 1990.

[3] J. R. Jain et al., "Displacement measurement and its application in interframe image coding," *IEEE Trans. Communications*, Vol. COM-29, Dec. 1981, pp. 1799-1808.

[4] D. LeGall, "MPEG: a video compression standard for multimedia applications," *Communications of the ACM*, Vol. 34, Apr. 1991, pp. 47-58.

[5] B. Liu et al., "New fast algorithms for the estimation of block motion vectors," *IEEE Trans. Circuits and Systems for Video Technology*, Vol. 3, Apr. 1993, pp. 148-157.

[6] J. R. Nickolls, "The design of the MasPar MP-1: a cost effective massively parallel computer," *IEEE Compcon*, Feb. 1990, pp. 25-28.

[7] H. J. Siegel, J. K. Antonio, R. C. Metzger, M. Tan, and Y. A. Li, "Heterogeneous computing," in *Handbook of Parallel and Distributed Computing*, A. Zomaya, ed., McGraw-Hill, 1995.

[8] H. J. Siegel, T. Schwederski, W. G. Nation, J. B. Armstrong, L. Wang, J. T. Kuehn, R. Gupta, M. D. Allemang, D. G. Meyer, and D. W. Watson, "The design and prototyping of the PASM reconfigurable parallel processing system," in *Parallel Computing: Paradigms and Applications*, A. Zomaya, ed., Chapman and Hall, London, U.K., 1995.

[9] M. Tan, J. M. Siegel, and H. J. Siegel, "Parallel implementations of block-based motion vector estimation for video compression on three machines," *Technical Report*, E.E. School, Purdue, in preparation.

Properties of Binomial Coefficients and Implications to Parallelizing Lagged Fibonacci Random Number Generators

Srinivas Aluru

Computer and Information Science

Syracuse University, Syracuse, NY 13244-4100

aluru@cis.syr.edu

Abstract

In this paper, we explore the relationship between binomial coefficients and lagged Fibonacci generators and use this relationship to investigate the possibility of parallelizing lagged Fibonacci generators using the leapfrog technique. We show that while lagged Fibonacci generators with the exclusive or operator can be efficiently parallelized without any communication when the number of processors is a power of two, it is not possible to parallelize other lagged Fibonacci generators efficiently in a communication-free manner.

1 Introduction

A number of computer applications require the use of random numbers. To parallelize such applications, parallel random number generators are required. Sequential random number generators have been studied in great depth and a handful of methods are acknowledged to be appropriate: linear congruential generators, lagged Fibonacci generators, combination generators and more recently, the add with carry and subtract with borrow generators [13]. The leapfrog technique has gained popularity as an acceptable way of parallelizing sequential generators [3, 6]. Using the leapfrog technique, Bowman et al. [3] parallelize linear congruential generators and Aluru et al. parallelize lagged fibonacci exclusive or generators when the number of processors is a power of two [2]. In this paper, we investigate the possibility of applying the leapfrog technique to lagged Fibonacci generators.

2 Parallel Random Number Generation

A sequential random number generator is expected to have good randomness properties and a constant generation time per random number. The generator is started by specifying the initial state called the 'seed'. Two runs of an application with the same seed produce the same result, ensuring reproducibility. The same requirements naturally extend to a parallel random number generator, making the following features desirable: good randomness properties for the sequence generated on each processor, absence of mutual correlations among any pair of sequences, reproducibility of results, generation time per random number and memory required similar to the generator on one processor and absence of communication.

The small number of good sequential generators precludes the possibility of using a separate generator per processor. We are thus forced to split the sequence of random numbers generated by a sequential generator to the many parallel processors. The leapfrog technique prescribes a way of splitting that preserves the reproducibility and randomness properties. Let $x_0, x_1, x_2, \ldots$ be the sequence generated by a sequential generator. In the leapfrog technique using N processors, processor i generates every N^{th} number in the sequence starting at x_i: i.e., it generates $x_i, x_{i+N}, x_{i+2N}, \ldots$. It is easily seen that the multiple streams generated are non-overlapping and together generate the original sequence. This allows the user to write code in such a way that the results are independent of the number of processors. This also ensures that the randomness properties of the sequential generator carry over to the parallel generator.

It is not straightforward to implement a parallel random number generator using the leapfrog technique and satisfying all the requirements outlined before. To satisfy the no-communication requirement, a way of computing x_k using only x_{k-mN} (the previously generated random numbers of the same processor) should be found. Furthermore, the generation of x_k should use only a constant number of operations. Since a random number generator is an equation describing the computation of the kth random number x_k, recursive application of the same equation should be studied to describe all possible ways of computing x_k from previous numbers. If a way of computing x_k is possible using a constant number of operations and random numbers previously generated on the same processor, we can use it to parallelize the generator with the leapfrog technique. Such a method is easy for the linear congruential generator [6].

The purpose of this paper is to investigate all possible ways of computing the kth random number x_k of the sequence generated by a lagged Fibonacci generator, to determine if lagged Fibonacci generators are amenable to the leapfrog technique.

3 Lagged Fibonacci Generators

Lagged fibonacci generators are specified by the recurrence $x_k = x_{k-p} \otimes x_{k-p+q} \bmod m$, where $\otimes$ denotes the operation which could be any of $+, -, \times$ or $\oplus$ (exclusive or). For generating l bit random numbers, $m = 2^l$. The seed for these generators is the first p random numbers. For the multiplicative generator, the random numbers are odd numbers $mod\ m$.

Linear congruential generators have been known to exhibit regularities that make them unsuitable for certain kinds of Monte Carlo applications [12, 11, 10]. The advantages of lagged Fibonacci generators over Linear congruential generators include multi-dimensional uniformity, large period, repeatability of the random numbers within a full period and independence of the period from the word size of the computer

Figure 1: First few rows of the Pascal's triangle.

Figure 2: Odd numbers in the Pascal's triangle.

on which it is implemented. Lagged Fibonacci generators with $+, -$ and $\times$ give good results on most of the stringent statistical tests. Lagged Fibonacci generators with $\oplus$ (exclusive or) are not good for small values of p, but for large values of p (607, for example), they pass all statistical tests [9].

4 Properties of Binomial Coefficients

Several properties of binomial coefficients modulo two have been known for a long time [5, 7, 8, 15, 16]. In this paper, we provide a self-contained treatment of the properties relevant to our discussion and offer independent proofs of these.

We begin with the Pascal's triangle, a pictorial way of representing all binomial coefficients. The nth row of the Pascal's triangle consists of all the binomial coefficients of n. A notable property of the diagram is that each number is formed by adding the two numbers above it in the previous row. This corresponds to the property that $(n, i) = (n - 1, i - 1) + (n - 1, i)$, where (n, i) represents the ith binomial coefficient of n. The first few rows of the Pascal's triangle are shown in Figure 1.

Figure 2 shows the structure obtained by representing each odd number in the Pascal's triangle by a black square. The diagram between row 2^k and row 2^{k+1} consists of two copies of the diagram between row 0 and row 2^k, placed next to each other. Thus, the structure of odd numbers in the 0th row determines the entire diagram. Theorem 1 formalizes and proves this property. Since $(n, r) = (n, n - r)$, only values of $r \leq \lfloor \frac{n}{2} \rfloor$ are considered.

Theorem 1 $(2^k + l, i)$ *is odd*, $0 \leq i \leq \lfloor \frac{2^k + l}{2} \rfloor$
$\Leftrightarrow i \leq l$ *and* (l, i) *is odd.* $\forall 0 \leq l < 2^k$

Proof By induction on k. If $k = 0$, $l = 0$ and both $(1, 0)$ and $(0, 0)$ are odd. Otherwise, for any k, consider induction on l. If $l = 0$, $(2^k, i) = \frac{2^k}{i}(2^k -$

$1, i - 1)$. If $i \neq 0$ or 2^k, i has less than k powers of 2 in its factorization and since both $(2^k, i)$ and $(2^k - 1, i - 1)$ are integers, $(2^k, i)$ is even. Therefore, for $i \leq \lfloor \frac{2^k + 0}{2} \rfloor$, $(2^k, i)$ is odd only when $i = 0$, corresponding to $(0, 0)$ being odd. If $l > 0$ and $i < l$,
$$(2^k + l, i) = (2^k + l - 1, i) + (2^k + l - 1, i - 1)$$
$$(l, i) = (l - 1, i) + (l - 1, i - 1)$$
By induction hypothesis, $(2^k + l - 1, i)$ and $(l - 1, i)$ are both even or both odd and $(2^k + l - 1, i - 1)$ and $(l - 1, i - 1)$ are both even or both odd. Therefore, the results of the summation are both even or both odd. If $i = l$, $(2^k + l, l) = \frac{2^k + l}{l}(2^k + l - 1, l - 1)$. Since $(l - 1, l - 1)$ is odd, $(2^k + l - 1, l - 1)$ is odd (by induction hypothesis) and hence $(2^k + l, l)$ is odd. This completes the proof. □

Corollary 2 *The number of odd terms in the $(2^k + l)^{th}$ row, $0 \leq l < 2^k$ is twice the number of odd terms in the lth row.*

Lemma 3 *Let k be the number of 1's in the binary representation of n. The number of odd terms in the nth row of the Pascal's triangle is 2^k.*

Proof By induction on k. If $k = 0$, the statement is obviously true. Otherwise, $n = 2^{m_1} + 2^{m_2} + \ldots + 2^{m_k}$, $m_1 > m_2 > \ldots > m_k$. By Corollary 2, the number of odd terms in the the nth row is twice the number of odd terms in the row corresponding to $n' = 2^{m_2} + 2^{m_3} + \ldots + 2^{m_k}$, which is 2^{k-1} by induction hypothesis. Therefore, there are 2^k odd terms in the nth row. □

The following lemma provides a simple test to determine if a binomial coefficient is odd or even. We say that r preserves the 0's of n iff for every bit position having a 0 in n, r has a 0 in the same bit position.

Lemma 4 (n, r) *is odd* $\Leftrightarrow r$ *preserves the 0's of n.*

Proof For the 'if' part, consider induction on the number of bits required to represent n. If $n = 0$ or 1, any r preserves the 0's of n, and all terms in the first two rows of Pascal's triangle are odd. Otherwise,
Case 1 $r \leq \lfloor \frac{n}{2} \rfloor$.
$n = 1........$
$r = 0........$
By Theorem 1, (n, r) is odd if (n', r') is odd, where n' and r' are obtained by removing the most significant bit from n and r. Clearly, r' preserves the 0's of n'. By hypothesis, (n', r') is odd, and hence (n, r) is odd.
Case 2 $r > \lfloor \frac{n}{2} \rfloor$.
It is easy to see that $n - r$ preserves the 0's of n and $n - r \leq \lfloor \frac{n}{2} \rfloor$. By Case 1, $(n, r) = (n, n - r)$ is odd.

The 'only if' part is by a counting argument. Let k be the number of 1's in the binary representation of n. There are 2^k choices of r preserving the 0's of n and these result in odd combinations. By Lemma 3, there are only 2^k odd combinations. Hence, any r not preserving the 0's of n results in even combination. □

Corollary 5 (n, r) *is odd* $\Leftrightarrow (n \vee r) = n$, *where* $\vee$ *is the bitwise logical or operator.*

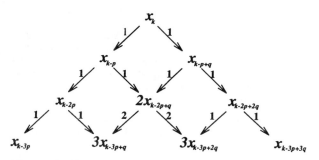

Figure 3: Recursive expansion of a Fibonacci generator.

5 Parallelizing Lagged Fibonacci Generators

Recursive expansion of $x_k = x_{k-p} \otimes x_{k-p+q} \bmod m$ is shown in Figure 3. The relationship with the Pascal's triangle is clear: Each row in Figure 3 represents a way of computing x_k. $(n, i-1)$ occurrences of $x_{k-np+(i-1)q}$ and (n,i) occurrences of $x_{k-np+iq}$ in the nth row contribute to $(n, i-1) + (n, i) = (n+1, i)$ occurrences of $x_{k-(n+1)p+iq}$ in the $(n+1)^{th}$ row. It follows that, $\forall n \ s.t. \ 1 \le n \le \lfloor \frac{k}{p} \rfloor$,

$$x_k = \bigotimes_{i=0}^{n} \bigotimes_{j=1}^{(n,i)} x_{k-np+iq} \bmod m \qquad (\otimes \ne -) \quad (1)$$

$$x_k = \sum_{i=0}^{n} (-1)^i \sum_{j=1}^{(n,i)} x_{k-np+iq} \bmod m \qquad (\otimes = -) \quad (2)$$

corresponding to the nth row of the Pascal's triangle. The notation $\bigotimes_{i=0}^{n}$ is much like the $\sum_{i=0}^{n}$, except that the operation involved is $\otimes$ instead of $+$.

Unfortunately, it is not immediately clear if this family of equations is of help in efficient parallel random number generation. The number of terms in the equation as a function of n is $\sum_{i=0}^{n}(n,i) = 2^n$.

Our goal is to compute x_k using only x_{k-mN} for $1 \le m \le \lfloor \frac{k}{N} \rfloor$. Therefore, equation (1) or (2) can be used if $x_{k-np+iq} = x_{k-(np-iq)}$ is available on the processor computing x_k for all $0 \le i \le n$. This implies that $(np - iq)$ should be a multiple of N for all $0 \le i \le n$. i.e. we require $(np - iq) \bmod N = 0$.

$$(np - iq) \bmod N = (np - (i+1)q) \bmod N$$
$$\Rightarrow \quad q \bmod N = 0 \Rightarrow q = kN$$

It is obviously impossible to satisfy this relation since it restricts the number of processors to be a factor of q. Even otherwise, substituting $i = 0$ in $(np - iq) \bmod N = 0$ implies that $np \bmod N = 0$ or that np is a multiple of N. This can be satisfied only by taking n to be a multiple of N. Since we want n to be as small as possible, choose $n = N$. This would imply that the equation for computing x_k has 2^N terms, where N is the number of processors. This cost is prohibitive and definitely non-scalable, even if all the terms were to be found on the same processor.

There is still one avenue left open: If we can choose n such that $\bigotimes_{j=1}^{(n,i)} x_{k-np+iq} \bmod m$ is 1 for the mul-

tiplicative generator and 0 for the other generators irrespective of $x_{k-np+iq}$ for most i, then it does not matter if $x_{k-np+iq}$ is not available on the same processor. Clearly, this can not be done for $i = 0$ and $i = n$ since $(n, 0) = (n, n) = 1$ and $x_{k-np+iq}$ is arbitrary. Therefore, we require that x_{k-np} and $x_{k-np+nq}$ be available on the same processor, which can be easily accomplished by choosing $n = N$. For the parallel generator to have the same speed per processor as the sequential generator, $\bigotimes_{j=1}^{(N,i)} x_{k-Np+iq} \bmod m$ should be equal to 0 for $+$, $-$ or $\oplus$ generators and 1 for $\times$ generator, for all i except 0 and N. For such values of N, we can have a parallel lagged fibonacci generator on N processors.

5.1 Lagged Fibonacci with $\oplus$

The equation $\bigotimes_{j=1}^{(N,i)} x_{k-Np+iq} \bmod m$ can be simplified in the particular case when the operation involved is exclusive or ($\oplus$), using the properties of exclusive or and binomial coefficients modulo two. Since for any binary vector $\mathbf{a}$ and integer m $\bigoplus_{i=1}^{m} \mathbf{a}$ is $\mathbf{0}$ for even m and $\mathbf{a}$ for odd m,
$$\bigoplus_{j=1}^{(N,i)} x_{k-Np+iq} = \{(N, i) \bmod 2\} \ x_{k-Np+iq}$$
By Corollary 5, $(N, i) \bmod 2$ is 1 iff $(N \vee r) = N$.

If N is a power of 2, by Lemma 3, there are only two possible values of i for which (N, i) is odd. By Lemma 4, these values are 0 and N. The equation is further simplified to $x_k = x_{k-Np} \oplus x_{k-Np+Nq} \bmod m$.

For generating l-bit random numbers, m is chosen to be 2^l. If $a < 2^l$ and $b < 2^l$, $a \oplus b < 2^l$, allowing us to drop $\bmod \ m$. Thus, from $x_k = x_{k-p} \oplus x_{k-p+q}$, we obtain $x_k = x_{k-Np} \oplus x_{k-Np+Nq}$, if N is a power of 2.

Consider the task of generating N parallel streams using the leapfrog technique. If each processor stores the last p random numbers it generated, the processor computing x_k will have $x_{k-N}, x_{k-2N}, \cdots,$ $x_{k-(p-q)N}, \cdots, x_{k-pN}$ in its local memory. In particular, $x_{k-(p-q)N}$ and x_{k-pN} are found in the local memory using which x_k can be computed.

There is a striking similarity between the sequential and parallel generators. Both generate the next random number by a simple exclusive or operation on two previous numbers found in the local memory. Both retain the last p random numbers as state information. In fact, the code is identical except for the initial generation of seeds for the parallel generator.

If N is not a power of 2, there are at least two 1's in the binary representation of N. By Lemma 3, there are at least four odd numbers in the Nth row of the Pascal's triangle, i.e. $\exists i(i \ne 0 \ , \ i \ne N)$ such that (N, i) is odd. For such i, $x_{k-Np+iq}$ can not be ignored. Hence, we can not parallelize the $\oplus$ generator when the number of processors is not a power of 2.

5.2 Lagged Fibonacci with $+$ or $-$

In this case, $\bigotimes_{j=1}^{(N,i)} x_{k-Np+iq} = (N, i)x_{k-Np+iq}$. Therefore, we require $(N, i) \bmod m$ to be 0 for all i such that $1 \le i < N$. i.e. m should divide (N, i) for all such i. Therefore, m should divide the greatest common divisor (gcd) of all (N, i) $(1 \le i < N)$. Since

m should be a power of 2, we are interested in the highest power of 2 that can divide this gcd.

If N is not a power of 2, we have already seen that $\exists i (i \neq 0 \text{ and } i \neq N)$ such that (N, i) is odd. Therefore, the gcd has to be an odd number, making it impossible to find a suitable m.

If N is a power of 2, consider the binomial coefficient $(N, N/2) = \frac{N}{N/2}(N-1, N/2-1) = 2(N-1, N/2-1)$. Since $N-1$ does not have any 0's in its binary representation, by Lemma 4, every binomial coefficient of $N-1$ is odd. In particular, $(N-1, N/2-1)$ is odd. Therefore, the highest power of 2 that can divide $(N, N/2)$ is 2. Since all binomial coefficients of N (except $(N, 0)$ and (N, N)) are even, it follows that the highest power of 2 that divides the gcd of all binomial coefficients of N except 0 and N is 2. Thus, we are forced to choose $m = 2$. If $m = 2$, we are generating 1-bit random numbers (random string of 0's and 1's), for which the additive generator is clearly equivalent to the 1-bit $\oplus$ generator.

From the above arguments, Lagged Fibonacci generators with $+$ or $-$ can not be efficiently parallelized in a communication-free manner using the leapfrog technique.

5.3 Lagged Fibonacci with $\times$

For the multiplicative lagged Fibonacci generator, $x_k = \prod_{i=0}^{N} (x_{k-Np+iq})^{(N,i)} \mod m$. Therefore, we want to find all N and m such that $(x_{k-Np+iq})^{(N,i)} \mod m = 1$ for arbitrary odd $x_{k-Np+iq}$, $1 \leq x_{k-Np+iq} \leq m-1$, $1 \leq i \leq N-1$.

Lemma 6 If m is a power of 2, a odd, $0 < a < m$, $a^n \mod m = 1 \Leftrightarrow m$ divides n
Proof For the 'if' part, let $a = 2j + 1$ from some j.
$$a^n = (1+2j)^n = 1 + \sum_{i=1}^{n} (n,i)(2j)^i$$
$$= 1 + \sum_{i=1}^{n} (n-1, i-1) \frac{n}{i} 2^i j^i$$
Since i has less than i powers of 2 in its prime factorization, each term in the summation has at least as many powers of 2 as n has in the respective prime factorizations. Since m divides n and m is a power of 2, $a^n \mod m = 1$.

For the 'only if' part, suppose m does not divide n. Let $n = mp + q$ where $q < m$.
$$a^n \mod m = a^{mp+q} \mod m$$
$$= ((a^{mp} \mod m)(a^q \mod m)) \mod m$$
$$= a^q \mod m$$
since $a^{mp} \mod m = 1$ by the 'if' part.

It is enough to show that $a^q \mod m \neq 1$ for some odd a. One can easily verify that when $a = m-1$, $a^q \mod m = m-1$, completing the proof. □

From Lemma 6, $(x_{k-Np+iq})^{(N,i)} \mod m = 1$ for all (N, i) $(1 \leq i \leq N-1)$ iff m divides all such (N, i). Therefore, m should divide the gcd of all such (N, i). This condition is the same as required for parallelizing the additive lagged Fibonacci generator. Therefore, multiplicative lagged Fibonacci generators can not be efficiently parallelized in a communication-free manner using the leapfrog technique.

6 Conclusions

To parallelize a random number generator using the leapfrog technique, the properties implied by the formula of the generator should be used to obtain an $O(1)$ time generation per random number per processor. In this paper, we studied the implications of the formula for lagged Fibonacci generators. There does not appear to be any way to parallelize the lagged Fibonacci generators in a communication-free manner except for the restrictive case when the operation is $\oplus$ and the number of processors is a power of 2.

Acknowledgements
I wish to thank Dr. G.M. Prabhu for his valuable suggestions during the development of this paper.

References

[1] S. Aluru, Random number generators for parallel computers, M.S. Thesis, Iowa State University, 1991.

[2] S. Aluru, G.M. Prabhu, and J. Gustafson, A Random Number Generator for Parallel Computers, *Parallel Comput.* 18 (1992) 839-847.

[3] K.O. Bowman, and M.T. Robinson, Studies of random number generators for parallel processing, *Proc. Second Conference on Hypercube Multiprocessors, M.T. Heath, ed., SIAM, Philadelphia* (1987) 445-453.

[4] I. Deak, Uniform random number generators for parallel computers, *Parallel Comput.* 15 (1990) 155-164.

[5] N.J. Fine. Binomial coefficients modulo a prime, *American Mathematical Monthly* 54 (1947) 589.

[6] G. Fox, et. al. *Solving Problems on Concurrent Processors: Volume I - General Techniques and Regular Problems.* Prentice Hall, Englewood Cliffs, NJ. (1988).

[7] J.W.L. Glaisher, On the residue of a binomial-theorem coefficient with respect to a prime modulus, *Quart. J. Math.* 30 (1899) 150.

[8] S.H. Kimball et. al., Odd binomial coefficients, *Amer. Math. Monthly* 65 (1958) 368.

[9] G. Marsaglia, A current view of random number generators, *XVIth Conference on: Computer Science and Statistics: The Interface, L. Billard, ed.* (1985) 3-10.

[10] G. Marsaglia, The structure of linear congruential sequences, *In Applications of Number Theory to Numerical Analysis, Z.K.Zaremba, Ed., Academic Press, New York* (1972).

[11] G. Marsaglia, Regularities in congruential random number generators, *Numer. Math.,* 16 (1970) 8-10.

[12] G. Marsaglia, Random numbers fall mainly in the planes, *Proc. Nat. Aca. Sci., 61* (1968), 25-28.

[13] G. Marsaglia, and A. Zaman, A new class of random number generators, *Annals of Applied Probability I, 3*(1991) 462-480.

[14] O.E. Percus and M.H. Kalos, Random number generators for MIMD parallel processors, *J. Par. and Dist. Comput., 6* (1987) 477-497.

[15] J.B. Roberts, On binomial coefficient residues, *Canadian J. Math 9* (1957) 363.

[16] S. Wolfram, Geometry of binomial coefficients, *Amer. Math. Monthly 91* (1984) 566-571.

*Anatomy of a Parallel Out-of-Core Dense Linear Solver

Kenneth Klimkowski
Texas Institute for Computational
and Applied Mathematics
The University of Texas at Austin
Austin, Texas 78712
ken@ticam.utexas.edu

Robert A. van de Geijn
Department of Computer Sciences
The University of Texas at Austin
Austin, Texas 78712
rvdg@cs.utexas.edu

Abstract − In this paper, we describe the design and implementation of the Platform Independent Parallel Solver (PIPSolver) package for the out-of-core (OOC) solution of complex dense linear systems. Our approach is unique in that it allows essentially all of RAM to be filled with the current portion of the matrix (slab) to be updated and factored, thereby greatly improving the computation to I/O ratio over previous approaches. Experiences and performance are reported for the Cray T3D system.

INTRODUCTION

Scalable, portable linear algebra libraries for distributed memory MIMD architectures that are currently being investigated put a limit on the size of the problem that can be solved, dictated by the aggregate RAM of the machine. A good example of this is the ScaLAPACK library for dense linear algebra [3]. This paper is concerned with the investigation of techniques for development of a dense linear solver package that is scalable both in machine and problem size.

A number of parallel out-of-core dense linear solver packages exist. These include Intel's Prosolver™ [4, 2] and a code developed at Sandia National Laboratories [9]. The current Prosolver™ package includes two OOC solvers: a block based solver and a slab based solver. The block solver brings essentially square blocks of the matrix incore, which improves the computation to I/O ratio, but limits pivoting to improve stability of the algorithm. Moreover, at any given time this approach juggles up to ten blocks in an effort to overlap computation and I/O. This limits the amount of memory that is dedicated to data on which useful computation is being performed. The current Prosolver slab solver works with column blocks and uses a full implementation of partial pivoting. It suffers from the fact that incore a linear data decomposition is used, which prevents scalability of the algorithm. Moreover, three to four slabs are being juggled at any given time, again in an effort to overlap computation and I/O. The Sandia approach is very similar to the Intel slab solver, except that it uses a two dimensional data decomposition for the incore computation, leading to a scalable incore algorithm. It also juggles four slabs simultaneously, with the same drawbacks.

Our code is based on a highly efficient incore solver, which was written for electromagnetics applications [5] and has also been used for acoustics applications [7]. We modified this code in a way that very naturally led to an OOC solver. This approach allows maximal code reuse, extending the robustness and manageability of the incore version to the OOC version. The incore computation uses a two dimensional data decomposition, yielding a scalable incore component, and does full partial pivoting, yielding the expected stability. While we also use a slab approach, we overlap computation and I/O at a finer level, which allows us to essential fill the entire incore memory with the current slab being updated and factored (the active slab). This gives us an edge by improving the computation to I/O ratio compared to previous approaches.

A PARALLEL INCORE DENSE LINEAR SOLVER

Our dense linear solver depends on the ability

*This work was performed in part at the Jet Propulsion Laboratory, California Institute of Technology under a contract with the National Aeronautics and Space Administration.

to perform an LU factorization of a given $n \times n$ matrix:

$$PA = LU$$

where L and U are unit lower and nonunit uppertriangular, respectively. Permutation matrix P represents the accumulation[1] of all pivots required for stability. Once this factorization is completed, the dense linear system $Ax = b$ can be solved by solving $Ly = Pb$ and $Ux = y$, the forwards and backward substitution phases.

Ignoring pivoting for now, a high performance version, known as a *blocked right-looking* variant of the algorithm can be generated by blocking the matrices:

$$\left(\begin{array}{c|c} A_{00} & A_{01} \\ \hline A_{10} & A_{11} \end{array} \right) = \left(\begin{array}{c|c} L_{00} & 0 \\ \hline L_{10} & L_{11} \end{array} \right) \left(\begin{array}{c|c} U_{00} & U_{01} \\ \hline 0 & U_{11} \end{array} \right)$$

where A_{00}, L_{00}, and U_{00} are $k \times k$ matrices. Given this partitioning, the algorithm becomes: Factor the first *panel* of k columns, overwriting this panel with the result; solve the triangular system with multiple right-hand-sides $L_{00}U_{01} = A_{01}$, overwriting A_{01}; update $A_{11} \leftarrow A_{11} - L_{10}U_{01}$; recursively compute $A_{11} = L_{11}U_{11}$.

The algorithm is formulated so that the bulk of the computation is now in matrix-matrix operations like matrix multiplication, which can inherently achieve higher performance by overcoming the memory bandwidth bottleneck of many current generation microprocessors. Pivoting can be easily added.

To parallelize the above algorithm, one must first decide upon a data distribution. For scalability reasons, a two dimensional data decomposition is required, as is described in [8, 6]: Given p processors, we view these as *logically* forming a $r \times c$ mesh, where $p = rc$. In this mesh, $\mathbf{P}_{ij}$ denotes the (i, j)th processor. The matrix is partitioned like:

$$A = \left(\begin{array}{c|c|c|c} \tilde{A}_{00} & \tilde{A}_{01} & \cdots & \tilde{A}_{0(N-1)} \\ \hline \tilde{A}_{10} & \tilde{A}_{11} & \cdots & \tilde{A}_{1(N-1)} \\ \hline \tilde{A}_{20} & \tilde{A}_{21} & \cdots & \tilde{A}_{2(N-1)} \\ \hline \vdots & \vdots & & \vdots \\ \hline \tilde{A}_{(N-1)0} & \tilde{A}_{(N-1)1} & \cdots & \tilde{A}_{(N-1)(N-1)} \end{array} \right) \quad (1)$$

Here all blocks are of dimension $k \times k$, except those in the last row and column, which may be smaller. These blocks are now wrapped onto the

logical mesh so that $\tilde{A}_{ij}$ is assigned to processor $\mathbf{P}_{(i \bmod r),(j \bmod c)}$.

Given this partitioning, the parallel algorithm becomes: Factor the first *panel* of k columns on the first column of nodes; broadcast the result within rows of nodes; solve the triangular system with multiple right-hand-sides $L_{00}U_{01} = A_{01}$; broadcast the result within columns of nodes; in parallel update $A_{11} \leftarrow A_{11} - L_{10}U_{01}$; recursively compute $A_{11} = L_{11}U_{11}$.

While the wrap mapping described above is necessary to maintain reasonable load balance throughout the computation, it is unreasonable to require the user to build the matrix, or the OOC solver to read it, as prescribed by that mapping. However, the matrix can be blocked into the alternative partitioning

$$A = \left(\begin{array}{c|c|c|c} B_{00} & B_{01} & \cdots & B_{0(N-1)} \\ \hline B_{10} & B_{11} & \cdots & B_{1(N-1)} \\ \hline \vdots & \vdots & & \vdots \\ \hline B_{(r-1)0} & B_{(r-1)1} & \cdots & B_{(r-1)(N-1)} \end{array} \right) \quad (2)$$

where B_{ij} is a matrix of dimensions *approximately* n/r by k. The matrix is now distributed among the processors (e.g. by reading it from a disk) so that all blocks B_{ij} are assigned to processor $\mathbf{P}_{i(j \bmod c)}$. I.e., the distribution uses blocking in the row direction, and wrapping in the column direction. It would appear that if the matrix is assigned to processors in this fashion, a redistribution of the matrix is necessary to be able to use the parallel incore solver routine that expects wrapping in both directions. However, notice that if each processor $\mathbf{P}_{ij}$ calls the incore solver without redistribution, a factorization and corresponding solve of a permuted matrix will be performed, which will not change the final answer if minimal care is taken. We can view this distribution as being physically blocked and logically wrapped in the row dimension, and physically and logically wrapped in the column dimension.

FROM IN-CORE TO OUT-OF-CORE

In this section, we describe how an incore factorization algorithm can be converted to an OOC version, motivating our implementation.

An alternative to the above described blocked right-looking LU factorization scheme is the

---

[1] The permutations are computed and applied as the algorithm progresses, but can logically be viewed as being applied upfront.

blocked left-looking variant. For this, one must perform an alternative partitioning of the matrix:

$$\left(\begin{array}{c|c|c} A_{00} & A_{01} & A_{02} \\ \hline A_{10} & A_{11} & A_{12} \\ \hline A_{20} & A_{21} & A_{22} \end{array} \right) =$$

$$\left(\begin{array}{c|c|c} L_{00} & & \\ \hline L_{10} & L_{11} & \\ \hline L_{20} & L_{21} & L_{22} \end{array} \right) \left(\begin{array}{c|c|c} U_{00} & U_{01} & U_{02} \\ \hline & U_{11} & U_{12} \\ \hline & & U_{22} \end{array} \right)$$

where A_{00}, L_{00}, and U_{00} are $n_1 \times n_1$ matrices, and A_{11}, L_{11}, and U_{11} are $k \times k$ matrices. The assumption is that the first n_1 columns have already been factored, i.e. L_{00}, L_{10}, L_{20}, and U_{00} are available, and have overwritten the corresponding parts of matrix A. From this, we derive the equalities

$$A_{01} = L_{00}U_{01} \quad (3)$$

$$\left(\begin{array}{c} A_{11} \\ A_{21} \end{array} \right) - \left(\begin{array}{c} L_{10} \\ L_{20} \end{array} \right) U_{01} = \left(\begin{array}{c} L_{11} \\ L_{21} \end{array} \right) U_{11} \quad (4)$$

The algorithm then becomes: If $n_1 > 0$ compute U_{01} from (3). and update according to the LHS of (4); factor the result to obtain the RHS of (4), overwriting the corresponding parts of A; repartition the matrix so that $n_1 \leftarrow n_1 + k$ and repeat. Initially, $n_1 = 0$.

The bulk of the computation is still in matrix-matrix operations. The left-looking variant has two advantages over the right-looking variant for OOC computation: it requires less I/O: each element of a matrix is only written to disk once, and in case of a system failure, it is easier to recover while salvaging computation already performed.

Assuming that the factorization of

$$\left(\begin{array}{c} A_{00} \\ \hline A_{10} \\ \hline A_{20} \end{array} \right) \quad (5)$$

was obtained using a right-looking variant, the *updating* of

$$\left(\begin{array}{c} A_{01} \\ \hline A_{11} \\ \hline A_{21} \end{array} \right) \quad (6)$$

w.r.t. that factorization is a matter of *recreating the history of the right-looking algorithm*. I.e., the same pivots and updates that were applied to factor (5) must be applied to (6). We will call the data in (6) a *slab*, which will typically be the data to be kept in-core during the current stage of the computation.

Once one has a parallel in-core blocked right-looking LU factorization, the implementation of an OOC version now becomes: Read in the current slab; create a virtual factorization by calling the in-core factorization, except that during the factorization of the part that is to the left of the current slab, the factored portions are read from disk, before being applied to the current slab; once the factorization reaches the current slab, the algorithm proceeds exactly like the in-core factorization; write the current slab; continue with next slab. We call this approach a *virtual* factorization.

DETAILS

Due to space limitations, we limit ourselves here to an enumeration of the issues.

The constraints in designing the I/O interface for the Cray T3D include the fact that each I/O request generates a system call to the Cray Y-MP with a comparatively large startup overhead, in addition to the cost per data item transferred. Reading of large blocks is hence essential. In addition, open files use resources on the Cray Y-MP agent, so they should be avoided to reduce communication traffic between the T3D and the Y-MP. Other parallel systems can be expected to have similar constraints.

We earlier made the observation that although wrapping requires small blocks, the matrix can be generated or read from disk in large blocks. This naturally accommodates the requirement that large blocks be read. In our implementation, we store the matrix so that each *row of nodes* shares one file. This limits the number of open files required.

Starting with N panels, the number of panels that comprise a slab, K, is solely determined by available memory. Naturally, K is an integer multiple of the number of columns of processors. In general, we will have $\lfloor N/K \rfloor$ full slabs, and one, possibly empty, slab of width $n - \lfloor N/K \rfloor Kk$ columns. A simple approach to managing these slabs will read a slab, perform a virtual factorization as described above, and write the slab. Notice that this leaves the reading and writing of slabs totally exposed (not overlapped with computation). Since this constitutes essentially the only exposed I/O, and is of lower order complexity, good efficiency can be achieved. It should be noted that even this I/O could be almost completely overlapped, requiring only slight modification of the current code.

After performing a factorization of a matrix, applications usually solve for one or several right hand sides (RHS) by forward and backward substitution. Since the forward substitution is similiar to a virtual factorization, we can treat the RHS as another slab logically appended to the matrix. For the backward substitution, we apply panels of the factored matrix to the RHS in the reverse order. Thus, forward and backward substitutions are created by modifying the virtual factorization.

The triangular solve is an $O(n^2)$ computation on $O(n^2)$ data items. Thus the forward and backward solves are inherently I/O bound. Where in the factorization we used $p = r \times c$ processors, we can reduce the size of the processor grid during these triangular solutions as long as we maintain the row dimension, r, of the blocked matrix. Explicit column wrapping and the row blocking facilitates this choice. As a tradeoff, larger numbers of processors can improve I/O by reading more of the matrix from disk.

Checkpointing is the process of monitoring the progress of a computation such that the computation could be restarted after system crash. In our implementation, the factored matrix is sequentially read, factored, and written, and hence our checkpointing requires only knowledge of how far we have progressed. For simplicity, we currently checkpoint after writing each slab of the matrix to disk. However, our granularity of I/O operations is a panel, so we could checkpoint after each panel when writing out a slab.

Checkpointing provides all the necessary information for restarting. Restarting is the ability to continue a computation after crashes such as system failure, exceeding time limits, or exceeding disk quotas. Our approach adds two features to checkpointing: restarting and the ability to continue with a different number of processors.

The user interface for our package is simple. An application builds the matrix piecewise and stores it on disk. The application builds a single logical file, which has properties much like a matrix: row and column dimensions and a leading dimension. To fill parts of this logical file, we provide an I/O interface, called the Platform Independent Parallel I/O (PIPIO) interface, which hides all details of how the logical file is physically mapped to disk(s), as well as system dependent details. The interface automatically splits the logical file into the required r physical files. The RHS is handled similarly.

An advantage of this approach is that during the building of the matrix and RHS, the processors need not be configured as the logical mesh required for the parallel OOC solver. For our primary application, an electromagnetics application developed by the Jet Propulsion Laboratory, it is more convenient and efficient to generate the matrix by columns.

RESULTS

This section describes preliminary results obtained using the PIPSolver on the Jet Propulsion Laboratory's Cray T3D system.

The performance of the virtual LU factorization is very respectable because much of the I/O during that portion of the code is overlapped with computation. The observed I/O rates can be greatly improved, which would improve overall performance to very acceptable levels. The I/O rates are highly influenced by the load on the Y-MP frontend. This can greatly affect the observed I/O rate from one run to the next.

The T3D provided reliable high performance. To achieve this performance, it is essential to use highly optimized compute kernels like the Basic Linear Algebra Subroutines. Moreover, although we achieved reasonable I/O rates, the very high I/O rates (200 MBPS) promised by the Cray T3D continue to elude us.

CONCLUSION

We have described a simple approach to converting a highly efficient and scalable incore complex dense linear solver to an OOC implementation. In doing so, we designed an OOC version which has an arguably better computation to I/O ratio than other implementations with similar stability properties. The simplicity of the approach inherently yields a more maintainable and flexible code. The simple application interface is another attractive feature.

ACKNOWLEDGEMENTS

We are indebted to Tom Cwik of the Jet Propulsion Laboratory and Dan Katz of Cray Research for their expertise with the Cray T3D. In addition, we would like to acknowledge the Jet Propulsion Laboratory for providing generous access to their 256 node Cray T3D during the development and evaluation of the codes.

Table 1: Performance data on Cray T3D. For each run, we report timing data for the factorization only, giving the exposed reading and writing of slabs, as well as the virtual factorization (which includes some I/O). We grouped the timings so that the *number of slabs* are approximately constant. This gives a better insight into the scalability of the method. MBPS are aggregate I/O rates, reported in megabytes per second. MFLOPS reported are *per processor*.

n	p	$r \times c$	# of Slabs	Read Time	Read MBPS	Write Time	Write MBPS	Virtual LU Time	Virtual LU MFLOPS	All Time	All MFLOPS
5000	8	2×4	0.96	7.4	54.4	18.1	22.1	4436.5	95.5	62.0	90.2
7000	16	4×4	0.94	16.7	47.0	37.9	20.7	598.0	95.6	652.6	87.6
10000	32	4×8	0.96	46.9	34.1	198.0	8.1	872.3	95.5	1117.2	74.6
14000	64	8×8	0.94	100.3	31.3	237.2	13.2	1210.3	94.5	1547.8	73.9
6000	8	2×4	1.39	10.7	54.0	21.2	27.2	762.0	94.5	793.9	90.7
8400	16	4×4	1.35	21.1	53.6	69.9	16.1	1067.5	92.5	1158.5	85.3
12000	32	4×8	1.39	68.7	33.5	139.0	16.6	1614.8	89.2	1822.5	79.0
16800	64	8×8	1.35	108.5	41.6	257.8	17.5	2311.4	85.5	2677.7	73.8
8000	8	2×4	2.50	43.0	23.8	55.7	18.4	1820.8	93.7	1919.5	88.9
11200	16	4×4	2.41	92.7	21.6	173.0	11.6	2679.0	87.4	2944.7	79.5
16000	32	4×8	2.50	141.7	28.9	267.9	15.3	3991.6	85.5	4401.2	77.6
22400	64	8×8	2.41	179.5	44.7	484.0	16.6	5855.8	80.0	6519.3	71.8

References

[1] E. Anderson, Z. Bai, J. Demmel, J. Dongarra, J. DuCroz, A. Greenbaum, S. Hammarling, A. McKenney, S. Ostrouchov, and D. Sorensen, *LAPACK Users' Guide,* SIAM, Philadelphia, 1992.

[2] E. Castro-Leon and M. L. Barton, "The Prosolver(TM) Libraries : Parallel Solvers for Intel High Performance Scalable Computers", Internal Document, Intel Scalable Systems Division.

[3] J. Choi, J. J. Dongarra, R. Pozo, and D. W. Walker, "Scalapack: A Scalable Linear Algebra Library for Distributed Memory Concurrent Computers, *Proceedings of the Fourth Symposium on the Frontiers of Massively Parallel Computation.* IEEE Comput. Soc. Press, 1992, pp. 120-127.

[4] T. Cwik, J. Patterson, and D. Scott, "Electomagnetic Scattering Calculations on the Intel Touchstone Delta," *Proceedings of Supercomputing '92,* IEEE Comput. Soc. Press, 1992, pp. 538-542.

[5] Tom Cwik, Robert van de Geijn, and Jean Patterson, "Application of Massively Parallel Computation to Integral Equation Models of Electromagnetic Scattering," *Journal of the Optical Society of America A,* Vol. 11, No. 4, April 1994, pp. 1538-1545

[6] Jack. J. Dongarra, Robert A. van de Geijn, and David W. Walker, "Scalability Issues Affecting the Design of a Dense Linear Algebra Library," *Journal of Parallel and Distributed Computing,* Vol. 22, No. 3, Sept. 1994, pp. 523-537.

[7] P. Geng, J. T. Oden, and R. A. van de Geijn, "Massively Parallel Computation for Acoustics Scattering Problems using Boundary Element Methods," *Journal of Sound and Vibration,* (to appear).

[8] B. A. Hendrickson and D. E. Womble, "The Torus-wrap Mapping for Dense Matrix Calculations on Massively Parallel Computers," *SIAM J. Sci. Comput.,* vol. 15 (1994), no. 5, pp. 1201-1226.

[9] D. E. Womble, D. S. Greenberg, R. E. Riesen, and S. R. Wheat, "Out of Core, Out of Mind: Practical Parallel I/O," *Proceedings of the Scalable Libraries Conference,* Oct. 6-8, 1993, Mississippi State University, pp.10-16.

BLOCK-ROW SPARSE MATRIX-VECTOR MULTIPLICATION ON SIMD MACHINES [a]

Nirav Kapadia and Jose A. B. Fortes
School of Electrical Engineering,
Purdue University,
West Lafayette, IN 47907-1285, USA
kapadia@ecn.purdue.edu fortes@ecn.purdue.edu

Abstract -- *The irregular nature of the data structures required to efficiently store arbitrary sparse matrices and the architectural constraints of a SIMD computer make it difficult to design an algorithm that can efficiently multiply an arbitrary sparse matrix by a vector. A new "block-row" algorithm is proposed. It allows the "regularity" of a data structure with a row-major mapping to be varied by changing a parameter (the "blocksize"); a heuristic to find a very good approximation of the optimal blocksize is also described. The block-row algorithm has been implemented on a 16,384 processor MasPar MP-1, and, for the matrices studied, the algorithm was found to be faster than any of the other algorithms considered.*

1. INTRODUCTION

This paper presents a new *block-row algorithm* for matrix-vector multiplication which was primarily developed for the large unstructured sparse matrices arising from a scattering-matrix approach to device simulation [1],[3]. The algorithm has been implemented on a 16,384 processor MasPar MP-1, and was also tested on four large matrices selected from the Harwell-Boeing Sparse Matrix Collection [4]. It was found to be faster than the randomized packing algorithms [10], the "segmented-scan" algorithm [5], and the "snake-like" method [8]. In addition, the block-row algorithm is memory-efficient; for the largest problem described here, involving 93,602 unknowns and 1,427,614 non-zero elements, the block-row algorithm used approximately 36 MBytes of memory (vs. 237MBytes for the randomized packing algorithm). The only assumption made about the matrix is that most of its columns are "sparse", and consequently, the advantages of this algorithm over competing approaches can be expected to hold for many problems for which this characteristic holds.

A brief listing of references is presented in Section 2. Section 3 covers some background information, and the block-row algorithm is presented in Section 4. Finally, experimental results are discussed in Section 5, and Section 6 concludes the paper.

2. PRIOR WORK

Most sparse matrix algorithms that are designed for SPMD or MIMD models do not work efficiently on SIMD architectures without extensive modifications; consequently, focus is restricted to SIMD algorithms. A brief overview of parallel sparse matrix computations along with some references can be found in [7]; additional techniques can be found in [5].

Several methods for sparse matrix-vector multiplication involve the partitioning of the matrix in some way. M. Morjaria and G. Makinson [9] proposed a block partitioning method for the storage of large sparse matrices on a two-dimensional mesh processor array; J. Anderson et. al. proposed a modification using a less compact data structure and a heuristic scheduling procedure that enabled them to exploit more parallelism and reduce the amount of interprocessor communication [2]. In [11], a sparse matrix is decomposed into submatrices with an equal number of non-zero elements (multiple recursive decomposition); however, the different sizes (in general) of the individual submatrices make this method unsuitable for SIMD computers. A randomized packing algorithm proposed in [10] divides the matrix into submatrices of equal size (except, possibly, the ones on the "boundaries" of the matrix). This algorithm is conceptually similar to (in the sense that it divides the matrix into submatrices) the block-row scatter method [11], and is designed for a SIMD machine. It is evaluated in Section 5, along with the "segmented-scan" method [5], and the "snake-like" method [8],[11]; these three algorithms were found to be the most competitive for the matrices associated with the intended application.

3. SPARSE MATRIX-VECTOR MULTIPLICATION

Consider the problem of matrix-vector multiplication with the notation $b = Ax$, where A is a sparse matrix of size N with 'N_{elts}' non-zero elements (a_{ij}). Then, the process of parallel sparse matrix-vector multiplication can be divided into several basic steps:

In parallel, for each a_{ij} in each processor:

  1. Fetch required vector element (x_j).
  2. Perform local multiply ($c_{ij} = a_{ij} \times x_j$).
  3. Add partial products ($b_i = \sum_j c_{ij}$).
  4. Put result-vector element in appropriate processor.

Observe that the actual order in which the operations are performed is not completely specified by the loop above - for example, all processors could execute step 1 for every a_{ij}, then execute step 2 for every a_{ij}, etc., or they could sequentially perform steps 1 through 4 for

(a) This research was partially funded by the National Science Foundation under grants MIP-9500673 and CDA-9015696.

each a_{ij}. Obviously, other alternatives are also possible; while designing an algorithm for parallel sparse matrix-vector multiplication, it would be beneficial to perform the above steps in some "optimal" order. If there are N_{proc} processors in the (SIMD) computer, then N_{elts} elements can be mapped into L_{elts} *layers*, where $L_{elts} = \lceil N_{elts}/N_{proc} \rceil$ (a layer can be viewed as a "row" in a two-dimensional array, where each "row" consists of elements in the same position in the memory of each processor). Then, the time required for one matrix-vector multiplication is given by:

$$t_{parallel} = t_{fetch} + t_{multiply} + t_{add} + t_{arrange}$$

$$= \sum_{k=1}^{L_{elts}} t_{fetch,\,k} + \sum_{k=1}^{L_{elts}} t_{multiply,\,k} + \sum_{k=1}^{L_{elts}} t_{add,\,k} + \sum_{k=1}^{L_{elts}} t_{arrange,\,k} \qquad (1)$$

Thus, the procedure of parallel sparse matrix-vector multiplication can be divided into four (possibly interleaved) *phases*: the *fetch phase*, the *multiply phase*, the *reduce phase*, and the *result phase*. Time t_{fetch} represents the time required to fetch the vector elements, and $t_{multiply}$ includes the *intra-processor* reduction time, in addition to the multiplication time. Time t_{add} represents the *inter-processor* reduction time, and $t_{arrange}$ represents the time required to execute the result phase. It can be inferred that the fetch phase and the reduce phase account for a large fraction of the total time [6].

The fetch phase can be carried out by using regular communication primitives if the elements of the matrix are distributed in a column-major format, while the reduce phase can be carried out by using (only) regular communication if the elements of the matrix are distributed in a row-major format (on most commercially available SIMD computers, regular communication is faster than communication among arbitrary processors). In addition, it does not appear to be possible to design a data-structure that simultaneously allows a row-major and a column-major mapping, and at the same time distributes the non-zero elements of the matrix evenly among the processors for arbitrary sparse matrices [10]. The matrices associated with the intended application have fewer (or about the same) non-zero elements per column as compared to the number of non-zero elements per row; consequently, a row-major data distribution is used for the block-row algorithm (because of this, the columns of the matrix must be "sparse"; else t_{fetch} will dominate $t_{parallel}$).

4. THE BLOCK-ROW ALGORITHM

4.1 Introduction

In this algorithm, elements from each row of a sparse matrix are grouped into *blocks*, with *blocksize* elements per block; zero elements are stored only if the number of non-zero elements in a given row is not a multiple of the blocksize. Then, each block (rather than each element) is processed as a basic unit, which facilitates the design of a data structure whose "regularity" can be varied by changing a parameter (the blocksize). The reduce and result phases are executed at the most once for every block because of the regularity of the associated data structure, which results in a faster algorithm.

4.2 The Block-Row Algorithm: An Example

Consider the "sparse" matrix and the (dense) vector shown in Figure 1, and a processor array with four processors. Assume the intermediate-stage representation of the matrix shown in Figure 2; this representation is obtained by "compressing" the non-zero elements in each row of the matrix - that is, by moving them to the left hand side of the zero elements.

$$A = \begin{bmatrix} a_{00} & 0 & a_{02} & 0 & 0 & a_{05} \\ 0 & a_{11} & 0 & a_{13} & a_{14} & a_{15} \\ 0 & a_{21} & a_{22} & 0 & a_{24} & a_{25} \\ a_{30} & 0 & 0 & a_{33} & 0 & 0 \\ 0 & 0 & 0 & 0 & a_{44} & 0 \\ a_{50} & 0 & 0 & a_{53} & 0 & a_{55} \end{bmatrix} \qquad x = \begin{bmatrix} x_0 \\ x_1 \\ x_2 \\ x_3 \\ x_4 \\ x_5 \end{bmatrix}$$

Figure 1: An example "sparse" matrix with $N = 6$ and 17 non-zero elements, and the corresponding vector.

Row 0	a_{00}	a_{02}	a_{05}	0	0	0
Row 1	a_{11}	a_{13}	a_{14}	a_{15}	0	0
Row 2	a_{21}	a_{22}	a_{24}	a_{25}	0	0
Row 3	a_{30}	a_{33}	0	0	0	0
Row 4	a_{44}	0	0	0	0	0
Row 5	a_{50}	a_{53}	a_{55}	0	0	0

Figure 2: The intermediate-stage representation of the matrix shown in Figure 1.

From this intermediate-stage representation, data structures with different amounts of "regularity" can be obtained by changing the value of a parameter (the blocksize). All blocks that contain at least one non-zero element are mapped on to the processor array, whereas blocks that have only zero elements are discarded. Let S_{blk} be the value of the blocksize. Then, N_{tot}/S_{blk} blocks are mapped onto the processor array, where N_{tot} is the total number of elements (including zero elements) in the blocks.

Consider a blocksize of one; each row is divided into blocks, with one element per block. This data structure is simply a row-major mapping of the *non-zero* elements of the matrix (Figure 2 walked row-wise). This data structure is then mapped into the processor array as shown in Figure 3. Each processor reads $\lceil N_{tot}/N_{proc}/S_{blk} \rceil$ (= 4, in this case) blocks from the data structure in a row-major format; thus, there are four *complete layers* of data in the memory of the processors. The remaining blocks are mapped into *incomplete layers* of data, with each processor (starting from the first one) being assigned one block. The vector is also distributed among the processors, as shown in the figure.

Usually, to multiply a sparse matrix by a vector, each of the four phases must be executed for each layer of data (in the processor memory). In the block-row algorithm, however, the reduce phase (consisting of inter-processor reduction operations) and the result phase are

PE 0	PE 1	PE 2	PE 3
a_{00}	a_{13}	a_{22}	$\underline{a_{33}}$
a_{02}	a_{14}	a_{24}	$\underline{a_{44}}$
$\underline{a_{05}}$	$\underline{a_{15}}$	$\underline{a_{25}}$	a_{50}
a_{11}	a_{21}	a_{30}	$\underline{a_{53}}$
$\underline{a_{55}}$	-	-	-
x_0	x_1	x_2	x_3
x_4	x_5	-	-

Figure 3: Distribution of the elements of the matrix shown in Figure 1 (processors: 4, blocksize: 1). An underscore indicates the execution of the reduce/result phases.

executed only once for each block (the multiply phase now consists of local multiply-add operations). In general, for a given processor, the reduce phase and the result phase do not need to be executed for layer 'i' (of data) if the element in layer '$i+1$' belongs to the same row as the element in layer 'i'. In Figure 3, an underscore below a particular element indicates that the reduce phase and the result phase need to be executed for that layer (by the corresponding processor).

Row 0	a_{00}	a_{02}	a_{05}	0
Row 1	a_{11}	a_{13}	a_{14}	a_{15}
Row 2	a_{21}	a_{22}	a_{24}	a_{25}
Row 3	a_{30}	a_{33}		
Row 4	a_{44}	0		
Row 5	a_{50}	a_{53}	a_{55}	0

Figure 4: The data structure for the matrix shown in Figure 1 with a blocksize of two; single vertical lines indicate block boundaries.

Now consider a blocksize value of two - the resulting data structure is shown in Figure 4. Notice that this data structure is more "regular" than the data structure for a blocksize of one (the number of elements in each row is forced to be a multiple of the blocksize); this regularity, however, is obtained at the cost of having to store zero elements. For the reduce phase, the zero elements are assumed to belong to a specific block (specified by the row index), whereas for the fetch phase, the zero entries are ignored (indicated by a ''*'' for the column index in Figure 5). The entries in the data structure are mapped on to the processor array as shown in Figure 5. Each processor first reads two (in this case) blocks corresponding to the complete layers of data; the remaining (two) blocks corresponding to the incomplete layers of data are distributed evenly among the processors in the processor array (by "flattening" them; Figure 5).

PE 0	PE 1	PE 2	PE 3
a_{00}	a_{11}	a_{21}	a_{30}
a_{02}	a_{13}	a_{22}	a_{33}
a_{05}	a_{14}	a_{24}	a_{44}
0_{0*}	a_{15}	a_{25}	0_{4*}
$\underline{a_{50}}$	$\underline{a_{53}}$	$\underline{a_{55}}$	$\underline{0_{5*}}$
x_0	x_1	x_2	x_3
x_4	x_5	-	-

Figure 5: Distribution of the elements of the matrix shown in Figure 1 (processors: 4, blocksize: 2). An underscore indicates the execution of the reduce/result phases, and no vector-element needs to be fetched for zero elements (0_{x*}).

PE 0	PE 1	PE 2	PE 3
a_{00}	a_{11}	a_{21}	a_{30}
a_{02}	a_{13}	a_{22}	a_{33}
a_{05}	a_{14}	a_{24}	0_{3*}
$\underline{0_{0*}}$	$\underline{a_{15}}$	$\underline{a_{25}}$	$\underline{0_{3*}}$
a_{44}	0_{4*}	a_{50}	a_{55}
$\underline{0_{4*}}$	$\underline{0_{4*}}$	$\underline{a_{53}}$	$\underline{0_{5*}}$
x_0	x_1	x_2	x_3
x_4	x_5	-	-

Figure 6: Distribution of the elements of the matrix shown in Figure 1 (processors: 4, blocksize: 4). An underscore indicates the execution of the reduce/result phases, and no vector-element needs to be fetched for zero elements (0_{x*}).

Finally, consider a blocksize of four: the corresponding data structure can be obtained by truncating the last two columns in Figure 2 (the number of elements in row is a multiple of four), and the data distribution in the processors is shown in Figure 6. As before, the two blocks in the incomplete layers of data are flattened to maximize the utilization of the processors. With this blocksize, the reduce/result phases are executed only twice (compared to five times for a blocksize of one) for each matrix-vector multiplication. Observe that a further increase in the blocksize will add only zero elements; a *meaningful increase* in the blocksize is *limited* by the maximum number of non-zero elements in any one row of the given matrix.

4.3 Description of the Block-Row Algorithm

Assume the notation described in Section 3 and Section 4.2, and the intermediate-stage matrix representation of Figure 2 (Section 4.2), with a blocksize of S_{blk}. Then, the blocks are distributed in the processor array as follows: each processor, starting with the first one, ini-

```
┌─────────────────────────────────────────────┐
│        Matrix-vector Multiplication y = Ax    │
├─────────────────────────────────────────────┤
│ initialize                                    │
│   in parallel in all processors:              │
│     partial_product = 0.                      │
│     for k = 0,..., L_vect − 1                  │
│       result_vect_elt[k] = 0.                 │
│     end for                                   │
│ for block = 0,..., L_tot/S_blk − 1             │
│   for layer = 0,..., S_blk − 1                 │
│     in parallel in all processors:            │
│       if a_ij in the current layer ≠ 0        │
│         /* Fetch */                           │
│         temp = processor[j % N_proc].x[j / N_proc], │
│         /* Multiply-add */                    │
│         partial_product = partial_product + a_ij × temp. │
│       end if                                  │
│   end for                                     │
│   in parallel in all processors:              │
│     /* "inter-processor" reduction */         │
│     reduced_result = 0.                       │
│     if next block does not belong to the same row as │
│          this block                           │
│       if adjacent processor(s) is enabled and has blocks │
│             from this row                     │
│         in parallel in the last processor of each │
│             reduction set:                    │
│           reduced_result = sum(partial_products │
│                 in that reduction set).       │
│       else                                    │
│         reduced_result = partial product.     │
│       end if                                  │
│       partial_product = 0.                    │
│     end if                                    │
│   in parallel in the last processor in each reduction set: │
│     /* result phase */                        │
│     processor[i % N_proc].result_vect_elt[i / N_proc] │
│       = processor[i % N_proc].result_vect_elt[i / N_proc] │
│           + reduced_result.                   │
│ end for                                       │
└─────────────────────────────────────────────┘
```

Figure 7: C-like pseudocode for the block-row algorithm; L_{tot} (L_{vect}) is the number of layers that the matrix (vector) elements map into, and N_{proc} represents the number of processors. '%' denotes the *mod* operator, and *processor[a].b* represents the variable *b* in processor *a*. A *reduction set* is defined as the set of processors with (current) partial products that belong to the same row.

tially reads $\lceil N_{tot}/N_{proc}/S_{blk} \rceil$ blocks corresponding to the complete layers of data. Then, the remaining blocks (e.g. *k*) are distributed among the first 'k' processors, with one block per processor. The number of layers of data that the elements are mapped into is given by

$$L_{tot} = \left\lceil \frac{N_{tot}}{N_{proc} \times S_{blk}} \right\rceil \times S_{blk} . \qquad (2)$$

Note that, for this analysis, the blocks in the incomplete layers of data are not flattened; this results in a clearer explanation, and "flattening" can be added to the algorithm with relatively minor modifications.

For the given setup, the pseudocode for sparse matrix-vector multiplication using the block-row algorithm is given in Figure 7. With reference to the figure, observe that the fetch phase and the multiplication phase are executed for all layers of data, whereas the reduce phase and the result phase are executed *only once* for each block. In practice, if a processor has more than one

block from a given row of the matrix, the reduce/result phases are only executed after the last block (in that processor) belonging to that row has been processed.

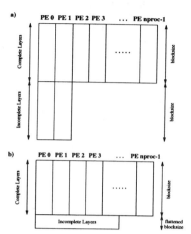

Figure 8: Distribution of the blocks in the memory before (a) and after (b) "flattening" the incomplete layers.

With reference to Figure 8a, "incomplete layers" can result in "sequential" execution - 'flattening' the blocks by reducing the blocksize can result in better processor utilization (Figure 8b). Flattening can, however, increase t_{add} and $t_{arrange}$. The performance improvement obtained by 'flattening' depends on both, the problem-size, and the blocksize.

4.4 Timing Analysis and Blocksize Selection

In theory, if the dependence of t_{fetch} on the data distribution is ignored, and the number of zeros stored is reasonably small, the fetch phase should be unaffected by the blocksize because no fetches are required for zero entries. In practice, however, because of the overhead of testing for zero elements, and the fact that, depending on the distribution of the zero elements, a smaller number of enabled processors (because of processors with zero elements being disabled) does not necessarily result in fewer communication conflicts, t_{fetch} is proportional to the number of layers required to store the elements of the matrix (including the zero elements that are stored).

The time required for the multiplication phase is directly proportional to the number of layers that are processed (L_{tot}). Time $t_{add, k}$ is proportional to the largest number of processors across which blocks from any one row are distributed; and $t_{arrange, k}$ is dependent on the distribution of the processors generating the results, and on the rows to which those results belong. However, because the number of times that the reduce and the result phases are executed depends on the (maximum) number of blocks per row, t_{add} and $t_{arrange}$ are lower for higher blocksizes (in general). In addition, increasing the block-size reduces the total number of communications required in the result phase (from a maximum of $2 \times N$ to a minimum of N).

On the other hand, a larger blocksize can result in more overhead (in this paper, *overhead* is defined as the number of zero elements of the given matrix that are explicitly stored in memory, which can be indirectly measured by L_{tot}; this value is a function of the blocksize), which, in turn, increases $t_{multiply}$, and can increase the value of t_{fetch} because of under-utilization of the communication bandwidth of the interprocessor communication network in the processor array. Thus, there is an optimal blocksize for which the sum of t_{fetch}, $t_{multiply}$, t_{add}, and $t_{arrange}$ is minimized.

As noted before, the blocksize is bounded on both sides - with a blocksize of one, the block-row algorithm is essentially the same as the segmented-scan algorithm considered in Section 5.3, though the resulting data distribution will be different, while a blocksize equal to the maximum number of non-zero elements in any row results in a data structure that is similar to the Ellpack-Itpack storage format described in [7]. A very good *approximation* to the optimal blocksize can found with ease. If S_{opt} is the optimal blocksize, then using a blocksize value that is a *submultiple* of S_{opt} results in a greater number of tests in the reduce phase (the actual number of reductions taking place do not increase; see Figure 7). Thus, if the effects of a possibly different distribution of data among the processors are ignored, a blocksize of S_{opt}/i, where i is small, will result in approximately ideal performance (relative to S_{opt}). Consequently, a very good "approximation" to the optimal blocksize can be found simply by "searching" through small blocksize values (say, those below twenty).

A potential problem with the above analysis occurs if S_{opt} is a prime number. The effectiveness of a specific blocksize, say S_{blk}, is dependent on the number of rows that have S_{blk} (or a multiple of S_{blk}) non-zero elements; for example, if most rows have exactly 'r' non-zero elements, then it should be obvious that S_{opt} is equal to 'r'. So, an *element distribution* plot (number of non-zero elements, say 'r' vs. the number of rows that have exactly 'r' non-zero elements), which can be easily obtained, can be used to predict the potential optimal blocksize values. A peak in the plot indicates the possibility of (but *does not* guarantee) that value (on the X-Axis) being the optimal blocksize (see Figures 12 - 15 for examples). If a peak occurs at a prime number, then simply "searching" through small blocksize values may not result in satisfactory performance, and the blocksize value(s) corresponding to the peak(s) will also have to be evaluated.

5. EXPERIMENTAL EVALUATION OF THE BLOCK-ROW ALGORITHM

5.1 Introduction

The block-row algorithm was coded on a 16,384 processor MasPar MP-1. The fetch phase is optimized for relatively small matrices ($N \leq N_{proc}$) by making additional copies of the vector, and interprocessor communication in the reduce phase is restricted to nearest-neighbor communication via the X-Net. Additionally, all frequently used variables are kept in registers (register operations are up to ten times faster than local memory operations on the MP-1). Additional details, along with code for the block-row algorithm (written in MPL) can be found in [6]. The data for a given sparse matrix is stored in multiple (binary) files, in order to utilize the parallel read capa-

bility of the MP-1; a matrix with approximately one million non-zero elements set up in this format (stored in double precision) can be read in about 1.5 seconds.

5.2 Experimental Results

The application for which the block-row algorithm was developed involves the simulation of carrier transport in modern semiconductor devices using the *Scattering Matrix Approach* [1],[3]. In this approach, the device is viewed as a set of interconnected thin slabs, where each slab is thin enough so that the electric field and the doping density can be considered constant within the slab. Carrier transport across each slab is described by a matrix equation which relates the incident fluxes to the emerging fluxes through transmission and reflection coefficients. The sparsity structure of a scattering matrix depends on how the energy space is discretized; the structure of a scattering matrix based on a k-space discretization, and evaluated at an electric field strength of 300kV/cm (Matrix 'A') is shown in Figure 9.

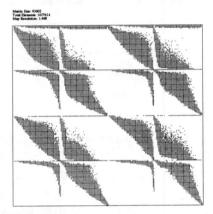

Figure 9: The sparsity structure of a scattering matrix evaluated at an electric field of 300kV/cm.

Matrix 'A' is a scattering matrix of size 93,602 (square) and 1,427,614 non-zero elements - for this matrix, $R_{min} = 0$, $R_{max} = 20,488$, and $R_{median} = 4$. The low value of the median suggests that the value of the optimal blocksize will be somewhere in this range (verified experimentally). As seen in Figure 10, the optimal value for the blocksize is seven. Observe that t_{fetch} and $t_{multiply}$ are proportional to the overhead, and t_{add} and $t_{arrange}$ decrease with increasing blocksizes. Also observe that the number of layers (and consequently overhead) do not always increase with an increase in the blocksize. The same data plotted with respect to the number of layers (Figure 11) verifies that t_{fetch} and $t_{multiply}$ increase with overhead, and shows that t_{add} and $t_{arrange}$ decrease with an increase in the blocksize in spite of higher overheads.

The block-row algorithm was also evaluated using the sparsity patterns of four test matrices (BCSSTK29, BCSSTK30, BCSSTK32, and BCSSTK33 - Figures 12 - 15, respectively) associated with the detailed modeling of structures, obtained from the Harwell-Boeing Sparse Matrix Collection [4]. The actual values of the non-zero entries were not available, and a random (non-zero) value was assigned to each non-zero entry (the vector entries

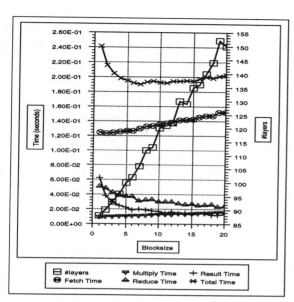

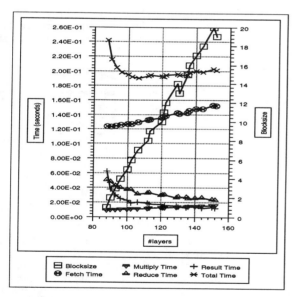

Figure 10: Times for the various phases as the blocksize is varied, for Matrix 'A'.

Figure 11: Times for the various phases with respect to the amount of data, for Matrix 'A'.

were also chosen randomly). These matrices are symmetric; however, for the purpose of this paper, the symmetric nature of the matrices was not exploited and they were treated as non-symmetric matrices. Note that the figures representing each of the matrices have different resolutions and scaling factors, and consequently, no attempt should be made to derive any information about the relative sparseness of the matrices (each dot represents a submatrix with at least one non-zero element).

$N = 13,992$ $N_{elts} = 619,488$

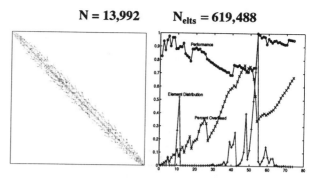

Figure 12: Matrix BCSSTK29; Y-Axis of graph is normalized; see text for details (Section 5.2). The element distribution plot is normalized to 2,543.

$N = 28,924$ $N_{elts} = 2,043,492$

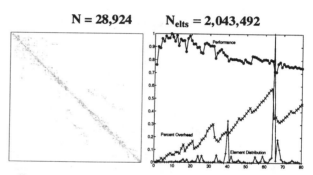

Figure 13: Matrix BCSSTK30; Y-Axis of graph is normalized; see text for details (Section 5.2). The element distribution plot is normalized to 8,084.

$N = 44,609$ $N_{elts} = 2,014,701$

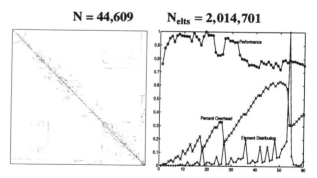

Figure 14: Matrix BCSSTK32; Y-Axis of graph is normalized; see text for details (Section 5.2). The element distribution plot is normalized to 11,249.

In Figures 12 - 15, the Y-Axis is normalized. The *performance* curve is normalized to data in Table 2 (1.0 on the axis corresponds to the time quoted for the block-row algorithm in Table 2, for the relevant matrix). The actual *percent overhead* is obtained by multiplying the Y-Axis values by 100, and the *element distribution* plot is normalized to the maximum number of rows with a given number of non-zero elements. The X-Axis represents the blocksize for the 'performance' curve and the 'percent overhead' graph; it represents the number of non-zero elements for the 'element distribution' plot. For example,

N = 8,738 N_{elts} = 591,904

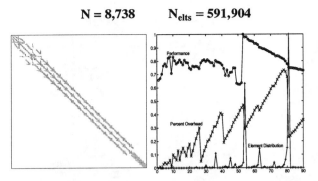

Figure 15: Matrix BCSSTK33; Y-Axis of graph is normalized; see text for details (Section 5.2). The element distribution plot is normalized to 3,440.

it can be seen from Figure 12 that a blocksize of 54 results in the best performance (0.043022 seconds, from Table 2), and the corresponding overhead is approximately 34%. The 'element distribution' plot shows that there were 2,543 rows with exactly 54 non-zero elements.

Observe that for all matrices except for BCSSTK33, the best performance, or performance that is within 5% of the best performance, is obtained for a blocksize that is less than twenty; this supports the claim that a very good approximation to the optimal blocksize can be obtained by exhaustively evaluating the performance of the algorithm for relatively small blocksize values. For BCSSTK33, the best performance is at a blocksize of 53, which is a prime number (consequently, no submultiple of that blocksize was possible). However, this fact can be easily determined *a priori* by plotting the element distribution graph.

5.3 Comparative Evaluation of the Algorithm

In this section, the performance of the block-row algorithm is compared with the performance of three other algorithms discussed in literature. A variation of the "snake-like" method [11], the "segmented-scan" method [5], and a randomized packing algorithm [10] were implemented on a MasPar MP-1, and compared with the block-row algorithm.

The "snake-like" method requires the non-zero elements from each column of the matrix are stored in connected (adjacent) processors of the processor array (optimized fetch phase). Conversely, for the "segmented scan" method, a row-major storage format, along with a "scan" primitive is used to optimize the reduce phase; each row is considered to be a "segment", and the reduction of partial products in all rows can be implemented in parallel. The randomized packing algorithm implemented here is the second algorithm (best of two) presented in [10]; the data structure for this algorithm requires both, the non-zero elements from a row and from a column, to be stored in adjacent processors (at the cost of allowing a load imbalance among processors).

The randomized packing algorithm, as described in [10], has five 'phases'; the vector-distribution phase and the scatter phase are grouped into the fetch phase, and the gather phase and the row-sum phase are grouped into the 'reduce + result' phase. The output vector generated by

the original randomized packing algorithm does not have the same format as the input vector - a consistent format is enforced here, and the conversion time is included in the 'reduce + result' phase (typically, the conversion time accounts for only a few percent of the total time for the phase). Additionally, the randomization changes the matrix 'A' to 'PAQ^T', and this effect must be reversed at the end of the computations; this time has been ignored here (the preprocessing took several minutes of CPU time for the larger matrices, as compared to tens of seconds for the block-row algorithm). This implementation of the randomized packing algorithm achieved approximately 110 MFLOPS for the largest dense matrix-vector multiplication problem that could be solved on a 16,384 processor MP-1 with 256 MBytes of memory - versus the approximately 116 MFLOPS achieved by the authors of [10].

The data presented in this section represents the "best" performance obtained for each algorithm; for the randomized packing algorithm, the best randomization (of ten, using two different random number generators) is used, and the best blocksize is used for the block-row algorithm. Detailed results for Matrix 'A' are presented in Table 1, while the results for the other matrices are summarized in Table 2; the values shown do not include the time required to read the matrices into memory.

Table 1: A comparison of the *best* times (in seconds) for the various algorithms for the scattering matrix problem (Matrix 'A'); normalized times are w.r.t. the block-row algorithm.

N: 93602		N_{elts}: 1427614		
	Snake-like	Seg-Scan	Randomized	Block-Row
t_{fetch}	4.01E-01	1.66E-01	7.40E-02	1.30E-01
$t_{multiply}$	6.09E-03	6.01E-03	2.40E-02	1.11E-02
t_{add}	9.08E-01	1.03E-01	4.83E-01	3.09E-02
$t_{arrange}$		4.93E-02		2.01E-02
t_{total}	1.31E-00	3.23E-01	5.75E-01	1.90E-01
t_{normal}	6.89	1.70	3.03	1.00

For matrix 'A', the (best) random permutation of the rows and columns of the matrix resulted in a maximum processor load of 275, and a minimum processor load of 44. The ideal load is 88, and without randomization, the maximum processor load would have been 4654 (minimum 0). In this case, the block-row algorithm is about 1.7 times faster than any of the other algorithms (Table 1).

The matrices represented in Table 2 are more dense and have a higher (maximum) number of non-zero elements per column (the distribution of the minimum/maximum number of non-zero elements per column for the different matrices is as follows: Matrix 'A': 1/20; BCSSTK29 5/71; BCSSTK30 4/219; BCSSTK32 2/216; and BCSSTK33 20/141). The randomized packing algorithm has a better performance as the matrices get more dense [10], while the segmented-scan algorithm and the block-row algorithm have higher fetch phase times because of the relatively dense columns; this explains the improved performance of the

Table 2: A summary of the *best* times (in seconds) for the various algorithms for the matrices BCSSTK: (BCS in Table) 29, 30, 32, 33; normalized times are with respect to the block-row algorithm.

Comparative Evaluation				
	BCS29	BCS30	BCS32	BCS33
Time (seconds)				
Block-Row (blocksize)	0.043022 (54)	0.174204 (09)	0.204133 (19)	0.045224 (53)
Normalized Time ((min/max) load)				
Block-Row	1.000 (54)	1.000 (135)	1.000 (143)	1.000 (53)
Randomized	2.160 (1/65)	1.146 (76/168)	1.388 (72/164)	1.418 (0/62)
Seg-Scan	4.231 (38)	5.819 (125)	4.883 (123)	3.510 (37)
Snake-like	20.035 (38)	23.455 (125)	15.940 (123)	20.653 (37)

randomized packing algorithm relative to the performance of the other algorithms, when compared to the results for Matrix 'A'. The performance of the snake-like method is affected the most because of the symmetric structure of the matrices, and the relatively dense columns (which, for a symmetric matrix implies that the rows are as populated). Note that the "load" shown in Table 2 represents the number of elements of the matrix that are stored in the memory of one processor; for the randomized packing algorithm, the load distribution is not even, and consequently, the algorithm requires more memory per processor than the other algorithms.

6. CONCLUSIONS AND FUTURE WORK

The block-row algorithm allows the "regularity" of a data structure that uses a row-major mapping to be varied by a changing a parameter (the "blocksize"). As confirmed by experiments, a very good approximation to the optimal blocksize can be obtained by performing a "search" in conjunction with information from an 'element distribution' plot. The information from the 'element distribution' plot allows the search to be narrowed to a few values; analytical determination of the optimal blocksize is difficult because the execution times depend on the data distribution - which changes with a change in the value of the blocksize. For the matrices considered here, the block-row algorithm out-performed the snake-like method, the segmented-scan method, and the randomized packing algorithm, when implemented on the MasPar MP-1. An *adaptive* block-row algorithm is currently being developed; the adaptive algorithm allows different blocksizes to be selected for different rows of a given matrix.

Acknowledgements

The authors are indebted to Professor Mark Lundstrom (Purdue University) and his students for providing and explaining the 'Scattering Matrix Approach' application used here. The authors would also like to thank Professor Valerie Taylor (Northwestern University) for bringing to their attention and facilitating access to the Harwell-Boeing Sparse Matrix Collection. The reviewers' comments were helpful in preparing the final version of the paper.

LIST OF REFERENCES

[1] M. A. Alam, M. A. Stettler, and M. S. Lundstrom, "Formulation of the Boltzmann Equation in terms of Scattering Matrices", *Solid-State Electronics*, (Feb, 1993), pp. 263-271.

[2] J. Anderson, G. Mitra, and D. Parkinson, "The Scheduling of Sparse Matrix-Vector Multiplication on a Massively Parallel DAP Computer", *Parallel Computing*, (Jun, 1992), pp. 675-697.

[3] Amitava Das, and Mark S. Lundstrom, "A Scattering Matrix Approach to Device Simulation", *Solid-State Electronics*, (Oct, 1990), pp. 1299-1307.

[4] Iain S. Duff, Roger G. Grimes, and John G. Lewis, "Sparse Matrix Test Problems", *ACM Transactions on Mathematical Software*", (Mar, 1989), pp. 1-14.

[5] Steven Warren Hammond, "Mapping Unstructured Grid Computations to Massively Parallel Computers", PhD. Thesis, Rensselear Polytechnic Institute, (1992), pp. 93-104; also RIACS Technical Report 92.14, Research Institute for Advanced Computer Science, NASA Ames Research Center, (1992).

[6] Nirav Kapadia, *A Sparse Matrix-Vector Multiplication Algorithm for Computational Electromagnetics and Scattering Matrix Models*, Technical Report TR-EE 94-31, Electrical Engineering, Purdue University, (Oct, 1994).

[7] Vipin Kumar, Ananth Grama, Anshul Gupta, and George Karypis, *Introduction to Parallel Computing: Design and Analysis of Algorithms*, The Benjamin/Cummings Publishing Company, Inc., (1994), Chapter 11.

[8] Manavendra Misra and V. K. Prasanna, "Efficient VLSI Implementation of Iterative Solutions to Sparse Linear Systems", *Proceedings of the International Conference on Systolic Arrays*, (May, 1989), pp. 52-61.

[9] M. Morjaria and G. J. Makinson, "Unstructured Sparse Matrix Vector Multiplication on the DAP", *Supercomputers and Parallel Computation*, (1984), pp. 157-166.

[10] Andrew T. Ogielski and William Aiello, "Sparse Matrix Computations on Parallel Processor Arrays", *SIAM Journal on Scientific Computing*, (May, 1993), pp. 519-530.

[11] L. F. Romero and E. L. Zapata, *Data Distributions for Sparse Matrix Vector Multiplication*, Technical report, Department of Computer Architecture, University of Malaga, Spain, 1993; also published in the *Proceedings of the Fourth International Workshop on Compilers for Parallel Computers*, (1993).

A DISTRIBUTED MEMORY, MULTIFRONTAL METHOD FOR SEQUENCES OF UNSYMMETRIC PATTERN MATRICES

Steven M. Hadfield
US Air Force Academy
2354 Fairchild Drive, Suite 6D2A, USAFA, Colorado 80840
email: hadfieldsm%dfms%usafa@dfmail.usafa.af.mil
and
Timothy A. Davis
University of Florida
E301 CSE, Univ. of Florida, Gainesville, Florida 32611
email: davis@cis.ufl.edu.

Abstract -- *This paper describes a parallel algorithm for the LU factorization of sequences of matrices with an identical, unsymmetric pattern. The algorithm is based on the unsymmetric pattern multifrontal method by Davis and Duff and repeatedly uses the computational structure defined by their method to factor each matrix in the sequence. A lost pivot recovery mechanism is built into the parallel algorithm to dynamically alter the computations to preserve numerical stability. Initial performance results from an nCUBE 2 are presented and show that the algorithm achieves significant parallelism and reduces overall factorization time even on a single processor.*

Introduction. The multifrontal method for the LU factorization of unsymmetric pattern sparse matrices developed by Davis and Duff [2] has demonstrated exceptional sequential performance [13] and significant potential parallelism [10]. This paper describes a parallel algorithm based on Davis and Duff's method that factors sequences of identically structured matrices such as those that arise when solving systems of differential algebraic equations [4,15]. The algorithm reuses the computational structure originally defined by Davis and Duff's method for each matrix in the sequence but employs lost pivot recovery techniques to maintain numerical stability. Performance evaluation of the algorithm's implementation indicates that substantial execution time savings are possible with both sequential factorization and via parallelism.

Multifrontal Concepts. Multifrontal methods for sparse matrix factorization decompose the sparse matrix into a set of overlapping dense submatrices called frontal matrices. Each frontal matrix is partially factored by one or more pivots. Entries in the unfactored portion of the frontal matrix (called the contribution block) must be uniquely assembled (added) into subsequent frontal matrices. An assembly directed acyclic graph (DAG) is used to define the computational structure with nodes representing frontal matrices and edges representing the passing of contribution block entries from one frontal matrix to another. With symmetric pattern multifrontal methods, the assembly DAG is a tree (or forest) as each frontal matrix's contribution block can be completely absorbed within a single subsequent frontal matrix [11]. With the unsymmetric pattern multifrontal method, the contribution block can be fragmented and portions must be passed to different subsequent frontal matrices which results in the more generalized DAG structure. Three levels of parallelism are available within this structure: between distinct connected components, between nodes within a component, and within the nodes representing frontal matrices.

Lost Pivot Recovery Issues. When anticipated pivot entries become no longer numerically acceptable for subsequent matrices in the sequence, changes in the pivot ordering are necessary (we call this Lost Pivot Recovery (LPR)). Multifrontal methods can address lost pivot recovery with three distinct strategies. First, avoidance tries to preclude the need for recovering lost pivots by relaxing the criteria used to judge potential pivots as being numerically acceptable through a technique called threshold pivoting [3]. The second strategy, intra-frontal matrix recovery, tries to recover from lost pivots using only permutations confined to the pivot block of the frontal matrix. Such permutations are advantageous because they have no effects on the assembly DAG or composition of other frontal matrices.

Since avoidance and intra-frontal matrix recovery are not capable of recovering from all lost pivots, there is a need for the third strategy which is inter-frontal matrix recovery. This strategy is similar to border casting in MCSPARSE [5], except that we consider the effects of pivot failure on our rectangular frontal matrices. Inter-frontal matrix recovery uses symmetric permutations to the entire matrix to shift lost pivot rows and columns to subsequent frontal matrices. In the case of a symmetric pattern multifrontal method, the impacts are limited to a single subsequent frontal matrix. With the unsymmetric pattern multifrontal method, multiple subsequent frontal matrices can be affected in a variety of ways depending on the type of relationship that exists between themselves and the frontal matrix possessing the failed pivots. A detailed theoretical development of lost pivot recovery is beyond the scope of this paper but can be found in [8,9]. Details of these LPR techniques include explicit definitions of the extent of fill-in, required synchronization between and within frontal matrices, handling multiple and repeated pivot failures, ordering the resolution of multiple failures, handling contribution block extensions, and minimizing required communications.

Parallel Algorithm with LPR. Our algorithm consists of a host-based portion does some necessary preprocessing using the matrix's assembly DAG and a parallel factorization component (referred to as PRF) that does the actual factorization on a hypercube topology multiprocessor.

The host-preprocessing takes an assembly DAG for the first matrix in the sequence from an implementation of Davis and Duff's unsymmetric pattern multifrontal method called UMFPACK [1]. The assembly DAG is the primary computational structure which is then used in scheduling, allocation, and assignment. Scheduling is done using a critical path priority scheme. Determination of the size of a subcube to be assigned to each frontal matrix is called allocation. Allocation is done based on a proportional scheme that looks at the number and sizes of frontal matrices ready for factorization and determines a particular frontal matrix's subcube size based on its portion of the available workload. Assignment of a frontal matrix's factorization to a particular subcube is done in a manner that attempts to minimize inter-subcube communications requirements.

The developed schedules and assignments are selectively passed to the multiprocessor's processing nodes and the parallel numerical factorization of the first matrix in the sequence starts. In the absence of the lost pivot recovery techniques, the basic process has each processing node factor each frontal matrix in its schedule. For each frontal matrix, the assigned processors allocate their portion of the frontal matrix's storage, assemble (add) in the original values and contributions from previous frontal matrices, and then partially factor the frontal matrix. Contribution block entries from other processors are passed using blocking message receives with special message typing conventions and known sender IDs. This provides for the necessary synchronization between subcubes assigned to frontal matrices. The partial factorization process is done with a column-oriented, pipelined fan-out routine similar to that found in [6]. Entries in the contribution block of each frontal matrix are forwarded to the appropriate subsequent frontal matrices. Message typing conventions allow these forwarded contributions to be selectively read by the receiving frontal matrices.

When lost pivot recovery is enabled in the parallel factorization code, additional synchronization and processing is required. Specifically, pivot failures in one frontal matrix require that subsequent frontal matrices be expanded to accommodate the lost pivot rows and/or columns. Such expansions affect the memory allocation and row/column patterns of subsequent frontal matrices. Special message passing between the root nodes of the frontal matrices' assigned subcubes is used to synchronize and coordinate lost pivot recovery between frontal matrices. Furthermore, a particular frontal matrix may be affected by pivot failures in several preceding frontal matrices and the recovery information may take several disjoint paths to a subsequent frontal matrix. Along the various paths some entries may be absorbed by intermediate frontal matrices and other entries may have additional contributions added due to fill-in. Thus, the resolution of lost pivot recoveries involves resolving multiple instances of specific recoveries as well as resolution of multiple distinct recoveries. After the effects of all the recoveries are resolved by the assigned subcube's root processor, they are forwarded to the rest of the subcube.

Once the frontal matrix's definition has been updated per any lost pivot recoveries, the factorization process can commence. In the original partial dense factorization routine, the cooperating processor that holds the next pivot's column will receive the multipliers from the current pivot (column of the L factor) and then update just the next pivot's column with these multipliers. The multipliers for this next pivot can then be computed and sent out after which the rest of columns are updated. In this way the next pivot's multipliers can be on their way to the other processors before those processors need them.

To adapt this algorithm for intra-frontal matrix lost pivot recovery, the processor owning the next pivot column will first update that column and check for a valid pivot. If a valid pivot cannot be found, the other pivot columns held by this processor are updated and checked for valid pivots. If one of these other columns has a valid pivot, its multipliers are computed and sent out with the necessary permutation information appended. If the processor has no columns with valid pivots, it sends out a failure status. When the other processors see the failure status, each will check its own potential pivot columns for a valid pivot. If a valid pivot is found, the processor will contribute the forward distance from the original pivot owner's processor ID to its own processor ID to a subcube synchronization operation. If no valid pivot was found, the size of the subcube will be forwarded. The synchronization operation uses a parallel prefix operation followed by a broadcast to determine and distribute the minimum of these contributions. Thus the entire subcube knows which processor now owns the next pivot or if no pivots are available. In the latter case, inter-frontal matrix recovery is necessary.

Following the factorization process, the contribution block must be forwarded to the subsequent frontal matrices. The original contribution block is passed in the same way as without LPR. Extensions to the contribution block from previous recoveries or from lost pivots in the current frontal matrix must be handled specially. All of these entries are passed to the subcube's root processor where they are integrated into the correct recovery structure and then forwarded via the LPR control messages.

Performance Results. The parallel numerical factorization was evaluated using the matrices and matrix sequences described in Table 1. Table 2 reports the achieved speedups on 2, 4, 8, 16, 32, and 64 processors. All the test matrices are highly unsymmetric in pattern. GEMAT11 is an electrical power problem [4]; the others are chemical engineering matrices [15]. These speedups were achieved with the lost pivot recovery mechanisms disabled. These results compare favorably with similar results using symmetric pattern multifrontal methods on matrices of like order [7,12,14].

Table 1: *Test Matrix Sequences*

Matrix	n	nonzeros	# in seq
RDIST1	4134	94408	41
EXTR1	2837	11407	1
GEMAT11	4929	33108	1
RDIST2	3198	56834	40
RDIST3A	2398	61896	37

Table 2: *Achieved Speedups*

Matrix	2	4	8	16	32	64
RDIST1	1.9	2.9	5.2	9.5	14.1	20.2
EXTR1	1.8	2.8	4.5	6.5	9.0	12.2
GEMAT11	1.9	3.2	5.1	8.7	13.4	16.8
RDIST2	1.8	2.6	4.4	7.5	11.9	17.3
RDIST3A	1.8	2.8	5.0	7.4	12.3	17.8

LPR was evaluated using the RDIST1, RDIST2, and RDIST3A sequences both with and without an avoidance strategy of relaxing the pivot threshold from 0.1 to 0.001. The total factorization time for all matrices in a sequence using the PRF code with LPR was compared to a corresponding factorization time estimate based on doing a full UMFPACK factorization for each matrix in the sequence. Since UMFPACK could not run on the nCUBE 2 due to memory limitations, an estimate of the UMFPACK factorization time was developed based on timings done on a CRAY YMP. The percentage time savings are provided in Table 3.

Table 3: *Sequential LPR Performance*

Matrix Name	Improvement w/o avoidance	Improvement w/ avoidance
RDIST1	58.8%	60.3%
RDIST2	48.3%	54.3%
RDIST3A	1.2%	31.3%

The parallel performance of PRF with LPR was evaluated using selected matrices in each of the test sequences on a 32 processors. Table 4 reports the range of speedups achieved both without and with the use of the avoidance strategy described earlier.

Conclusion. The parallel algorithm described in this paper builds on Davis and Duff's work by providing a robust lost pivot recovery mechanism and also demonstrates the availability of significant parallelism in their approach. Future efforts will likely focus on a parallel shared memory version of the algorithm as many of the parallelism-limiting factors result directly from the distributed memory environment

Table 4: *Parallel LPR Performance (P=32)*

Matrix Name	Speedup range w/o avoidance	Speedup range w/ avoidance
RDIST1	12.2-12.4	12.3-12.4
RDIST2	7.3-10.6	8.1-10.6
RDIST3A	5.2-12.0	7.8-12.0

REFERENCES

[1] T.A. Davis, *Users' Guide for the Unsymmetric-Pattern Multifrontal Package (UMFPACK, Version 1.1)*, CISE Department, University of Florida, Tech. Report TR-95-004, (January, 1995).

[2] T.A. Davis and I.S. Duff, *An Unsymmetric-Pattern Multifrontal Method for Sparse LU Factorization*, CISE Department, University of Florida, Tech Report TR-94-038, (November, 1994), 21 pp.

[3] I.S. Duff, A.M. Erisman, and J.K. Reid, *Direct Methods for Sparse Matrices*, Oxford Science Publications, New York, NY, (1989).

[4] I.S. Duff, R.G. Grimes, and J.G. Lewis, *User's Guide for Harwell-Boeing Sparse Matrix Collection (Release I)*, CSSD, Harwell Laboratory, Tech Report TR/PA/92/86, (October, 1992).

[5] K.A. Gallivan, *Solving Large Nonsymmetric Sparse Linear Systems Using Mcsparse*, CSRD, University of Illinois, Urbana-Champaign, CSRD Tech Report 1424, (April, 1995).

[6] G.A. Geist and M.T. Heath, "Matrix Factorization on a Hypercube", *Hypercube Multiprocessors 1986*, Society for Industrial and Applied Mathematics, Philadelphia, PA, (1986), pp. 161-180.

[7] A. George, M.T. Heath, J.W.-H. Liu, and E.G.-Y. Ng, "Solution of Sparse Positive Definite Systems on a Hypercube", *Journal of Computational Applied Mathematics*, Vol. 27, (1989), pp 129-156.

[8] S.M. Hadfield, *On the LU Factorization of Sequences of Identically Structured Sparse Matrices within a Distributed Memory Environment*, CISE Department, University of Florida, Tech Report TR-94-019, (April,1994).

[9] S.M. Hadfield and T.A. Davis, *Lost Pivot Recovery for an Unsymmetric Pattern Multifrontal Method*, CISE Department, University of Florida, Tech Report TR-94-029, (October,1994).

[10] S.M. Hadfield and T.A. Davis, "Potential and Achievable Parallelism in the Unsymmetric-Pattern Multifrontal LU Factorization Method for Sparse Matrices" in *Proceeding of the Fifth SIAM Conference on Applied Linear Algebra*, (1994), pp 387-391.

[11] J.W.-H. Liu, "The Multifrontal Method for Sparse Matrix Solution: Theory and Practice", *SIAM Review*, Vol. 34, (1992), pp. 82-109.

[12] R. Lucas, T. Blank, and J. Tiemann, "A Parallel Solution Method for Large Sparse Systems of Equations", *IEEE Transactions on Computer-Aided Design*, CAD-6, (1987), pp. 981-991.

[13] E.G.-Y. Ng, "Comparison of Some Direct Methods for Solving Sparse Nonsymmetric Linear Systems", in *Proceeding of the Fifth SIAM Conference on Applied Linear Algebra*, (1994), p. 140.

[14] A. Pothen and C. Sun, "A Mapping Algorithm for Parallel Sparse Cholesky Factorization", *SIAM Journal of Scientific Computing*, Vol. 14, (1993), pp. 1253-1257.

[15] S.E. Zitney and M.A. Stadtherr, "Supercomputing Strategies for the Design and Analysis of Complex Separation Systems" in *Proceeding of Conference on Industrial Engineering and Chemical Research*, (1993), pp. 604-612.

NOTE: All University of Florida Technical Reports are available via anonymous FTP to ftp.cis.ufl.edu:cis/tech-reports, or via the web at http://www.cis.ufl.edu

A PARALLEL ADAPTIVE FAST MULTIPOLE ALGORITHM FOR N-BODY PROBLEMS*

Sanjeev Krishnan and Laxmikant V. Kalé
Department of Computer Science
University of Illinois
Urbana, IL 61801.
E-mail: {sanjeev,kale}@cs.uiuc.edu

Abstract

We describe the design and implementation of a parallel adaptive fast multipole algorithm (AFMA) for N-body problems. Our AFMA algorithm can organize particles in cells of arbitrary shape. This simplifies its parallelization, so that good locality and load balance are both easily achieved. We describe a tighter well-separatedness criterion, and improved techniques for constructing the AFMA tree. We describe how to avoid redundant computation of pair-wise interactions while maintaining load balance, using a fast edge-partitioning algorithm. The AFMA algorithm is designed in an object oriented, message-driven manner, allowing latency tolerance by overlapping computation and communication easily. It also incorporates several optimizations for message prioritization and communication reduction. Preliminary performance results of our implementation using the Charm++ parallel programming system are presented.

1 Introduction

The N-body problem is an important core problem arising in several simulation applications in astrophysics, molecular dynamics and fluid dynamics. The computation in the N-body problem consists of calculating interactions between all pairs of bodies (particles) in the system. The order of complexity for a simple formulation of this problem would thus be $O(N^2)$, for a system consisting of N particles. There exist many good algorithms for solving the N-body problem, of which the Fast Multipole Algorithm due to Greengard and Rokhlin [5] has $O(N)$ time complexity for N particles, for any desired bound on error. Other algorithms include the $O(NlogN)$ Barnes-Hut method, particle-in-cell methods, and the Distance Class methods. Although the problem sizes at which the Fast Multipole method would outperform the Barnes-Hut method is not clear, there has recently been a lot of interest in implementing both algorithms on parallel machines.

In the FMA, the effect due to a *well-separated* (sufficiently far away) group of particles is approximated by a *multipole expansion*, which is a refined formalization of the

center-of-mass, leading to provable error bounds. Interactions are computed between particles and groups of particles, as well as between different groups of particles. In order to partition the particle set into groups, the FMA recursively divides the computational space into cubical "cells", which are ordered hierarchically in a tree. For the three dimensional problem an octtree is generated. In the *non-adaptive FMA*, a uniform grid is imposed on the computational space, resulting in a complete tree whose leaves all have the same depth. However, this is unsuitable for non-uniform particle distributions. Hence the *adaptive FMA* [3] divides cells until the number of particles in a leaf cell is less than some specified *grain-size*, leading to an irregular tree.

The adaptive FMA (AFMA) is difficult to efficiently program on parallel computers, especially for private memory ones. This is because of the difficulty of achieving good load balance and locality simultaneously, distributing shared data structures such as the tree among processors, overcoming latencies arising due to the inherent irregularity and unpredictability of the computation, and reducing communication volume. There are parallel implementations of the non-adaptive FMA on shared and distributed memory computers, including the work by Board and others [1, 6], and of the adaptive FMA on shared memory computers [10]. Our work is one of the first parallel implementations of the AFMA on message passing computers.

2 Modifications to the original AFMA

In the original AFMA [3], tree construction proceeds by recursively subdividing the computational space (which is initially cubical at the outermost level) into smaller cubical boxes of the same size (dimensions) : 4 boxes for the two-dimensional case and 8 in three dimensions. All boxes at the same level in the tree are the same size. For adaptiveness, the decision to subdivide a box depends on the number of particles it contains, so that the resulting tree is irregular.

The primary difficulty the original AFMA poses, especially for parallel implementations, is that the *partitioning of space is dictated by the AFMA algorithm*, and cannot be determined by either the parallel partitioning strategy or by the application using the AFMA code. While there is a

---

*This research was supported in part by the National Science Foundation grants CCR-90-07195 and CCR-91-06608.

lot of research into good libraries for parallel partitioning techniques and heuristics (which achieve good load balance and minimal communication), none of these techniques can be used because of the cubical subdivision of space the original AFMA requires. Instead new constrained techniques have to be developed, such as the costzones scheme [10]. These special techniques can give good load balance, but degrade locality since they cannot ensure that the regions mapped to processors are convex in shape.

Again, the application itself often has its own requirements for partitioning : e.g. in a molecular dynamics application NAMD [2] used in a Grand Challenge applications group at the University of Illinois, the partitioning of atoms into cells is determined by cutoff distances for other force calculations. In order to use the AFMA for long range Coulomb force calculations, the atoms have to be redistributed before AFMA and brought back to their original cells after the AFMA step, incurring significant overhead both in terms of performance and programmability.

To overcome these problems, we have developed a modified version of the FMA which does not require subdivision into cubical boxes (note that cubical subdivision is not theoretically required). In the modified algorithm, the computational space can be divided into *regions of any shape*, as determined by the partitioning algorithm or the application itself. In our implementation we have used orthogonal bisection for partitioning[1], in which each processor gets a contiguous, convex (rectangular) region of space, which ensures good load balance and low inter-processor communication.

In the original AFMA, two boxes of the same size are said to be well-separated if they are separated by a distance greater than their width. Implementations in three dimensions [1, 5] require the distance between well-separated boxes to be twice their width. Well-separation is required in order to maintain error bounds for the translation operators for multipole expansions. To handle boxes of arbitrary shapes and sizes, we now define a slightly different, more general well-separatedness criterion using Greengard's original theorem [5]. C_1 and C_2 are two cells, having r_1 and r_2 as their radii (distance from center to farthest particle), and z_1 and z_2 as their centers. We say that C_1 is *partially well separated* from C_2 if $|z_1 - z_2| > cr_1 + r_2$, where c is a constant greater than 1. This corresponds to a distance of cr_1 from the center of C_1 to the nearest particle in C_2. Similarly, C_2 is partially well separated from C_1 if $|z_1 - z_2| > cr_2 + r_1$. We say that C_1 and C_2 are well-separated if they are both partially well-separated from each other, i.e. $|z_1 - z_2| > max\{cr_1 + r_2, cr_2 + r_1\}$.

Using the above definition of well-separatedness, the four types of interaction lists for C_1 in the AFMA [3] can be defined as :

- V list : A cell C_2 is in the V list of C_1 if C_1 and C_2 are well-separated. Then C_2's multipole expansion can be converted to a *local expansion* at C_1's center.
- W list (leaf cells only) : if C_1 and C_2 are not well-

separated, but C_2 is partially well-separated from C_1, and C_1 is a leaf cell, then C_2's multipole expansion needs to be evaluated directly at C_1's particles.
- X list : if C_1 and C_2 are not well-separated, but C_1 is partially well-separated from C_2, and C_2 is a leaf cell, then individual particle fields from C_2 need to be converted to a local expansion at C_1's center.
- U list : (leaf cells only) if none of the above three conditions apply between C_1 and C_2, then interactions have to be computed directly, between each pair of their particles.

In the original non-adaptive FMA, the well-separation criterion between spherical regions is applied to the cubical boxes case by requiring two intervening boxes (for 3-D) between the boxes being compared [5]. This is clearly an overkill, since for example, corner boxes in a cubical region are more distant from the center than other boxes. Intuitively, the volume enclosed by a cube is greater than the the volume enclosed by a sphere inscribed in the cube. By comparing boxes for well-separation using a tighter spherical criterion, there are fewer of the expensive pair-wise interactions. E.g in the three dimensional case, instead of computing 125 interactions with neighboring cells at the leaf level with the cubical criterion, there are only 93 interactions with our tighter criterion.

We have also developed an optimization for formation of interaction lists for cells. In the original FMA, the interaction list for a cell C consists of those cells *at the same level in the tree* which are well separated from C, but whose parents are not well separated from C's parent. Thus comparisons for well-separatedness are always between cells of the same size and hence at the same level of the tree. Instead, if we use our more general well-separatedness criterion, we can compare cells at different levels in the tree. E.g. C and D are two non-leaf cells, with 8 children each. If they are not well separated, there will be totally 64 interactions (for each pair of children). However, if C is well separated from 6 of D's children, then those 6 interactions can be computed directly, and the remaining pair-wise, resulting in only $6 + 8 * 2 = 22$ interactions. Thus we can further reduce the number of interactions that need to be computed.

3 Partitioning and tree construction

We first describe a fast parallel partitioning algorithm. We use orthogonal recursive bisection (ORB), which is a well known technique which recursively partitions the computational space by planes parallel to the coordinate axes. The load measure for partitioning is obtained from the previous AFMA iteration by a scheme similar to [10]. The partitions resulting from ORB are rectangular, resulting in less communication with neighboring partitions. Also, since partitions are not restricted to be cubical, much better load balance can be achieved. If there are P processors, the number of parallel bisection operations needed is usually $P - 1$. However, all these bisections need not be serialized. In fact, the critical path involves only $log_2 P$ bisections, corresponding to the height of the binary tree formed by hierarchically ordering the partitions. Hence we

---

[1] We emphasize, however, that in our modified AFMA, the partitioning strategy used can be quite general.

overlap the parallel bisection operations to significantly reduce idle times of all processors. This is done with little programming effort because of the message-driven model of our design (see section 5).

At the end of partitioning, processors exchange particles so that each processor has its own set of particles. A copy of the top log_2P levels of the AFMA tree is also made on all processors, so that all processors have the root cell of all other processors' local trees. Now each processor proceeds to construct its own local part of the tree by recursively dividing the space assigned to it by ORB. The division continues till the load on each leaf cell is less than some pre-specified grainsize. After tree construction is completed, interaction lists for the shared (top log_2P levels) and local trees are computed.

After each processor builds its local tree, it needs to get parts of other processors' trees in order to compute the remote members of the interaction lists required for the AFMA. The purpose of this step is similar to the Locally Essential Tree (LET) construction step in [12], however, we avoid the overheads associated with their implementation by two optimizations. First, we assemble only the structure of the LET and the coordinates for each cell; the particles and multipoles for each cell are not transferred because they are not required for interaction list formation. Second, instead of fine-grained receiver initiated communication for expanding the tree one level at a time [12], each processor sends to its neighbors (the processors it needs to interact with, as determined by the list-formation algorithm) the *exact* part of its own local tree that they will require, resulting in just one message per neighbor. The key insight that allows us to do this for the AFMA is the following : if a local cell C is partially well separated from the root cell for processor P, then from the definition, C must appear in the V or W lists of all cells in P, and hence C's children do not need to be sent to P. Thus we identify all cells that need to be sent to neighboring processors using just the shared top levels of the AFMA tree.

Once each processor has received parts of the LET from other processors (no exchange of the LET is needed with well-separated processors), and attached them to its copy of the shared top levels of the tree, it has the entire LET, and can start computing members of the four lists for each of its cells. The list forming algorithm consists of a recursive depth first traversal of the tree starting at the root cell of the tree. Each cell has a list of "candidate cells" which is built while traversing its parents. For internal cells in the tree, the algorithm tries to classify candidate cells into either the V or X lists. Those cells that cannot be classified are added to the candidate-lists of all child cells, so that they can be classified at a lower (finer) level in the tree. Finally, at the leaf cells, all remaining candidate cells are added to the U or W lists. Details of this algorithm are not described here due to space limitations.

4   Balancing pair-wise interactions

Pair-wise (U list) interactions take up a major portion of the time for computing interactions between cells. To elim-
inate redundant pair-wise interactions, Newton's third law is usually used. For a pair of cells A and B, particle data is sent from A to B, where the forces are computed and sent back to A. Although this insight for removing redundant computation is well known, in the parallel context we are faced with the problem of deciding, for every U list computation, which processor it should be computed on. If U list computations are not assigned to processors carefully, it is possible that some processors will get overloaded.

We can arrive at a heuristic solution to this load balancing problem by formulating it as an *edge-partitioning problem* on a graph. The vertices of the graph are processors. An edge is drawn between a pair of processors P and Q if there is a U list interaction between a cell on P and a cell on Q. The *weight* of this edge is the total computation cost of all cross-processor U list interactions between cells on P and cells on Q. The problem now reduces to that of partitioning the load of each edge among the processors it connects such that all processors have approximately equal loads after all edges have been assigned. Although this edge-partitioning problem can be solved optimally, we require a fast, easily parallelizable algorithm.

We have developed a heuristic algorithm by making use of *spatial information* which is already present in the AFMA tree. This spatial information is in the form of the repeated bisections performed while constructing the shared levels of the tree. The main idea in our edge-partitioning algorithm is hence to recursively divide the edge loads between sibling cells (which represent adjacent partitions) in the hierarchical tree. The heuristic for dividing cross-partition edges between two partitions takes into account local as well as global criteria. The global requirement is that the loads of the two partitions should be balanced. The local requirement is that the edge should not be assigned to a processor which is already heavily loaded. We have obtained excellent results by using the local criterion when processor loads are close to the average load (as would be the case towards the end of this algorithm), and using the global criterion otherwise. Each processor recursively bisects *only the subtrees it belongs to*. If E is the total number of edges and V is the number of processors, then in the best case each processor has to examine $E + E/2 + E/4 + ...$ edges, which means that the algorithm has a best case complexity of $O(E)$ and a worst case complexity of $O(E * logV)^2$. In practice the running time is close to the best case because the initial ORB partitioning of particles ensures that the number of edges across partitions is balanced to some extent. Together with the ORB partitioning, this edge-partitioning heuristic results in good load balance, as well as good locality.

5   Overlapping communication latency

We have implemented our algorithm in the Charm++ parallel programming language [8, 7], which is an extension of C++. Charm++ has a *message driven model*, in which functions inside objects are invoked in response to

---

[2]Optimal algorithms for this have a complexity of $O(V * E)$ or more.

the receipt of messages; there are no explicit blocking "receive" calls. Cells in the AFMA tree are modeled as objects having functions for computing the different interactions. Thus the design consists of concurrently executing objects which interact by asynchronous function invocations.

As discussed in section 3, the partitioning by ORB can be optimized by performing bisections concurrently. This allows idle times during one bisection to be overlapped with computation from the other. Again, during the construction of the Locally Essential Tree, it is not necessary for a processor to wait for the parts of the tree from other processors ; it can continue with the construction of its own local tree. When a message containing tree-data from another processor arrives, the code for processing it is scheduled and that part of the tree is attached to the local tree.

Each of the four types of interactions (corresponding to the four lists) involve highly parallel and irregular interactions between cells in the AFMA tree. For each of the interactions, all processors iterate over their cells, computing local interactions and sending/receiving messages for remote ones. Interactions with remote processors are computed completely asynchronously: when a message containing remote data arrives, the object to which it is directed is automatically scheduled by the Charm runtime system, and the proper function for computing the interaction is invoked. Thus no time is wasted in waiting for remote data, and an almost ideal overlap of communication and computation is achieved.

Although there is much parallelism within each stage of the AFMA, executing them sequentially (as has been done in almost all previous implementations) can lead to serious imbalances and processor idling. This is because it is difficult to balance the load in each stage by itself. In [4] an attempt has been made to explicitly overlap two stages (in the context of the Barnes-Hut method) using a "non-synchronizing" global communication protocol. However, this requires some programming effort, and is difficult to achieve when there are many stages with complex dependences.

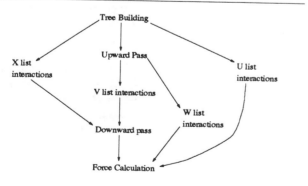

Figure 1: Dependences between stages of the AFMA.

Figure 1 shows the dependences across the various stages of the AFMA. It can be seen that there is a lot of scope for overlapping the idle times in one stage with computation from other stages. This is achieved naturally, and with *no extra programming effort*, because of the asynchronous, message driven nature of our design. Each processor just starts off all the interactions, and then dependences between stages are enforced *at the granularity of the cell objects*, thus avoiding any need for global synchronization between stages. This overlap across stages enables us to prevent processors from idling when there is work in other stages left.

6   Priorities and other optimizations

From Figure 1, it is clear that the dependence between stages is complex, and needs to be exploited carefully. There is a significant critical path consisting of tree construction, the upward pass in the AFMA tree, the V-list interactions, the downward pass in the tree, and finally the force evaluations. On the other hand, U-list interactions can proceed independently for all cells, and depend on only the tree construction to complete. Hence they can be thought of as "background computations" running at lower priority, which are used to fill up idle times during the other stages. We use the *message prioritization* feature of the Charm run-time system [7] to give the highest priority to messages traversing up and down the tree, while the V-, X-, W- and U-list interactions are given successively lower priorities. This enables the scheduler on each processor to always schedule computations on the critical path ahead of the others.

The communication patterns in the parallel AFMA are unpredictable and irregular. This is especially a problem for distributed memory computers which have high message latencies. Thus it is not possible to get good performance with fine-grained, receiver initiated communication, as is possible for shared memory machines [11]. We have extensively used a sender-initiated *advance-send* protocol to reduce communication overhead in our implementation. In [4, 9] a form of advance-send is used in the context of the Barnes-Hut method by sending particle data to remote nodes instead of requesting for their particles. For the AFMA, the basic problem to be solved before using advance-send is for each processor to determine which other processors will be consumers of its particle and multipole data. The key insight is to find, for each interaction list, the *dual interaction list*. From the definition of the four lists, we find that the U list is its own dual, the V list is its own dual, and the W and X lists are duals of each other. Thus each cell simply has to advance-send its particles to all cells in its U list, its multipole expansions to all V and X list cells, and a converted local expansion to all W list cells. In the AFMA, sending multipole expansions instead of particles has the advantage that a single multipole expansion is usually much less voluminous than the particle data itself, for cells containing tens or hundreds of particles. As described earlier, when a message carrying this data arrives at a processor, the Charm run-time automatically schedules the correct interactions in the appropriate cell objects. Thus advance-send is achieved with almost no programming effort.

To increase the granularity of communication, it is necessary to combine or aggregate messages going to the same destination processor. All-to-all personalized communication is also required during the partitioning stage. We have implemented library classes for these operations, which are used by simply inheriting a message class from them.

7 Preliminary performance results

We have implemented our parallel AFMA algorithm using the Charm++ portable object-oriented parallel programming system. Table 1 presents preliminary results on the TMC CM-5 and Intel Paragon for our 3-D AFMA implementation. The timings are for a nonuniform distribution[3] of 20,000 particles, for one time-step. The number of multipole expansion terms is 8, corresponding to the high-accuracy simulations in the work by Board [1]. The results do not include the parallel tree formation step. It was observed that several computation steps can be carried out before a tree formation is required; also a new tree-partitioning every time-step is not required for the correctness of the algorithm, as long as particles are moved to their new cells after a computation step. A detailed analysis of performance is not presented here due to lack of space.

Processors	16	32	64	128
Paragon	138.3	72.8	37.8	19.7
CM-5	315.4	147.1	74.9	42.2

Table 1: Time (in seconds) to simulate one time-step for 20,000 particles on the Intel Paragon and TMC CM-5.

8 Summary

In summary, the contributions of this work are :

- Our work is one of the first parallel implementations of the adaptive fast multipole algorithm on distributed memory computers.

- The algorithm is based on a modified version of the Greengard-Rokhlin AFMA; it allows particles to be organized in cells of arbitrary shape, instead of restricting them to be cubical. This simplifies parallel partitioning, so that good locality and load balance are both easily achieved.

- We have developed a tighter well-separatedness criterion which reduces the number of expensive pair-wise interactions, and an improved technique for constructing the locally essential tree.

- We have designed a fast edge-partitioning heuristic to solve the load-balancing problem caused when redundant pair-wise interactions are eliminated.

[3] Each coordinate value in this distribution was generated by doing a bitwise AND of two random integers, and then normalizing it within the computational box. Thus most particles are concentrated at one of the corners of the box.

- Our implementation automatically overlaps computation and communication within and across stages of the algorithm.

- Our implementation is optimized using message prioritization, advance-sends, and other communication optimizations.

Acknowledgements

We would like to thank John Board at Duke University for the sequential FMA code on which our implementation is based.

References

[1] J. A. Board et al. Scalable implementations of multipole accelerated algorithms for molecular dynamics. In *Proceedings of the Scalable High Performance Computing Conference*, May 1994.

[2] J. A. Board, L. V. Kale, K. Schulten, R. Skeel, and T. Schlick. Modeling biomolecules: Larger scales, longer durations. *IEEE Computational Science and Engineering*, 1(4), 1994.

[3] J. Carrier, L. Greengard, and V. Rokhlin. A fast adaptive multipole algorithm for particle simulations. *SIAM Journal of Scientific and Statistical Computing*, 9, July 1988.

[4] A. Grama, V. Kumar, and A. Sameh. Scalable parallel formulations of the Barnes-Hut method for n-body simulations. In *Proceedings of Supercomputing 94*, Nov. 1994.

[5] L. Greengard. *The Rapid Evaluation of Potential Fields in Particle Systems*. MIT Press, 1988.

[6] L. Greengard and W. D. Gropp. A parallel version of the fast multipole method. *Computers Math Applica*, 20(7), 1990.

[7] L. Kale. The Chare Kernel parallel programming language and system. In *Proceedings of the International Conference on Parallel Processing*, Aug. 1990.

[8] L. Kale and S. Krishnan. Charm++ : A portable concurrent object oriented system based on C++. In *Proceedings of the Conference on Object Oriented Programming Systems, Languages and Applications*, September 1993.

[9] P. Liu and S. Bhatt. Experiences with parallel n-body simulation. In *6th Annual ACM Symposium on Parallel Algorithms and Architectures*, 1994.

[10] J. Singh, C. Holt, J. Hennessy, and A. Gupta. A parallel adaptive fast multipole method. In *Proceedings of Supercomputing 93*, Nov. 1993.

[11] J. P. Singh, J. L. Hennessy, and A. Gupta. Implications of hierarchical n-body methods for multiprocessor architecture. *ACM Transactions on Computer Systems*. To Appear.

[12] M. S. Warren and J. K. Salmon. A parallel hashed oct-tree n-body algorithm. In *Proceedings of Supercomputing 93*, Nov. 1993.

PROGRAMMING REGULAR GRID-BASED WEATHER SIMULATION MODELS FOR PORTABLE AND FAST EXECUTION

Bernardo Rodriguez, Leslie Hart, and Tom Henderson
NOAA Forecast Systems Laboratory
R/E/FS5, 325 Broadway, Boulder, CO 80303 USA
e-mail: bernardo, hart, and hender @fsl.noaa.gov

Abstract -- *We have developed a high-level library, the Nearest Neighbor Tool (NNT), that can be used to code regular grid-based weather prediction models for portable execution on sequential, vector, and parallel computers. NNT is being used to parallelize operational weather forecast models, like the Eta and Rapid Update Cycle (RUC) models currently run at the National Meteorological Center (NMC). A user can gain three fundamental benefits from NNT: ease of programming, portability, and high performance. NNT provides a set of routines with Fortran 77 binding to decompose data arrays, execute I/O and communication operations, and simplify the parallelization of sequential loops and the optimization of their execution. The source codes written using NNT are fully portable, as are the data files. Since NNT is implemented as a layered set of routines, machine-dependent optimizations are possible and invisible to the user.*

1. INTRODUCTION

Operational weather forecast offices, like the National Meteorological Center, run numerical weather simulation programs to produce daily forecasts. These programs require very high computation and I/O speeds. Massively parallel computers are being investigated as a way of meeting current and future demands [4]. Besides performance requirements, "soft" requirements are also very important, such as portability, reduced development cost, and programmability. It is necessary that the code and data files port among computing platforms (sequential and parallel) without source changes so that new machines can be used and research can be conducted on different platforms. It is also important to maintain a single version of the code. The transfer from current sequential programs (sometimes with

directives for vector Supercomputers, like Crays) to the new programs that execute on parallel and sequential architectures should occur with minimal cost. Moreover, the changes to the code should have a minimum impact on the code "appearance," so that model developers can recognize and modify sections of code in the new program.

We have designed NNT to be used in the development of codes that comply with the requirements described above. NNT provides a local address space programming environment. It is a set of Fortran 77 callable routines built over a layer of lower level routines that form the interface with the specific computing platform where the codes must be run. NNT has been ported to single processor computing environments, such as workstations and single-processor Crays, and multiple-processor computing environments like the Intel Paragon, SGI Challenge, Cray T3D, IBM SP2, workstation clusters, and nCUBE. A port to the MPI message passing standard [14] is under way. Programs written exclusively using NNT and Fortran 77 port among platforms without any source code change, as do the input and output data files. Machine dependent code, like explicit message passing or synchronization operations, may be used with NNT but inhibit portability. NNT allows for optimization in two areas: at the user level and at the implementation level. At the user level, NNT provides routines for the run-time decomposition of data to processors, the overlap of computation and communication, and the overlap of processor activities with I/O operations. These are described in Section 3. At the implementation level, NNT allows machine-specific optimizations that are invisible to the user. These are usually related to message passing, I/O, global or local address space architectures, and single or multiple process execution. For example,

NNT implementations may use machine-dependent I/O calls that enhance I/O throughput, and avoid multilevel buffering in machines with global address space.

As a sequential code is modified for parallel execution using NNT, the changes to the code are minimal. For example, in the NNT version of NMC's Eta model [6], only 10% of the new code contained program lines that were added or modified (half of those were for I/O). Not only it is important that the number of changes introduced by NNT be minimal, but that the resulting loop structures and computation lines be identical to the original versions, as is shown in the examples in Section 3.

Several libraries have been proposed to parallelize weather prediction models. Among them, the GMD Grid Communication Library [5] and Argonne National Laboratory's RSL [9] seem to be the most promising [3][10]. Other efforts to parallelize weather prediction models have been based on high-level libraries [2][8]. The GMD library is of more general scope than NNT, since they allow for non regular domains. The cost of this generality is an added complexity in their use. RSL schedules individual columns of volume to each processor. This makes it simple to implement dynamic load balancing procedures but forces the programmer to change the original sequential code considerably. The advantages of NNT are the simplicity of its use, the treatment of I/O operations, and the possibility of overlapping computation, communication, and I/O operations. In Section 2 we present a general description of NNT. Using examples from the RUC and Eta models, we describe in Section 3 a method for the parallelization of a sequential code using NNT. In Section 3 we also present performance results of different applications coded using NNT, and show how fully portable code that includes performance optimizations can be generated and still maintain a simple programming interface. In Section 4 we show performance results of a parallel Laplace solver coded using NNT which was run on the SGI Challenge multiprocessor. We also present the performance of the same model coded using the native programming tool of the machine (DOACROSS directive) to examine the cost of using a local address space paradigm on multiprocessors with hardware support for global address space. We show that under certain conditions the NNT code executes faster than the native SGI code. We also present results of the RUC model coded using NNT on the Intel Paragon.

2. GENERAL DESCRIPTION OF NNT

NNT was designed for the parallelization of regular grid-based weather prediction models. In this type of model, the volume of the forecast area is mapped into a regular three-dimensional grid, and a set of partial differential equations is solved for each point in the grid. The computations found in weather prediction models have characteristic stencils that define the spatial data dependencies for the computation of a particular grid point. For example, the equation

$$df(i,j) = \frac{1}{4} \cdot (f(i-1,j) + f(i+1,j) + f(i,j-1) + f(i,j+1) - f(i,j))$$

has a 5 point stencil. We define the half-width of the stencil, in each dimension, as the maximum displacement in the index for that dimension. In the above equation, for example, the stencil half-width is 1 in the i and j dimension. The equation indicates that the computation of df in any grid point depends on the values of array f on the four neighboring points.

To solve this type of problem on a parallel computer, a subset of points (or region) is assigned to each particular process (data decomposition). Let V_n be the volume assigned to process n, defined by $V_n(i\_start)$, $V_n(i\_end)$, $V_n(j\_start)$, $V_n(j\_end)$, $V_n(k\_start)$, $V_n(k\_end)$. Let $W(i)$, $W(j)$, and $W(k)$ be the maximum stencil half-widths in each dimension. Process n will need to get (either through explicit message passing or through shared access) the data necessary to compute the points inside V_n. The mapping of processes to data defines a virtual process array, where neighboring processes have adjacent points in the three-dimensional grid. If f_n is the set of processes that are neighbors to n, NNT can be used to move data for the computation of V_n if

$$V_m(x\_end) - V_m(x\_start) > W(x)$$

for all

$$m \in f_n$$

and

$$x \in \{i, j, k\} \ .$$

That is, if the data needed for the computation of V_n are completely inside all of the regions assigned to the processes neighboring n.

NNT provides a routine to map processes to data (data decomposition). Each processor will have locally the set of data that corresponds to the region assigned to the process executing on the processor. This data can be input or output using the NNT I/O calls. Enough storage has to be allocated in each array to contain the region assigned for computation and the values from neighboring processes necessary for that computation. We will refer to the data in the region assigned for computation as interior data, and the neighboring data necessary for computations in the edge of the interior as the boundary data. NNT communicates boundary data between neighboring processes using boundary exchanges. The amount of data exchanged is given by the exchange *thickness*, which can be varied dynamically throughout the program execution. The maximum value of the exchange thickness is defined at the time of the data decomposition, and should be large enough to include the largest stencil related to a particular array. The boundary thickness should be set by the user according to the maximum value of exchange thickness.

The parallelization of a sequential code using NNT is based on data decomposition. Each process will execute the same program on different data. The first step in the parallelization process is to study the data dependencies in the code. These dependencies will define, for each array, the allowed decomposition types and the maximum thickness (stencil width) associated with it. The user is responsible for providing arrays large enough to hold the interior and boundary data. The specification of decomposition types and thickness can occur at run time. A data decomposition defines the regions assigned to each processor and also defines a virtual array of processors, creating neighboring relations between them. When the data decomposition routine is called, a decomposition structure is created and a structure handle is returned (a handle is a Fortran 77 integer). This handle will be used as an argument to subsequent NNT calls to specify the decomposition that is the context to the specific NNT operation. Many arrays can be associated with the same decomposition.

To execute I/O or exchange operations, NNT uses structures that must be defined by the user (at run time). These structures can be shared by many arrays, and allow for array aggregation so that multiple arrays can be input, output, or exchanged simultaneously. Handles to these structures are returned from routines that define the above operations. In Figure 2.1 we show the flow of creation and usage of NNT structures (accessed by handles). To define each operation the user must specify a decomposition handle and the number of arrays aggregated in the operation. To define an exchange operation a data type must be

specified. The parameters for the I/O or exchange operations can also be modified at run time. This allows to change the selection of array operands and the exchange patterns and thickness for boundary exchanges.

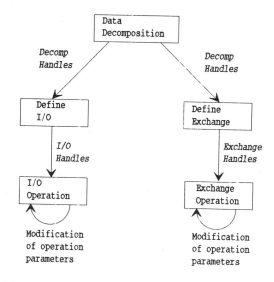

Figure 2.1. Dependence flow in the creation and usage on NNT handles

Once the decomposition, I/O, and exchange structures have been defined, the programmer inserts calls to NNT I/O and exchange functions where they are necessary. The remainder of the work is the loop index conversion. Since sequential codes are programmed in single address space, and NNT presents a local address view, the *start* and *end* parameters of loop structures must be changed to *start* and *end* values in the local space of each processor. NNT provides calls to do this translation.

3. PARALLELIZING A WEATHER FORECAST MODEL: EXAMPLES

To show how a weather prediction model is parallelized using NNT, we describe different aspects of this process using examples from the parallelization of the Eta and the RUC models [6][7]. A typical sequential program has I/O phases and computation phases. Figure 3.1 shows the typical flow of execution of an NNT program. In the NNT setup procedure, decomposition, I/O and exchange handles are created to define the NNT operations that will occur during execution. For large programs, it has been found useful to build a separate "NNT Setup" module that will call the NNT "define" routines. The handles can be passed to the rest of the program modules using common blocks. After the I/O and exchange handles are created, calls to NNT I/O and NNT exchange operations can be made. The computation part of a weather prediction program is primarily a set of loops. The *start* and *end* parameters of these loops can be modified using the NNT index translation routines (which can also be called from the NNT Setup module).

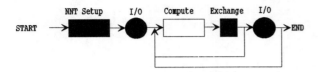

Figure 3.!. Typical flow of execution of an NNT program.

In the remainder of this section, we show examples of the usage of NNT. These examples were taken from the parallel versions of the Eta and RUC models. Details of the NNT calls are available in the NNT User's Guide [13].

3.1. Creation of NNT handles

In the RUC model a common block in an include file was used to declare the handles (decomposition, I/O, and exchange handles) used by NNT and the loop index translation vectors used to translate loop *start* and *end* parameters. This include file was initialized by calls to NNT define functions. The following code defines a data decomposition structure (the necessary declarations are omitted):

```
1    IDECOMPTYPE= NNT_DECOMP_1_2
2    IBDYTYPE(1) = NNT_DECOMP_BDY_NONPER
3    IBDYTYPE(2) = NNT_DECOMP_BDY_NONPER
4    IBDYTYPE(3) = NNT_DECOMP_BDY_NONPER
5    ITHICK(1) = I_THICKNESS
6    ITHICK(2) = J_THICKNESS
7    ITHICK(3) = 0
8    IDATASIZE(1) = mix
9    IDATASIZE(2) = mjx
10   IDATASIZE(3) = 1
11   CALL NNT_decomp(IDECOMPTYPE,
     & IBDYTYPE, IDATASIZE, ITHICK,
     & IDECOMP_2D, ISTAT)
```

Line 1 sets the decomposition type to a two-dimensional decomposition in the first two dimensions. Lines 2-4 set boundary types to nonperiodic. With periodic boundary conditions, the processors in the boundaries of the global domain exchange data with the processors in the opposite boundary. In nonperiodic boundaries, processors in the boundary of the global domain do not exchange data outside of the boundary. Lines 5-7 set the

boundary thickness, lines 8-10 set the size of the global array, and line 11 calls the NNT function that returns the decomposition handle IDECOMP_2D. Since Fortran 77 arrays are declared statically by the programmer, the information on the declared size is passed to NNT using the NNT_declared_size routine:

```
1    ISTORAGESIZE(1) = mix_A
2    ISTORAGESIZE(2) = mjx_A
3    ISTORAGESIZE(3) = 1
4    CALL NNT_declared_size(IDECOMP_2D,
     & ISTORAGESIZE, ISTAT)
```

Lines 1-3 sets the size used by the programmer in the array declaration, and line 4 calls the NNT function to set the declared size to the appropriate value. Since the size of the regions assigned to a processor decreases as the number of processors participating in the run increases, it is possible and generally beneficial to use the minimum declared array size for a given number of processors. This will optimize performance by improving the usage of the machine's memory hierarchy.

To define a three-dimensional exchange the nnt_define_exch call is used:

```
1    CALL NNT_define_exch(IDECOMP_3D,
     & IEXCH_F_3D_5VAR,
     & NNT_REAL,5,ISTAT)
```

The decomposition handle IDECOMP_3D is from a three-dimensional decomposition previously defined. The nnt_define_exch call creates an exchange handle (IEXCH_F_3D_5VAR) for an exchange of up to 5 real arrays. Assignment of model variables to an exchange handle is shown in Section 3.3.

The following are examples of the calculation of index translation vectors for translation of the *start* and *end* parameters in loop structures:

```
1    CALL NNT_loops(IDECOMP_3D,1,
     & I_START,I_END,1,NREGIONS,ISTAT)
2    CALL NNT_loops(IDECOMP_3D,2,
     & J_START,J_END,1,NREGIONS,ISTAT)
3    CALL NNT_loops_nd(IDECOMP_3D,1,
     & I_START_ND,I_END_ND,1,
     & NREGIONS,ISTAT)
4    CALL NNT_loops_nd(IDECOMP_3D,2,
     & J_START_ND,J_END_ND,1,
     & NREGIONS,ISTAT)
```

Lines 1 and 2 calculate the *start* and *end* parameters of the *I* (first) and *J* (second) dimensions of a loop structure that operates on arrays decomposed according to decomposition described by the handle IDECOMP_3D. The resulting vectors (I_START, I_END, J_START, J_END) will provide *start* and *end* indices to loop over the interior data of the processors. Lines 3 and 4 call `nnt_loops_nd`, which provides index translation vectors to loop over the interior and boundary data. NNT also provides calls to get index translation vectors to loop over the interior data and a limited area of the boundary data. The ability to perform computation in all or portions of boundary regions has proven very useful in optimizing parallel codes [7]. The computations in a processor's boundary region are redundant, since those computations are also executed in a neighbor's interior region. If the data dependencies allow it, redundant computations can be traded for communication, since data in the boundaries can be computed by several processors without need for communication. We show the use of the loop index translation vectors in Section 3.4. The parameters NREGIONS and 1 are only used for compatibility with future versions of NNT. The following is the definition of an I/O operation taken from Eta:

```
1   call NNT_make_io_handle(IO_2D_OUTUV,
    & IDECOMP_2D,2,ISTAT)
```

This call creates an I/O handle (IO_2D_OUTUV) that defines the input and output operations of any array decomposed with a two-dimensional decomposition identified by the handle IDECOMP_2D. A total of 2 arrays can be read or written simultaneously per operation per process using the handle IO_2D_OUTUV.

3.2. I/O operations

NNT provides a full set of I/O operations. These include string and binary I/O, and decomposed and non-decomposed array I/O. We only describe input and output operations on decomposed arrays. To execute an I/O operation, the arrays that will be the operands to the operation have to be assigned to the I/O handle. The following are examples taken from Eta:

```
1   CALL NNT_assign_io_var(IO_2D_OUTUV,
    & EGRID1_2d,1,ETAFLOATTYPE,ISTAT)
2   CALL NNT_assign_io_var(IO_2D_OUTUV,
    & EGRID2_2d,2,ETAFLOATTYPE,ISTAT)
```

Lines 1 and 2 assign arrays EGRID1_2d and EGRID2_2d as the first and second arrays to be operands of the I/O operation. Lines 1 and 2 are only necessary once, unless other arrays use the same I/O handle, and the array assignment of those lines is overwritten. I/O is performed using the following calls:

```
1   CALL NNT_write_multi(IFIL_OUT,
    & IO_2D_OUTUV,ISTAT)
```

```
2   CALL NNT_flush_all(ISTAT)
```

Line 1 calls the NNT write routine, which writes the arrays into the file IFIL_OUT. The read routine has the same syntax. Line 2 is an optional call to the NNT flush routine. NNT I/O write operations buffer the data in separate nodes (*write buffers*) that serve as an interface to the storage system (a future version of NNT will provide I/O buffering for read operations). This buffered data can be flushed at the time of the NNT_flush_all call, or automatically by the system when the buffers are full. The call to NNT_flush_all (that returns immediately) avoids having to wait for an automatic flush. The NNT write flush operation was used on the Eta and RUC models to overlap disk write time with processor activity. The number of write buffers active in a particular run is specified at run time, using the following syntax:

nntex NP executable NWB write_buffs

where NP and NWB are the number of processors and number of write I/O buffer processors for the run, respectively. I/O buffers have proven to reduce considerably the impact of I/O operations on program execution time. For example, in the RUC model with 40 KM resolution over the continental US, the execution time on 144 processors of the Intel Paragon was reduced by an order of magnitude by adding 9 I/O buffer processors [15]. Experiments show that the disk transmission speed component to total execution time can be completely overlapped using NNT write operations, given, as is the case of the Eta and RUC model, there is enough processor activity to occur during the write time.

3.3. Exchange Operations

Once an exchange operation has been defined, the arrays to be exchanged have to be assigned to the exchange structure and the exchange pattern must be set. The following is an example taken from the parallel version of the RUC model:

```
1   CALL NNT_assign_var(IEXCH_F_3D_5VAR,
    & 1,ua,NNT_DO_ALL, ISTAT)
2   CALL NNT_assign_var(IEXCH_F_3D_5VAR,
    & 2,va,NNT_DO_ALL,ISTAT)
3   CALL NNT_assign_var(IEXCH_F_3D_5VAR,
    & 3,ta,NNT_DO_ALL,ISTAT)
4   CALL NNT_assign_var(IEXCH_F_3D_5VAR,
    & 4,dfpa,NNT_DO_ALL,ISTAT)
5   CALL NNT_assign_var(IEXCH_F_3D_5VAR,
    & 5,qva,NNT_DO_ALL,ISTAT)
```

Lines 1-5 assign arrays ua, va, ta, dfpa, and qva as arrays to be exchanged. Typically, these lines are necessary only once in a program. The exchange pattern (NNT_DO_ALL) is set to exchange boundaries between all neighbors, including the corner neighbors (a total of 8 neighbors in a two-dimensional decomposition, for example). The exchange operation is executed with the following call:

```
1    CALL NNT_exchange(IEXCH_F_3D_5VAR,
     & ISTAT)
```

The NNT exchange has as parameter the exchange handle. NNT also provides subroutines to overlap exchange operations with computation: NNT_start_exch and NNT_end_exch. NNT_start_exch returns immediately, effectively overlapping the exchange time with the computations that precede the call to NNT_end_exch. The syntax for these calls is the same as for NNT_exchange.

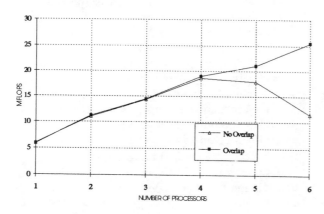

Figure 3.2. Performance of the Well-Posed Topographic model on a network of Sun Sparc2 workstations with Ethernet connection. Data size: 130x98x31. 32-bit precision.

NNT segmented exchanges were used to improve the scalability of the Well-Posed Topographic model running on a network of Sun Sparc2 workstations linked through Ethernet [11] (Figure 3.2). The boundary exchange was overlapped with the computation of the interior region assigned to each processor. The results show that performance optimizations were possible without sacrificing a simple programming interface.

3.4. Array Index Translation

Once the calls that compute the loop index translation vectors are made, the resulting vectors can be used to substitute *start* and *end* parameters in loop structures. The index translation vectors define the three-dimensional local space on which each

processor should execute the loop operations. The following examples are from the RUC forecast model:

Original sequential code:

```
1    do 122 j = 2,jlx
2    do 122 i = 2,ilx
3    do 122 k = 1,klx
4      ka = max(k-1,1)
5      qva(i,j,k) = qva(i,j,k)-.5*
     &  ((sdot(i,j,k)+abs(sdot(i,j,k)))
     &  *(qvb(i,j,k)*qvb(i,j,k+1))
     &  -(sdot(i,j,k)-abs(sdot(i,j,k)))*
     &  (qvb(i,j,k)-qvb(i,j,ka)))/
     &  (p(i,j,ka )-p(i,j,k+1))
122  continue
```

Parallel NNT code:

```
1    do 122 j = J_START(2),J_END(jlx)
2    do 122 i = I_START(2),I_END(ilx)
3    do 122 k = 1,klx
4      ka = max(k-1,1)
5      qva(i,j,k) = qva(i,j,k)-.5*
     &  ((sdot(i,j,k)+abs(sdot(i,j,k)))
     &  *(qvb(i,j,k)*qvb(i,j,k+1))
     &  -(sdot(i,j,k)-abs(sdot(i,j,k)))*
     &  (qvb(i,j,k)-qvb(i,j,ka)))/
     &  (p(i,j,ka )-p(i,j,k+1))
122  continue
```

Notice that the only changes to the code occur in lines 1 and 2, where the *start* and *end* parameters were replaced by the index translation vectors indexed with those start and end values. This greatly simplifies the translation process, which could be done by a preprocessor.

In summary, the parallelization of an application using NNT involves four steps: the creation of the NNT setup module, the substitution of loop index parameters, the insertion of exchange operations, and the substitution of I/O calls. The substitution of loop parameters is trivial and could be done by a preprocessor, the insertion of exchange operations is guided by data dependence analysis necessary in any parallelization effort, and the substitution of I/O calls should be easy given the simple interface to the NNT I/O calls. The NNT setup module contains the basis for the parallelization of the application and it defines the way data will be decomposed, exchanged, input, or output. This is where most of the calls to NNT will occur. It is our experience, however, that the structure of an NNT setup module can be reused between applications.

4. THE COST OF PORTABILITY IN NNT

One of the main reasons for designing NNT over a local address paradigm is that it allowed us to port between local and global address space architectures. The cost of portability is the inefficient use of global address space architectures if send and receive functions are called to move data between processors. We ported NNT to the SGI Challenge using System V Interprocess Communication (IPC) shared memory operations, which allow efficient movement of data during boundary exchanges. The data is gathered and scattered into a single shared buffer for communication, avoiding the high cost associated with the multi-level buffering used in message passing libraries [11].

We use a 2 dimensional Laplace solver that uses a five point stencil to probe the effectiveness of NNT on the SGI Challenge shared memory multiprocessor. We parallelized the Laplace solver using the DOACROSS directive and using NNT. The SGI DOACROSS directive only allows decomposition in one dimension. With NNT, however, the user can decompose in two dimensions. Figure 4.1 shows the execution time of the native SGI implementation (1D), the NNT implementations (1D and 2D), and the ideal execution time assuming perfect speed-up.

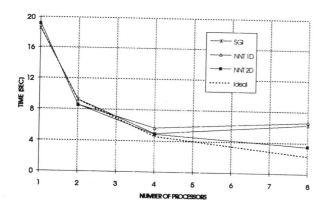

Figure 4.1. Execution time for a Laplace Solver on the SGI Challenge. The problem size was 512x512 and the algorithm used a five-point stencil. 32-bit precision.

We can see in Fig. 4.1 that the execution times are very similar between the native SGI and the 1D NNT implementation. The 2D NNT implementation,

on the other hand, executes faster. (Remember that the difference between 1D and 2D versions of an NNT program is simply the selection of the decomposition type, which can be done at run time.) It is not possible to implement a two dimensional decomposition using native SGI directives.

In general, global address space programs that decompose arrays on the dimension that is stored contiguously in memory will not use caches with large blocks effectively. Let us consider, for example, a two dimensional problem that is decomposed in both dimensions. If the cache block contains b words, then the cache hit ratio will be $b-1/b$ during the access to the boundaries in the non-contiguous storage dimension. However, during the access to the boundaries in the contiguous dimension, the cache hit ratio will be $W-1/b$, where W is the stencil half-width in the contiguous dimension. NNT (and in general message passing) offers an "unexpected" benefit in programming shared memory multiprocessors with caches, since it improves cache utilization on multidimensional decompositions. In message passing a processor copies data into a one-dimensional buffer before a send. The processor that later reads the buffer (receive operation), gets the full benefit of multiple-word fetch on a cache line load. Using message passing, a two dimensional decomposition of the two dimensional Laplace solver will be ideal, since it minimizes the ratio of communication to computation. Note that in a global address space multiprocessor with caches that have one-word blocks, a two dimensional decomposition would also minimize the number of coherence misses. Figure 4.1 shows that NNT is suitable for developing portable codes that run effectively on global address space architectures with small number of processors. We have not tested the global address port for large numbers of processors. The cost of using NNT on message passing architectures is negligible, related only to the overhead of two extra levels of subroutine calls on a boundary exchange.

Figure 4.2 shows the execution time for a 12 hour forecast of the RUC model. The model resolution was 40 KM over the continental US. The time includes the input of the initial model conditions (~30 MB) at the start of a run and the output of the atmosphere state (~18 MB) every hour of forecast, or thirteen times per run. The number of processors indicated in Fig. 4.2 includes the I/O buffers (see Section 3.2). Out of 153 processors, for example, 144 processors were compute processors and 9 were I/O buffer processors. For 144 processors, the addition of I/O buffers increased performance by an order of magnitude. The ideal execution time shown as a dotted line in the figure is obtained from perfect speed-up after 18 processors. The program was not run below 18 processors because paging to disk would have been necessary, resulting in very high execution times. The 40 Km RUC model scaled well up to 153 processors. Early results indicate that higher resolution models (20 Km) will have even better scaling.

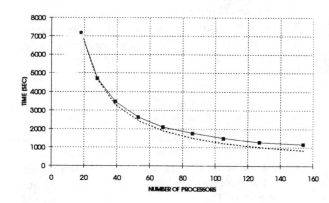

Figure 4.2. Execution time on the Intel Paragon for the RUC model, 40 KM resolution, 12 hour forecast over the continental US. Dotted line indicates ideal scaling after 18 processors. 32-bit precision.

5. CONCLUSIONS

NNT provides a set of subroutines that can be used for the parallelization of regular grid based weather prediction models. The Eta and the RUC operational models are being parallelized using NNT [6]. The definition of data decompositions, I/O operations, and exchanges, can be localized in a single program module. The model code appearance is minimally affected, since only I/O operations and *start* and *end* parameters in loop structures need to be modified. Boundary exchanges need to be placed according to data dependence analysis. The code developed using NNT runs without source code change on any computing platform where NNT is ported. The data files also port between computers without changes. For the types of problems investigated and small number of processors, the performance of the codes that use NNT is comparable to the native implementation on machines with global address space supported in hardware. Moreover, NNT provides additional benefits on global address space multiprocessors with caches, since it allows for efficient cache use on multidimensional decompositions.

NNT overlaps data movement with local computation, during exchange and I/O operations. We have shown that the scalability and performance of a parallel code can be increased considerably by using these optimizations, while still maintaining a simple programming interface.

Acknowledgment. Funding for this project is being provided by the FAA/ARD-80, Aviation Weather Development Program Office. This work is supported by the Aviation Development Program. The authors would like to thank George Carr, of Intel's Scalable Systems Division, for reviewing this paper.

6. REFERENCES

[1] C. Baillie, A. E. MacDonald, "Porting the Well-Posed Topographical Meteorological Model to the KSR Parallel Supercomputer," *Proceedings of the Fifth Workshop on Use of Parallel Processors in Meteorology*, World Scientific, (1992).

[2] A. Dickinson, P. Burton, E. Hibling, J. Parker, and R. Baxter, "Implementation and Initial Results from a Parallel Version of the Meteorological Office Atmosphere Prediction Model," *Proceedings of the Sixth Workshop on Use of Parallel Processors in Meteorology*, World Scientific, (1994).

[3] U.Gartel, W. Joppich, and A. Schuller, "Parallelizing the ECMWF's Weather Forecast Program: The 2D Case," *Arbeitspapiere der GMD 740*, GMD, (3,1993).

[4] L. Hart, T. Henderson, and B. Rodriguez, "Evaluation of Parallel Processing Technology for Operational Capability at the Forecast Systems Laboratory," *Proceedings of the Sixth Workshop on Use of Parallel Processors in Meteorology*, World Scientific, (1994).

[5] R. Hempel, and H. Ritzdorf, "The GMD Communications Library for Grid-Oriented Problems," *Technical Report 589*, GMD, Sankt Augustin, Germany, (1991).

[6] T. Henderson, C. Baillie, G. Carr, L. Hart, A. Marroquin, and B. Rodriguez, "Parallelizing the Eta Weather Forecast Model: Initial Results," *Proceedings of High Performance Computing '94.* Society for Computer Simulation, La Joya, California, (1994).

[7] T. Henderson, C. Baillie, S. Benjamin, T. Black, R. Bleck, G. Carr, L. Hart, M. Govett, A. Marroquin, J. Middlecoff, and B. Rodriguez, "Progress Towards Demonstrating Operational Capability of Massively Parallel Processors at the Forecast Systems Laboratory," *Proceedings of the Sixth Workshop on Use of Parallel Processors in Meteorology,* World Scientific, (1994).

[8] T. Kauranne, "The Operational HIRLAM 2 Model on Parallel Computers," *Proceedings of the Sixth Workshop on Use of Parallel Processors in Meteorology,* World Scientific, (1994).

[9] J. Michalakes, "RSL: A Parallel Runtime System Library for Regular Grid Finite Difference Models Using Multiple Nests," *Tech. Rep. ANL/MCS-TM-197,* Argonne National Laboratory, Argonne. Illinois, (12,1994).

[10] J. Michalakes, G. Grell, Parallel "Implementation, Validation and Performance of MM5," *Proceedings of the Sixth Workshop on Use of Parallel Processors in Meteorology,* World Scientific, (1994).

[11] B. Rodriguez, L. Hart, T. Henderson, "Performance and Portability in Parallel Computing: a Weather Forecast View," *High Performance Computing in the Geosciences,* Kluwer, (1995).

[12] B. Rodriguez, L. Hart, and T. Henderson, "A Library for the Portable Parallelization of Operational Weather Forecast Models," *Proceedings of the Sixth Workshop on Use of Parallel Processors in Meteorology,* World Scientific, (1994).

[13] B. Rodriguez, L. Hart, T. Henderson, "NNT 1.0 User's Guide," *Forecast Systems Laboratory Technical Memorandum ,* in preparation.

[14] Message Passing Interface Forum, "MPI: A Message-Passing Interface Standard," *International Journal of Supercomputing Applications,* Vol. 8, Number 3/4, (1994).

[15] T. Henderson, L. Hart, B. Rodriguez, "High Performance Computing at FSL: Demonstrating Operational Capability of Massively Parallel Processors," *FSL Forum,* Forecast Systems Laboratory, (3,1995).

PARALLEL MULTIPLE SEQUENCE ALIGNMENT USING SPECULATIVE COMPUTATION

Tieng K. Yap[1], Peter J. Munson[1], Ophir Frieder[2], and Robert L. Martino[1]

[1]Division of Computer Research and Technology, National Institutes of Health
12 South Dr. ● Bldg 2A ● Rm 2033 ● Bethestha, MD 20892-5624
{yap | munson | martino}@alw.nih.gov
[2]Department of Computer Science, George Mason University
Fairfax, VA 22030-4444
ophir@cs.gmu.edu

Abstract -- *Many different methods have been presented for aligning multiple biological sequences. These methods can be classified into three categories: rigorous, tree-based, and iterative. The rigorous method, which always generates the optimal alignment, requires memory space and computation time proportional to the product of the sequence lengths. Even for a modest number of sequences, this method becomes impractical. As a result, the other two methods were introduced. The iterative methods were shown to generate better alignments than the tree-based methods. However, these methods require as much as 100 times longer computation time. A number of days may be required to align a large number of sequences (e.g., over 100 sequences) sequentially. We present a parallel speculative computational method which reduces the computation time of the iterative methods from days to minutes. To evaluate our parallel method, we implemented a speculative computation version of the iterative improvement method of Berger and Munson on an Intel iPSC/860 parallel computer. The empirical results demonstrate that our parallel method obtained a significant speed up in comparison to the sequential method.*

INTRODUCTION

Pairwise sequence alignment is an important tool for locating similarity patterns between two biological (DNA and protein) sequences. This analytical tool has been used successfully to predict the function, structure, and evolution of biological sequences. The first biological sequence alignment algorithm, referred to as dynamic programming, was introduced by Needleman and Wunsch [21]. This algorithm was later improved by a number of researchers through an improved formulation [24], a reduced time complexity [8], and a reduced space complexity [18]. As more sequences were generated by the biomedical community, researchers began to use multiple sequence alignment to obtain a better understanding of biological sequences. Although the rigorous dynamic programming algorithm can be extended to align more than two sequences, it is impractical to extend this algorithm to the alignment of more than three sequences [19, 20] since memory space and computation time is proportional to the product of the sequence lengths. By restricting the number of possible alignments, the algorithm can be extended to align six to eight sequences [4, 16].

To practically align a large number of sequences, many researchers use the tree-based methods which generate a multiple sequence alignment by combining a number of pairwise alignments in a particular order. For a binary tree where there is only one branch at each level, the order is linear [1, 17, 25]. These linear ordering strategies start with the most similar sequence pair and continue to add sequences to the alignment in order of decreasing similarity. The linear ordering strategies produce a good multiple alignment if all the sequences belong to a single homologous family. However, as pointed out by Taylor [26], these strategies can produce a poor alignment if the sequences belong to two or more distinct subfamilies.

To improve the tree-based methods, researchers introduced sophisticated ordering strategies [2, 5, 7, 9, 10, 25]. These strategies apply various clustering techniques to order groups of related sequences in a hierarchical tree. Then, the final multiple alignment is obtained by combining clusters of sequences to the most

[2]This author was partially supported by the National Science Foundation under contract number IRI-9357785, by the Virginia Center for Innovative Technology under contract number INF-94-002, and by XPAND Corporation.

similar cluster in decreasing order of relatedness. Using tree-based methods, the order used to align and combine the sequences has a great effect on the final multiple sequence alignment. Consequently, a great deal of effort has been spent on designing new ordering strategies that would generate better alignments.

A few researchers [3, 12, 14, 15] have taken the opposite approach by not ordering the sequences in a systematic way. Instead, they applied randomized techniques with optimization functions to iteratively improve the multiple sequence alignment. These iterative methods produce better alignments than the tree-based methods and can be used to improve alignments that were generated from a tree-based method. However, they require a much longer computation time. In this paper, we use the Berger-Munson algorithm [3] to illustrate that the computation times of these iterative methods can be reduced significantly by using a parallel speculative computation technique.

SEQUENCE ALIGNMENT ALGORITHMS

Needleman and Wunsch [21] were the first to introduce a heuristic algorithm for aligning two biological sequences. Smith and Waterman [24] then formulated a more rigorous dynamic programming representation of this algorithm. Gotoh [8] followed by improving the time complexity of the algorithm. Let two sequences be $A=a_1a_2a_3...a_M$, and $B=b_1b_2b_3..b_N$ and let $sub(a_i,a_j)$ be a given similarity score for substituting residue a_i by a_j. The penalty for introducing a gap into a sequence is defined by the penalty function $w_k=-uk-v$, $(u,v \geq 0)$ where k is the gap length. The Gotoh algorithm is given as follows:

$$S_{i,j} = MAX \begin{cases} P_{i,j} \\ S_{i-1,j-1} + sub(a_i,b_j) \\ Q_{i,j} \end{cases}$$

$$P_{i,j} = MAX \begin{cases} S_{i-1,j} + w_1 \\ P_{i-1,j} + u \end{cases}$$

$$Q_{i,j} = MAX \begin{cases} S_{i,j-1} + w_1 \\ Q_{i,j-1} + u \end{cases}$$

$S_{i,j}$ is the cumulative alignment score between two sequences A and B up to the *ith* and *jth* positions. The initial conditions are defined as follows: $S_{i,0}=P_{i,0}=Q_{i,0}=0$ and $S_{0,j},P_{0,j}=Q_{0,j}=0$, for $0 \leq i \leq M$ and $0 \leq j \leq N$. As can be seen, the similarity scores $sub(a,b)$ between all possible

pairs of residues must first be defined before $S_{i,j}$ can be calculated. For our application, we use the PAM250 [6] substitution scoring matrix to define the similarity score between two residues.

To obtain the multiple sequence alignment, many algorithms (including tree-based and iterative) align two groups of sequences against each other a number of times. To align two groups, X and Y, of sequences, the algorithm of two-sequence alignment can be extended as follows:

$$S_{i,j} = MAX \begin{cases} P_{i,j} \\ S_{i-1,j-1} + sub'(X_i,Y_j) \\ Q_{i,j} \end{cases}$$

$$sub'(X_i,Y_j) = \sum_{k=1}^{K}\sum_{l=1}^{L} sub(X_{k,i},Y_{l,j}) +$$

$$\sum_{k=1}^{K}\sum_{m=k+1}^{K} sub(X_{k,i},X_{m,i}) + \sum_{l=1}^{L}\sum_{m=l+1}^{L} sub(Y_{l,j},Y_{m,j})$$

K is the number of sequences in the X group and L in the Y group. $P_{i,j}$, $Q_{i,j}$, and the initial conditions remain the same. If the sequences have already been aligned, the alignment score can be calculated using the following formula.

$$Score(A_N) = \sum_{i=1}^{N}\sum_{j=i+1}^{N}\sum_{k=1}^{L} sub(A_{i,k},A_{j,k})$$

where N is the number of sequences and L is the number of aligned positions.

BERGER-MUNSON ITERATIVE ALGORITHM

The Berger-Munson iterative improvement algorithm has been successfully used to perform multiple sequence alignment. Figure 1 shows the core part of this algorithm. The C language implementation contains approximately 1900 statements. To implement a parallel version of this algorithm, we separated the computational process into three steps. In step 1, the *n* input sequences are first randomly partitioned into two groups. Then, the alignment score between these two groups of sequences is calculated. In this step, the new gap positions are also saved for performing the alignment in step 3. In step 2, a decision flag is set to A (accepted) if the new resulting alignment is accepted; otherwise, it is set to R (rejected). A new alignment is accepted if the current score is higher than the current best score. If the decision flag in step 2 is set to A, the gap positions determined in step 1 are used to modify the current alignment in step 3 and the best score is updated. Then, the modified or unmodified alignment is used as the input for the next iteration. This iterative improvement algorithm continues until the stop criterion is met. We define the stop criterion as follows. After q consecutive iterations of rejections, the process is stopped where q is the number of all possible partitions.

```
best_score=initial_score();
While (stop criteria is not met){
    1 current_score = calculate(seq, gap_positions);
    2 flag = decide(current_score, best_score);
    3 seq = modify(seq, flag, gap_positions);
}
```

Figure 1. Berger-Munson Sequential Algorithm.

The Berger-Munson algorithm is highly sequential due to a loop-carried dependence between iterations. Iteration i depends on iteration $(i-1)$ since step 3 may modify the alignment during the $(i-1)th$ iteration and the modified alignment must be used by the ith iteration. In addition, the three steps within each iteration are also dependent on each other. Step 1 uses *seq* which may be modified by step 3. Step 2 uses *current_score* which produces by step 1. Step 3 uses *flag* variable which is set in step 2 and gap positions which are generated in step 1. These dependencies make it difficult to implement a parallel version of this algorithm while preserving the behavior of the original sequential version.

REVIEW OF A PRIOR BERGER-MUNSON PARALLELIZATION EFFORT

Ishikawa *et al.* [11] previously implemented a parallel version of the Berger-Munson algorithm on a parallel inference machine (PIM) using a parallel logic programming language (KL1). This parallel approach can be described as follows. All $(2^{n-1}-1)$ possible partitions or $n + \dfrac{n(n-1)}{2}$ restricted partitions, which are defined later in the Alignment Search Space section, are evaluated simultaneously in parallel. The resultant alignment which has the best score is selected as the input for the next iteration. One processor is used as the manager and the remaining processors as workers. Initially, the manager distributes each possible partition to a worker. When a worker finishes its calculation, it sends its alignment to the manager. Based on all the alignments collected from the workers, the manager selects the best alignment to be used as the

input for the next iteration.

The approach taken by Ishikawa *et al.* [11] has a few drawbacks. First, it becomes impractical for a large number of sequences. For example, approximately 10^{90} processors are needed to align 300 sequences if the unrestricted search space is used or 45,150 processors for the restricted space. Their implementation can be modified so that a large number of sequences can be aligned by dividing the number of partitions among the available processors as evenly as possible. However, it is still too costly to evaluate all partitions at each iteration. In the sequential version, only one random partition is evaluated at each iteration. Second, the parallel version is no longer a randomized process and its resultant alignment is not guaranteed to be as good as the one that is obtained from the original sequential version. That is, the quality of the derived alignment is unpredictable. Therefore, it is difficult to evaluate its performance. Third, the communication cost of the Ishikawa version can be reduced significantly.

In their approach, the sequences were sent back and forth between the manager and workers twice per parallel iteration. The manager sends the input sequences with the partition information to all the workers at the beginning of each iteration and all the workers send their alignments to the manager when they are done. As a result, the communication cost per iteration is approximately $2pnm$, where p is the number of processors, n is the number of sequences, and m is the length of the longest aligned sequence (original residues plus gaps). In our approach, we do not send sequences between processors. Only the gap and partition information are sent. As a result, our communication cost is approximately $2m$ since only one array of length m is used to hold the gap positions of each group. For large p and n, our communication cost is negligible in comparison to their approach.

PARALLEL SPECULATIVE BERGER-MUNSON ALGORITHM

Speculative computation [22, 23, 27] has been applied efficiently to parallelize sequential algorithms like simulated annealing, an algorithm similar to the Berger-Munson algorithm. By applying speculative computation to the parallelization of the Berger-Munson algorithm, we were able to achieve a higher speedup and a more scalable implementation than the prior effort mentioned above. In addition, our parallel alignment is guaranteed to be the same as the sequential one. The basic concept of speculative computation is to speculate the future solutions based on the current input parameters. Therefore, we can

Sequential Iteration number 1 2 3 4 5 6 7 . . .
Decision sequence AAAAARAAARRARRRARRRARRRARRRRR...

Figure 2. A Possible Sequential Decision Sequence.

speculate $(p-1)$ future solutions if we have p processors. In this application, we can speculate the alignments for the next $(p-1)$ iterations based on the current alignment.

In the original Berger-Munson algorithm, the final alignment is obtained by performing a sequence of alignments between two groups of sequences. Each iteration is accepted (A) if its alignment score is higher than the current best score. Otherwise, it is rejected (R). An example of a corresponding sequence of decisions is shown in Figure 2. Initially, every new alignment is accepted (e.g., iteration numbers 1-5). However, fewer and fewer are accepted as the alignment progresses. We stated earlier that the ith iteration may depend on the $(i-1)$th iteration. To be exact, the ith iteration depends on $(i-1)$th iteration only if the $(i-1)$th iteration has accepted a new alignment; otherwise, it only depends on the last accepted iteration.

Our parallel speculative computation approach is based on the recognition of the fact that a consecutive sequence of rejected iterations are not dependent on each other and can be done in parallel. Therefore, we can speculate that the $(p-1)$ previous iterations will be rejected so that they can be done in parallel. If the speculations are correct, the computation time could be reduced by a factor of p.

In the decision sequence of Figure 2, the first 28 sequential iterations can be reduced to 13 parallel steps if 4 processors are used. The parallel computation steps are shown in Figure 3. Three iterations $(p-1)$ are speculated at each parallel step where P_1 speculates that P_0 will reject its new alignment, P_2 speculates that P_0 and P_1 will reject their new alignments, and P_3 speculates that P_0 to P_2 will reject their new alginments. P_0 does not speculate. The numbers in the boxes of each row represent the speculated sequential iteration

numbers for the processor in that row at each parallel step. The iteration numbers that are speculated correctly, which also correspond to the sequential iteration number, are shown in bold. After each parallel step, the alignment of the last iteration that was speculated correctly is used as the input for the next step as shown by the lines leading from one parallel step to the next.

As the above illustration shows, we parallelized the Berger-Munson algorithm while preserving its sequential algorithmic behavior. The parallel alignment is guaranteed to be the same as the sequential one. For a large number of sequences, this algorithm can benefit significantly from parallel computation. Our parallel algorithm, which is implemented on every processor, is summarized in Figure 4 as C pseudocode with minor details omitted to improve clarity.

The variable gi is the global or sequential iteration number; bgi is the iteration number when the best score was obtained; q is the number of all possible partitions; $partn$ is a selected partition number for each individual processor; p is the number of processors; pid is the processor id ranging from 0 to $(p-1)$ and ap is the id of the processor that has accepted the best alignment.

To reduce the I/O time, only processor 0 reads the input sequences and then broadcasts them to the other processors since inter-processor data transfer is much faster than the I/O data transfer. Initially, every new alignment is usually accepted. As a result, we do not start to speculate until we encounter a rejection, (see lines 4 to 11). Every processor evaluates the same partition by initializing the same random seed. This strategy avoids the communication cost associated with the parallel speculative computation. The iteration

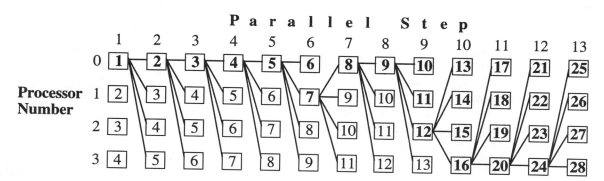

Figure 3. An Illustration of the Parallel Speculative Computation Process.

number is used as the random seed so that we can easily backtrack our steps when we make an incorrect speculation. This technique is also used to guarantee that the sequence of pairwise alignments is the same for both the parallel and sequential implementations.

```
1    processor 0 reads input sequences and broadcasts them to
       other processors.
2    gi = 0;
3    best_score = initial_score();
4    Do {
5      seed(gi); /*all processors set the same iteration seed gi*/
6      partn = select_partition();
7      current_score = calculate(seq, partn, gap_positions);
8      flag = decide(current_score, best_score);
9      seq = modify(seq, partn, flag, gap_positions);
10     gi = gi + 1;
11   }While (flag == A);
12   clear_partitions();
13   While((gi - bgi) < q){
14   for(i = 0; i < p; i++){
15     seed(gi + i);
16     itemp = select_partition();
17     set_partition(itemp);
18     if (i == pid) partn = itemp;
19   }
20   current_score = calculate(seq, partn, gap_positions);
21   flag = decide(current_score, best_score);
22   global_operation(ap, flag, best_score, partn, gap_positions);
23   seq = modify(seq, partn, flag, gap_positions);
24   if(flag == A){
25       gi = gi + ap + 1;
26       clear_partitions();
27   }
28   else gi = gi + p;
```

Figure 4. Parallel Speculative Berger-Munson Algorithm.

After a rejection is encountered, we start to speculate and continue until q (number of all possible partitions) rejections have been encountered. A random partition is selected only if it has not already been selected since the last accepted partition. That is, no partition is selected more than once by any processor or by different processors simultaneously. To ensure that no partition is selected more than once, each processor must know two pieces of information: the partitions that have already been selected and the partitions that are currently being selected by other processors. To avoid the costly inter-processor communication, these two pieces of information are obtained as follows. Each processor manages an array of q bits which correspond to the q possible partitions. Initially, these bits are cleared by a function in line 12. Then, the ith bit is set when the ith partition is selected. Therefore, each processor knows that a particular partition has already been selected if its corresponding bit is set. When this situation occurs, it simply selects another random partition. All q bits are cleared every time a new

partition is accepted. To determine the partitions that are being selected by other processors, each processor generates p random selectable partitions instead of just one and then selects the (pid)th one as show in lines 14-19. The remaining partitions are being selected by the other processors. All p bits that are corresponding to the p selectable partitions are set.

The global operation (line 22) is performed after each processor makes its decision. The accepted alignment with the smallest iteration number is selected as the input for the next iteration since the alignments with higher iteration numbers are invalid. That is, they were based on incorrect speculations. When a new partition is accepted, the contents of variables (ap, $flag$, $best\_score$, $partn$, $gap\_posititons$) are copied from the accepted processor to the other processors.

In lines 24-28, we determine the number of sequential iterations which were correctly speculated for skipping. If there is a global accepted partition (iteration) among the p partitions evaluated, only the iterations smaller than or equal to the accepted iteration are skipped (line 25). Otherwise, all p iterations are skipped (line 28).

ALIGNMENT SEARCH SPACE

We adapted the restricted search space as presented by Ishikawa *et al.* [11] who observed that the number of sequences in the divided groups had a great effect on the final alignment. They observed that if only one or two sequences were allowed in the first group, a better alignment was obtained. Our experiments yield the same observation. If only one or two sequences are allowed in one of the two groups, the number of possible partitions is reduced to $n + \dfrac{n(n-1)}{2}$ from a total of $2^{n-1} - 1$. We represent a restricted partition by a single random number. This number is then used to generate a sequence of 2 or 3 unique numbers. The first number is the number of sequences in the first group and the following one or two numbers representing the sequence numbers in this group. The remaining sequences are placed into the second group.

RESULTS

To evaluate the performance of our approach, we have used it to improve the alignments generated manually by experts, Kabat *et al.* [13], and automatically by a popular program, CLUSTALV [9, 10], which uses a tree-based method. Three different groups of immunoglobulin sequences with varying lengths and numbers of sequences were selected from the Kabat Database (Beta Release 5.0) which is maintained by Kabat *et al* [13]. Their statistical summaries are shown in Table 1. The average sequence length is about the same for all three groups. However, the number of sequences in the third group is about twice the second one which is about twice the first one. MKL5 is the largest group in this database. CLLC is the chicken immunoglobulin lambda light chains V-region group. HHC3 is the human immunoglobulin heavy chains subgroup III V-region group. MKL5 is the mouse immunoglobulin kappa light chains V V-region group. The initial score is the score before any alignment is performed.

Table 1. Statistical Summaries of the Three Groups of Test Sequences.

Group Name	Number of Sequences	Average Length	Initial Score (10^3)
CLLC	93	62	1,575
HHC3	185	65	6,651
MKL5	324	83	31,394

The scores of the alignments manually generated by experts, Kabat *et al.* [13] and their improved scores performed by the sequential Berger-Munson program, MUSEQAL, are shown in Table 2. The number of iterations and the sequential computation times taken by MUSEQAL are also shown in this table. These sequential computation times were obtained from executing a sequential program on a single processor. Similarly, we used MUSEQAL to improve the alignments generated by CLUSTALV. The corresponding information is presented in Table 3. Comparing Table 2 and Table 3, we can see that MUSEQAL improved the alignments generated by both Kabat *et al.* and CLUSTALV significantly. The sequential computation times in these tables are used to calculate the speedup factors of the parallel speculative Berger-Munson algorithm in the next table.

Table 4 shows the speedup factors for the three groups of sequences on varying numbers of processors. The speedup factor is defined as the ratio of the total run time of the sequential version of the program to the total run time of the parallel version. Table 4 demonstrates that significant speedups were obtained for all three groups of sequences. From this table, we can make three observations. First, we obtained the best speedup factors with the largest group, MKL5. Second, we obtained better efficiencies by using a smaller number of processors where efficiency is defined as the ratio of the speedup factor to the number of processors. Third, we achieved higher speedup

factors when improving the Kabat alignments compared to CLUSTALV alignment improvement. The results of the first two observations are as expected. The first observation was due to a larger number of partitions (search space) and the second to less communication and lower rates of incorrect speculations. For a larger number of processors, we had to speculate further into the future which resulted in a higher error rate. The third observation is due to the fact that the Kabat alignments were already better than the CLUSTALV alignments. As a result, there were more rejections in improving the Kabat alignments than the CLUSTALV alignments.

Table 2. Kabat and MUSEQAL Alignment Score Comparison

Group Name	Kabat Score (10^3)	MUSEQAL Score (10^3)	Number of Iterations	Sequential Run Time (10^3 s)
CLLC	1,827	1,957	16,193	55
HHC3	7,505	7,681	56,232	567
MKL5	38,569	38,766	112,374	2,535

Table 3. CLUSTALV and MUSEQAL Alignment Score Comparison.

Group Name	CLUSTALV Score (10^3)	MUSEQAL Score (10^3)	Number of Iterations	Sequential Run Time (10^3 s)
CLLC	1,809	1,957	12,716	36
HHC3	7,366	7,655	58,285	564
MKL5	37,778	38,749	112,390	2,611

Table 4. Parallel Speculative Berger-Munson Algorithm Speedup Factors.

Number of Processors	Speedup w.r.t Kabat Alignment			Speedup w.r.t CLUSTAL Alignment		
	CLLC	HHC3	MKL5	CLLC	HHC3	MKL5
1	1.0	1.0	1.0	1.0	1.0	1.0
2	1.8	1.9	1.9	1.8	1.8	1.9
4	3.4	3.4	3.8	3.3	3.3	3.7
8	6.4	6.2	7.3	6.1	5.9	7.1
16	11.6	11.4	14.1	10.7	10.7	13.6
32	19.5	20.8	28.3	17.0	19.4	26.3
64	29.5	38.1	53.1	23.8	35.1	50.7

DISCUSSION

The Berger-Munson iterative method is a good tool for improving the alignments generated by other methods. As an improvement tool, it can never generate a worse alignment than other methods. Its computation time is significantly reduced by using a parallel speculative computation technique. We also think that the computation of other iterative methods [3, 12, 14, 15] can also be reduced by using a parallel speculative computation approach.

It is difficult to accurately compare our speedup factors with that obtained by Ishikawa *et al.* [11] since their parallel algorithm does not always generate the same alignment as the sequential one. To evaluate their parallel algorithm, they aligned 7 sequences which has 63 possible partitions that were assigned to 63 processors. Their parallel implementation stops after no improvement was obtained. Their sequential (single processor) implementation stops after 32 iterations of no improvements, about one half of the number of partitions. For the sequential execution, they used different random seeds to generate different alignments. On average, they obtained a speedup factor of ten. We obtained a speedup range of 23 to 53 with our speculative method.

In spite of the above difficulty, our results clearly showed that our speedup factors are about three to five times higher than those obtained by Ishikawa *et al.* In addition, we were able to achieve higher speedup factors without changing the algorithmic behavior of the original sequential algorithm.

REFERENCES

[1] G.J. Barton, and M.J.E. Sternberg, "A Strategy For The Rapid Multiple Alignment Of Protein Sequences," *J. Mol. Bio.*, (1987), pp. 327-337.

[2] G.J. Barton, "Protein Multiple Sequence Alignment And Flexible Pattern Matching," *Methods Enzymol.*, (1990), pp. 403-427.

[3] M.P. Berger, and P.J. Munson, "A Novel Randomized Iterative Strategy For Aligning Multiple Protein Sequences," *Comput. Appl. Biosci.*, (1991), pp. 479-484.

[4] H. Carillo, and D. Lipman, "The Multiple Sequence Alignment Problem In Biology," *SIAM J. Appl. Math.*, (1988), pp. 197-209.

[5] F. Corpet, "Multiple Sequence Alignment With Hierarchical Clustering," *Nucleic Acids Research*, (1988), pp. 10881-10891.

[6] M.O. Dayhoff, R.M. Schwartz, and B.O. Orcutt, "A Model Of Evolutionary Change In Proteins," In Dayhoff (ed) , *Atlas of Protein Sequence and Structure Vol. 5, Suppl. 3, Nat. Biomed. Res. Found.*, Washington, D.C., (1978), pp.345-352.

[7] D.F. Feng, and R.F. Doolittle, "Progressive Alignment And Phylogenetic Tree Construction Of Protein Sequences," *Methods Enzymol.*, (1990), pp. 375-387.

[8] O. Gotoh, "An Improved Algorithm For Matching Biological Sequences," *J. Mol. Biol.*, (1982), pp. 705-708.

[9] D.G. Higgins, and P.M. Sharp, "CLUSTAL: A Package For Performing Multiple Sequence Alignment On A Microcomputer," *Gene*, (1988), pp. 237-244.

[10] D.G. Higgins, and P.M. Sharp, "Fast And Sensitive Multiple Sequence Alignments On A Microcomputer," *Comput. Appl. Biosci*, (1989), pp. 151-153.

[11] M. Ishikawa, M. Hoshida, M. Hirosawa, T. Toya, K. Onizuka, and K. Nitta, "Protein Sequence Analysis By Parallel Inference Machine," *Proceedings of the international conference on fifth generation computer systems,* (June, 1992), pp. 57-62 and 294-299.

[12] M. Ishikawa, T. Toya, M. Hoshida, K. Nitta, A. Ogiwara, and M. Kanehisa, "Multiple Sequence Alignment By Parallel Simulated Annealing," *Comput. Appl. Biosci.*, (1993), pp. 267-273.

[13] E.A. Kabat, T.T. Wu, H.M. Perry, K.S. Gottesman, and C. Foeller, "Sequence Of Proteins Of Immunological Interest," *U.S. Dept. of Health and Human Services, Public Health Service, National Institutes of Health*, NIH Publication No. 91-3242, (1991).

[14] J. Kim, S. Pramanik, and M.J. Chung, "Multiple Sequence Alignment Using Simulated Annealing," *Comput. Appl. Biosci.*, (1994), pp. 419-426.

[15] C.E Lawrence, S.F. Altschul, M.S. Boguski, J.S. Liu, A.F. Neuwald, J.C. Wootton, "Detecting Subtle Sequence Signals: A Gibbs Sampling Strategy For Multiple Alignment," *Science*, (1993), pp. 208-214.

[16] D.J. Lipman, S.F. Altschul, and J.D. Kececioglu, "A Tool For Multiple Sequence Alignment," *Proc. Natl. Acad. Sci.* USA, (1989), pp. 4412-4415.

[17] H.M. Martinez, "A Flex Multiple Sequence Alignment Program," *Nucleic Acids Research,*

(1988), pp. 1683-1691.

[18] E. Myers, and W. Miller, "Optimal Alignments in Linear Space," *Comput. Appl. Biosci.*, (1988), pp. 11-17.

[19] M. Murata, J.S. Richardson, and J.L. Sussman, "Simultaneous Comparison Of Three Protein Sequences," *Proc. Natl. Acad. Sci. USA*, (1985), pp. 3073-3077.

[20] M. Murata, "Three-way Needleman-Wunsch algorithm," *Methods Enzymol.*, (1990), pp. 365-375.

[21] S.B. Needleman, and C.D. Wunsch, "A General Method Applicable To The Search For Similarities In The Amino Acid Sequences Of Two Proteins," *J. Mol. Biol.*, (1970), pp. 443-453.

[22] A. Sohn, Z. Wu, and X. Jin, "Parallel Simulated Annealing By Generalized Speculative Computation," *Proceedings of the Fifth IEEE Symposium on Parallel and Distributed Processing*, Dallas, Texas, (December 1993).

[23] A. Sohn, "Parallel Speculative Computation Of Simulated Annealing," *Proceedings of the International Conference on Parallel Processing*, August, (1994), pp. III8-11.

[24] T.F. Smith, and M.S. Waterman, "Identification Of Common Molecular Subsequence," *J. Mol. Biol.*, (1981), pp. 195-197.

[25] W.R. Taylor, "Multiple Sequence Alignment By Pairwise Algorithm," *Comput. Appl. Biosci.*, (1987), pp. 81-87.

[26] W.R. Taylor, "A Flexible Method To Align Large Numbers Of Biological Sequences," *J. Mol. Evol.*, (1988), pp. 161-169.

[27] E.E. Witte, R.D. Chamberlain, and M.A. Flanklin, "Parallel Simulated Annealing Using Speculative Computation," *IEEE Transactions on Parallel and Distributed Systems*, (1991), pp. 483-494.

RELAXATION AND HYBRID APPROACHES TO GRÖBNER BASIS COMPUTATION ON DISTRIBUTED MEMORY MACHINES[a]

Hemal V. Shah and José A. B. Fortes
School of Electrical Engineering, Purdue University, W. Lafayette, In 47907.
{hvs,fortes}@ecn.purdue.edu

Abstract. *Two new techniques to compute an important polynomial basis (Gröbner basis) on distributed memory machines are presented. The first approach is based on relaxation of dependencies present in the sequential computation. In the hybrid approach, a Gröbner basis of a set of polynomials is computed in a tree-structured fashion. The upper level nodes compute the Gröbner basis using the relaxation approach. This hybrid approach takes advantages of both the tree-structured computation and dependency relaxation at upper level nodes in the tree. Both approaches exploit parallelism in an irregularly structured Gröbner basis computation. A comparative study of both approaches is provided. The performance data in this paper results from experiments performed on an Intel Paragon machine.*

1 INTRODUCTION

Gröbner basis is an important symbolic polynomial manipulation concept developed by Buchberger in the 1960s [3]. Among other applications, it can be used in checking ideal membership of a polynomial, solving a system of polynomial equations, proving geometric theorems, finding an implicit equation of a parametric surface, and checking state reachability of a Petri net model. A well-known algorithm for the computation of a Gröbner basis is also due to Buchberger [3]. Given the set of all possible pairs of an initial basis set of polynomials, it can be briefly described as follows:

1. An S-polynomial (defined in Section 2) of a selected pair of polynomials is reduced.

2. Every nonzero polynomial after reduction is added to the current basis set.

3. The set of pairs is updated for every nonzero polynomial added to the basis set.

4. Steps 1, 2, and 3 are repeated till the set of pairs of polynomials is empty.

The algorithm has the following interesting properties.

- The computational structure is unpredictable. The execution time depends on the order in which the pairs of polynomials are reduced and the order in which the polynomials are added to the basis set.

- The memory demands of the polynomials generated during the computation are not known before the execution of the algorithm.

- The dependence of each step on the results of previous steps makes the algorithm difficult to parallelize.

- The dynamic input dependent behavior of the algorithm makes it difficult to predict its performance for a given set of polynomials.

The main features of the two approaches developed for computing Gröbner basis in this paper are

- the relaxation approach exploits the parallelism available by relaxing the dependency structure of Buchberger's algorithm,

- the hybrid approach performs tree-structured computation with relaxation of dependencies at computationally intensive nodes of the tree.

First, basic definitions related to polynomials and Gröbner basis are provided in Section 2. Related work is summarized in Section 3. In Section 4, the relaxation approach is developed and the hybrid approach is developed in Section 5. A comparison of both approaches is provided in Section 6 along with some case studies. Finally, conclusions and directions for future research are provided.

2 POLYNOMIALS AND GRÖBNER BASIS

A polynomial ring is denoted by $k[x_1, \cdots, x_n]$, where k is a field and $x_1, \cdots, x_n$ are variables. A *monomial* in $x_1, \cdots, x_n$ is a product of the form $x^\alpha = x_1^{\alpha_1} \cdots x_n^{\alpha_n}$, where $x = (x_1, \cdots, x_n), \alpha = (\alpha_1, \cdots, \alpha_n)$ and all exponents $\alpha_1, \cdots, \alpha_n$ are nonnegative integers. A *polynomial* f in $x_1, \cdots, x_n$ with coefficients in field k is a linear combination of monomials. Let

$$f = \sum_\alpha a_\alpha x^\alpha \text{ be a polynomial in } k[x_1, \cdots, x_n].$$

a_α is the *coefficient* of the monomial x^α. If $a_\alpha \neq 0$, then $a_\alpha x^\alpha$ is a *term* of f. The *multidegree* of f is $multideg(f) = max(\alpha : a_\alpha \neq 0)$ with respect to some monomial ordering. The *leading coefficient* of f is $LC(f) = a_{multideg(f)} \in k$. The *leading monomial* of f is $LM(f) = x^{multideg(f)}$. The *leading term* of f is $LT(f) = LC(f)LM(f)$. In practice, the terms of a polynomial are ordered using lexicographical (lex) or total degree lexicographical (tdeg) ordering. A subset $I \subset k[x_1, \cdots, x_n]$ is an *ideal* if it contains 0 and it is closed under polynomial addition and

(a) This research was partially funded by the National Science Foundation under grants MIP-9500673 and CDA-9015696.

polynomial multiplication. The *S-polynomial* of two polynomials f and g is

$$Spoly(f,g) = \frac{LC(g)x^\gamma f}{LM(f)} - \frac{LC(f)x^\gamma g}{LM(g)}$$

where $x^\gamma = LCM(LM(f), LM(g))$ is the least common multiple of $LM(f)$ and $LM(g)$.

Definitions related to polynomial and S-polynomial are now illustrated for two polynomials, $f_1 = 2x_1^2x_2 - 1$ and $f_2 = x_1x_2^2 - x_1$ with $x_1 \succ_{lex} x_2$. The multidegrees of f_1 and f_2 are $multideg(f_1) = (2,1)$, $multideg(f_2) = (1,2)$, their leading coefficients are $LC(f_1) = 2, LC(f_2) = 1$, their leading monomials are $LM(f_1) = x_1^2x_2, LM(f_2) = x_1x_2^2$, their leading terms are $LT(f_1) = 2x_1^2x_2, LT(f_2) = x_1x_2^2$, the least common multiple of the leading monomials is, $LCM(LM(f_1), LM(f_2)) = LCM(x_1^2x_2, x_1x_2^2) = x_1^2x_2^2$. The S-polynomial of f_1 and f_2 is

$$\frac{1.x_1^2x_2^2.(2x_1^2x_2 - 1)}{x_1^2x_2} - \frac{2.x_1^2x_2^2.(x_1x_2^2 - x_1)}{x_1x_2^2} = 2x_1^2 - x_2.$$

A *normal form* of a polynomial h with respect to a set of polynomials F is a fully reduced form of a polynomial with respect to F and is denoted by $NF(h,F)$. It can be computed as follows:

1. Set $h_0 = h$ and $i = 0$.

2. For $i = 0, 1, 2, \cdots$ repeat step 3 until h_i cannot be rewritten; then output h_i and stop.

3. If there is a polynomial f in F such that the leading monomial of f divides a term $a_\alpha x^\alpha$ in h_i, then rewrite h_i as $h_{i+1} = h_i - \frac{a_\alpha x^\alpha f}{LT(f)}$.

Let $h_0 = h = x_1^2 - x_2$ and $F = \{f\}$, where $f = -x_1 + x_2^2$ and $x_1 \succ_{lex} x_2$. Then the sequence of computations $h_1 = h_0 + x_1.f = x_1x_2^2 - x_2, h_2 = h_1 + x_2^2.f = x_2^4 - x_2$ gives $NF(h,F) = h_2 = x_2^4 - x_2$. Using the normal form algorithm, a set of polynomials F can be reduced on itself by replacing each polynomial $f \in F$ with its normal form with respect to set $F - \{f\}$. The algorithm for reduction of a set of polynomials can be found in [3].

The definition of Gröbner basis and Buchberger's algorithm to compute it are provided next. Let F be a set of polynomials. A basis G for $Id(F)$, the ideal generated by F, is a Gröbner basis iff

1. for all pairs $(g_i, g_j), g_i, g_j \in G, g_i \neq g_j$, the remainder of the division of $Spoly(g_i, g_j)$ by G is zero, in other words $NF(Spoly(g_i, g_j), G) = 0$, and

2. the ideals generated by F and G are identical.

Algorithm 1 Buchberger's Algorithm (Gröbner)

Input: $F = \{f_1, \cdots, f_m\}$.
Output: A Gröbner basis $G = \{g_1, \cdots, g_n\}$.

begin
    $G \leftarrow F$
    $B \leftarrow \{(g_i, g_j)|g_i, g_j \in G, g_i \neq g_j\}$

    While $B \neq \phi$ **do**
        Select (g_i, g_j) from B
        $B \leftarrow B - \{(g_i, g_j)\}$
        $h \leftarrow Spoly(g_i, g_j)$
        $h_0 \leftarrow NF(h, G)$
        if $h_0 \neq 0$ **then**
            $B \leftarrow B \cup \{(g, h_0)|g \in G\}$
            $G \leftarrow G \cup \{h_0\}$
        end
    end
end

Buchberger's algorithm can be resource-intensive and time-consuming for large polynomial computations. Furthermore, it is difficult to parallelize. The next section discusses some related work on parallel Gröbner basis computation.

3 RELATED WORK ON PARALLELIZATION OF GRÖBNER BASIS COMPUTATION

Buchberger [4] first proposed the parallel L-Machine to compute Gröbner basis in parallel. It was based on manager-worker approach. This scheme was not implemented but the pair set manager can become a bottleneck in this type of computation as discussed in [18]. Ponder [9] proposed two parallel algorithms which were inherently "less parallel" because of serialization in keeping the basis set reduced. Senechaud [13, 14] developed parallel algorithms for a ring of processors and a hypercube of processors. For large enough data set of a small class of problems (sequential execution time of about 10 seconds), a ring with several processors provided good speedups. The second approach, based on "divide-and-conquer" strategy, was very close to the sequential algorithm and did not provide significant speedups.

Vidal and Schwab [12, 18] both used shared memory machines to perform Gröbner basis computation. Vidal used synchronization and lock primitives to keep the Gröbner basis set shared. Schwab extended Vidal's technique to take advantage of two-level parallelism in Gröbner basis computation. A pipeline of processors was used to reduce a pair (fine-grain parallelism) and many pairs were reduced in parallel (coarse-grain parallelism). Superlinear speedups were achieved for some examples but not properly explained. On shared memory machines Schreiner and Hong [10, 11] showed that by invoking optimized library routines of PACLIB, a maximum speedup of 10 could be achieved for Gröbner basis computation.

Recently researchers have started to address the problem of computing Gröbner basis on distributed memory machines. In Siegl's [16, 17] approach, every pair was reduced by a bidirectional pipeline of processes. For every nonzero polynomial after reduction, a new process was created and appended to the pipeline of processes. Good speedups were reported for some examples along with insights on cases with superlinear speedups. According to [5], this approach suffers from poor parallelism, pipeline imbalance, and communication overheads with large number of large polynomials. Hawley [8] concentrated on performing reduction on polynomials in parallel, but this fails to take advantage of "magic polynomials" to achieve superlinear speedups. Chakrabarti and Yelick [5] proposed a distributed approach based on non-deterministically scheduled generation and reduction of polynomial pairs and on replication of basis across all processors. "Magic polynomials" generated in parallel computation of Gröbner basis also led to superlinear speedup. The algorithm efficiency compared favorably with shared memory solutions and scaled better than previous approaches to a large number of processors. Chakrabarti and Yelick's algorithm bears some resemblance to one of the two approaches (relaxation) introduced in this paper. The differences are detailed in the next section.

4   RELAXATION APPROACH

The following are the key points in the design of a new parallel algorithm based on the relaxation approach:

- In Buchberger's algorithm, a selected pair is reduced and if the reduced polynomial is nonzero, it is added to the basis set and the set of pairs is updated. When the set of pairs is empty, the algorithm terminates. By relaxing the dependence of every reduction step on the previous reduction steps, one can achieve good parallelism at the cost of convergence rate. The relaxation approach reduces S-polynomials in parallel, thus achieving fine-grain parallelism on a distributed memory machine.

- Let the initial set of polynomials be F and a Gröbner basis of F be G. Let $|F| = m$, $|G| = n$, $n \geq m$. So, the number of added polynomials during Gröbner basis computation is $n - m$. But, the total number of pairs pr being processed by the algorithm is

$$\sum_{i=1}^{n}(i-1) = \frac{n(n-1)}{2} = \theta(n^2) \qquad (1)$$

Thus, polynomials generated by the S-polynomials of the pairs that were reduced to zero are $\theta(n^2)$. This suggests that most of the polynomials generated in the intermediate computation are reduced to zero. By replicating the basis, every processor can independently reduce a pair in parallel. Reduction of a pair to a zero polynomial results in less communication after a significant computation. On the other hand, if the basis is distributed across processors, every pair reduction results in a ring like communication. Therefore, replicating the basis on all processors will result in better computation/communication ratio than distributing the basis across processors. This was also observed in [5].

- Early generation of important basis polynomials can result in speculative parallelism.

- In the relaxation approach, the kth pair is assigned the number $(k \bmod nproc) + 1$. This guarantees equal distribution of the number of pairs to each processor. But, this does not guarantee equal distribution of computation to processors. The computation time taken by each pair depends on the problem structure. This dynamic input dependent behavior will be considered in future research.

In words, the parallel relaxation algorithm can be briefly described as:

1. Initially, the set of polynomials is replicated and the set of pairs is uniformly distributed across processors.

2. Every processor

   (a) selects a pair from its local set of pairs,

   (b) reduces the selected pair,

   (c) sends nonzero reduced polynomial to coordinator processor.

3. The coordinator processor broadcasts all nonzero polynomials to all processors.

4. Each processor updates its local set of pairs and adds nonzero polynomials to current basis set.

5. Steps 2, 3, 4 are repeated until each processor's local set of pairs is empty.

Unlike Chakrabarti and Yelick's approach tasks do not migrate in the relaxation approach. Each processor reduces only those pairs that reside in its local set. This results in a simple implementation without overheads of scheduling, validation/invalidation protocols, distributed task queue but at the cost of load balance. In the initial experiments, synchronous communication without any computation overlap is used to make the implementation as simple as possible. Steps 2(c) and 3 of the algorithm are potentially non-scalable. They can be replaced by distinct non-simultaneous broadcasts, possibly overlapped with computation. The relative advantages of these alternative implementations are under investigation.

The following primitives are used in the parallel relaxation algorithm $Par\_Relax\_Gröbner$. Here f, g and h represent polynomials and F and G represent sets of polynomials.

$nproc$ - the number of processors used in computation,
$myid$ - a unique number assigned to each processor,
  where $1 \leq myid \leq nproc$,
$Send(f, i)$ - send f to processor i,
$Send(F, i)$ - send F to processor i,
$Send(f, all)$ - send f to all processors,
$Send(F, all)$ - send F to all processors,
$Recv(f)$ - receive a polynomial and store it in f,
$Recv(F)$ - receive a set of polynomials and store it in F,
$Spoly(f, g)$ - S-polynomial of f and g,
$NF(h, F)$ - a normal form of h with respect to F,
$op(i, j)$ - assign a pair (i, j) a number in $[1, nproc]$,
$criteria(f, g)$ - it returns 0 if $NF(Spoly(f, g), G)$ can be
  detected as 0 based on some criteria ([2]).

Algorithm 2 A Parallel Relaxation Algorithm

(Par_Relax_Gröbner)
Input: $F = \{f_1, \cdots, f_m\}$.
Output: A Gröbner basis $G = \{g_1, \cdots, g_n\}$.

(Here, G and B are the local sets of each processor.)
begin
    for $p = 1$ to $nproc$ pardo
        $G \leftarrow F$

$$B \leftarrow \{(g_i, g_j)|g_i, g_j \in G, g_i \neq g_j,$$
$$op(i,j) = myid, criteria(g_i, g_j) \neq 0\}$$
end
While (in any processor set B is nonempty) **pardo**
$\quad H \leftarrow \phi$
$\quad$ **if** $B \neq \phi$ **then**
$\quad\quad$ select a pair $pr = (g_i, g_j)$ from B
$\quad\quad B \leftarrow B - \{pr\}, h \leftarrow Spoly(g_i, g_j)$
$\quad\quad h_0 \leftarrow NF(h, G), Send(h_0, coordinator\_id)$
$\quad$ **end**
$\quad$ **if** $(myid == coordinator\_id)$ **then**
$\quad\quad$ **for** $p = 1$ to $nproc$ **do**
$\quad\quad\quad Recv(h)$
$\quad\quad\quad$ **if** $h \neq 0$ **then** $H \leftarrow H \cup \{h\}$ **end**
$\quad\quad$ **end**
$\quad\quad Send(H, all)$
$\quad$ **else**
$\quad\quad Recv(H)$
$\quad$ **end**
$\quad G \leftarrow G \cup H$
$\quad B \leftarrow B \cup \{(g_i, g_j)| g_i \in G, g_j \in H, g_i \neq g_j,$
$\quad\quad op(i,j) = myid, criteria(g_i, g_j) \neq 0\}$
$\quad$ **end**
end

Performance of Parallel Relaxation Algorithm

Two examples are considered:

1. <u>Arnborg4:</u> [1] (The variables are ordered $x_1 \succ_{lex} x_2 \succ_{lex} x_3 \succ_{lex} x_4$.)

$$x_1 + x_2 + x_3 + x_4 = 0$$
$$x_1 x_2 + x_2 x_3 + x_3 x_4 + x_4 x_1 = 0$$
$$x_1 x_2 x_3 + x_2 x_3 x_4 + x_3 x_4 x_1 + x_4 x_1 x_2 = 0$$
$$x_1 x_2 x_3 x_4 = 1$$

2. <u>Mora:</u> [7] (The variables are ordered $x \succ_{lex} y \succ_{lex} z \succ_{lex} t$.)

$$\{x^{d+1} - yz^{d-1}t, xy^{d-1} - z^d, x^d z - y^d t\}$$

The final Gröbner basis of this example is guaranteed to have a polynomial of degree d^2. The number of polynomials added to the basis is linearly proportional to d.

Pseudo-speedup is defined here as the ratio of the total execution time of a parallel algorithm on one processor to solve a problem to the total execution time (computation time + communication time + initialization time) of the same parallel algorithm to solve the same problem on p processors. Speedups, which are computed with respect to the best sequential algorithm, are provided later. The pseudo-speedups achieved by the parallel relaxation algorithm for two examples described in this section are as shown in Figure 1.

From the performance curves presented in the Figure 1, the following observations and conclusions can be made:

1. The relaxation approach achieves good pseudo-speedups for both the examples considered in this section.

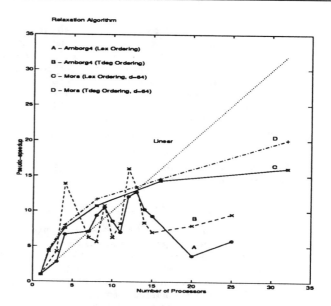

Figure 1: Pseudo-speedup achieved for Arnborg and Mora's Example Using *Par_Relax_Gröbner*

2. The "magic polynomials" are the basis polynomials that are generated early in the computation thus reducing the computation time and the number of polynomials added to the basis. The reason for the superlinear pseudo-speedup achieved for Arnborg's example is due to "magic polynomials" which are added to the Gröbner basis earlier than when they are added in the sequential execution of the Buchberger's algorithm. The oscillating behavior in the pseudo-speedup for Arnborg's example is due to the varying number of polynomials generated during the computation. An explanation for this behavior is provided in the next subsection.

3. In Mora's example, the pseudo-speedup increases with the number of processors.

4. For Arnborg4, pseudo-speedup scales down with the number of processors (≥ 14) due to the overheads.

The contributing sources of the overhead are the following.

- **Extra Computation:** At each step, the pairs are reduced in parallel. The reduced nonzero polynomials at each step are added to the basis. Some of these polynomials might have been reduced to zero by using other polynomials generated during that step. The addition of unnecessary polynomials to the basis results in extra computation.

- **Interprocessor Communication:** At every step, after reducing many pairs in parallel, the reduced nonzero polynomials need to be communicated to all processors. The time required for communication is one source of overhead. The addition of unnecessary polynomials to the basis not only results in extra computation, but also results in additional interprocessor communication.

• Load Imbalance: When reducing many pairs in parallel, if the time taken in reducing each pair is not the same, then some processors will be idle waiting for other processors to finish their computation. The other source of load imbalance is limited parallelism available for a specific problem. At any step, if the number of pairs that can be reduced in parallel is less than the number of processors used, then some processors will be idle during that step.

The above overheads become significant (relative to computation time) when the problem size is not large enough for the given number of processors.

An Explanation for Superlinear Pseudo-speedups

The dynamic nature of Gröbner basis computation makes it hard to provide a general analysis that explains the reasons for pseudo-speedups. Simplified assumptions are therefore made here which, nevertheless, enable an insightful analysis. Let the initial number of polynomials in the basis be m and the number of polynomials in the final basis be n_p when using p processors in the parallel relaxation algorithm. Let the processing time t for a pair be equal for all pairs and the number of pairs reduced in parallel be p. For the parallel relaxation algorithm, the computation time using p processors is

$$T_p = \frac{t \sum_{i=1}^{n_p}(i-1)}{p} = t\frac{n_p(n_p-1)}{2p}. \qquad (2)$$

So, the pseudo-speedup achieved using p processors (ignoring communication and initialization time) is

$$Pseudo-speedup = \frac{T_1}{T_p} = pR, \qquad (3)$$

where R is a Relaxation factor that is given by

$$R = \frac{n_1(n_1-1)}{n_p(n_p-1)} \approx \frac{n_1^2}{n_p^2} \text{(for } n_1, n_p \gg 1). \qquad (4)$$

So, when "magic polynomials" are added early to the basis, the number of added polynomials to the basis are reduced ($n_p < n_1$). Thus, the relaxation factor R becomes greater than 1, leading to superlinear pseudo-speedups.

In the next section, a tree-structured Gröbner basis computation is presented which can be faster than computation performed by Buchberger's algorithm for the sequential case for some large problems. Parallelization of this tree-structured Gröbner basis computation is discussed in the next section.

5 HYBRID APPROACH

In this section, a new variant of Buchberger's algorithm and a hybrid approach for its parallelization is presented. This hybrid approach is more efficient than the relaxation approach when the number of polynomials in a set is large and the set can be divided into almost independent subsets. The main idea behind the hybrid approach is to compute Gröbner basis of a given set of polynomials from the Gröbner basis of its subsets. *Gröbner_Tree* is the algorithm which sequentially computes the Gröbner basis of a given set in a tree structure

by using an algorithm similar to Buchberger's (called *Gröbner_Combine*) at each node. *Gröbner_Combine* is almost the same as Buchberger's algorithm. But, here pairs of the same set are not considered because they are from the same Gröbner basis set, thus they possess the first property of Gröbner basis definition. So, in *Gröbner_Combine* with input Gröbner basis sets G_1 and G_2, initially the set of pairs B is $\{(g_i, g_j)|g_i \in G_1, g_j \in G_2, criteria(g_i, g_j) \neq 0\}$ and the initial basis set G is $G_1 \cup G_2$. To maintain the brevity of this section, the description of these algorithms is not provided.

The tree-structured computation with the root node at the highest level can be easily parallelized by assigning each node computation to a processor and by executing the nodes at the same level in parallel. The amount of pseudo-speedup achieved in this approach is limited after a certain number of processors (for more details see [13, 15]). The main reason for this behavior is the unbalanced load distribution across processors and computation granularity that increases with the level of the nodes in the tree. The nodes at high levels in the computational tree tend to dominate the execution time of the whole computation. So, by executing each node of the tree on a processor where nodes at the same level are executed in parallel fails to take advantage of parallelism available inside each node. The shortcoming of this approach due to unbalanced tree-structured computation can be overcome by taking a hybrid approach. Here, the relaxation algorithm is used to exploit parallelism available inside each upper level node of the computational tree.

Algorithm 3 Parallel Hybrid Algorithm

(Par_Hybrid_Gröbner)

Input: $F = \{f_1, \cdots, f_m\}$ given in 2^q subsets $F_1, \cdots, F_{2^q}$ such that $F = \bigcup_{i=1}^{2^q} F_i$.
Output: A Gröbner basis $G = \{g_1, \cdots, g_n\}$.

```
begin
    for i = 1 to 2^q pardo
        G_i ← Gröbner_Tree(F_i)
    end
    for i = 1 to level do
        for j = 1 to 2^q step 2^i pardo
            G_j ← Gröbner_Combine(G_j, G_{j+2^{i-1}})
        end
    for i = level + 1 to q do
        for j = 1 to 2^q step 2^i do
            G_j ← Par_Relax_Gröbner_Combine(G_j, G_{j+2^{i-1}})
        end
    end
    G ← G_1
end
```

The only difference between the hybrid computation and pure tree-structured computation is that at the upper level nodes in the tree the Gröbner basis can be computed by using the parallel relaxation algorithm. The algorithm *Par_Relax_Gröbner_Combine* is similar to *Par_Relax_Gröbner* in all steps except that initially the pairs of polynomials from the same basis set are not considered. The pseudo-speedups achieved for both examples using the hybrid approach are shown in Figure 2. For

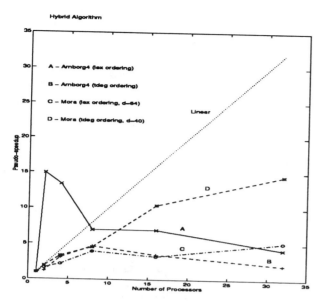

Figure 2: Pseudo-speedup achieved for Arnborg and Mora's Example Using *Par_Hybrid_Gröbner*

these two examples the root node computation is performed using relaxation approach by setting the value of *level* in *Par_Hybrid_Gröbner* to $q - 1$. No attempt should be made to compare the curves of Figures 1 and 2 as the definition of pseudo-speedup implies that different sequential times are used for each figure. A comparison of the relaxation and hybrid approaches is done in the next section.

From Figure 2 the following conclusions can be drawn:

- superlinear pseudo-speedup is achieved for Arnborg's example using the hybrid approach due to generation of "magic polynomials" early in the Gröbner basis computation,

- in Mora's example, even though reasonable pseudo-speedups are achieved, a large number of unnecessary polynomials were generated during the Gröbner basis computation of the subsets; thus, the execution times of the hybrid algorithm for Mora's example are longer than the execution times of the relaxation algorithm for the same problem with the same number of processors,

The hybrid approach developed in this section can be improved by grouping the polynomials from the initial set into subsets such that computation of Gröbner basis of the subsets takes much less time than the upper level node computation. Though this grouping is not always possible, it can be done in many real applications. Finding Gröbner basis of nonlinear equations arising in applications with many loosely coupled systems can benefit from the hybrid approach. Though this approach presented in this section uses a binary tree, one can also have a nonbinary tree-structured computation which might benefit even more from the hybrid approach.

6 COMPARISON OF RELAXATION AND HYBRID APPROACHES

Several examples of Grobner basis computations arising in geometric modeling, geometric theorem proving and Petri net modeling are used to compare the speedups achievable by relaxation and hybrid approaches. Unlike pseudo-speedup, speedup is defined as the ratio of execution time of the *best* sequential algorithm and the execution time of the parallel algorithm used to solve the same problem. Given a set of polynomials, it is not clear how to choose the sequential algorithm that yields the best execution time. In this comparative evaluation, the best sequential time is chosen as the shortest of the execution time of Buchberger's algorithm and the execution time of the sequential tree algorithm. In this sense, the speedups reported in Tables 1 and 4 are conservative (by at least a factor of two). The quantities in the parenthesis in Tables 3, 4 represent the time in seconds.

In Mora's example, the number of polynomials added to the basis is $\theta(d)$ and the number of intermediate polynomials reduced to zero is $\theta(d^2)$. So, for large d (64 in this case), the relaxation approach is able to achieve far better speedup than the hybrid approach. In this case, the Gröbner basis computation of subsets requires considerable time and generates many unnecessary polynomials which could have been reduced to zero if the relaxation approach is used instead. Furthermore, for Mora's example, the computation time at the root node in the hybrid approach increases because of these unnecessary polynomials. This explains the below unity speedups achieved by the hybrid approach for Mora's example. For Arnborg's example, the number of polynomials added to the basis in the hybrid approach is less than the number of polynomials added to the basis in the relaxation approach using the same number of processors. Thus, the hybrid approach performs better than the relaxation approach for Arnborg's example. The initial set of polynomials for Petri net model example can be decomposed into loosely coupled subsets. So, for Petri net model example, the hybrid approach provided slightly better performance than the relaxation approach. For Arnborg4 and Petri net model examples, both approaches are able to exploit a reasonable amount of parallelism. For Apollonios's circle theorem, moderate speedups are achieved by both approaches. The reasons for this behavior are explained at the end of this section.

The relaxation approach can be faster than the hybrid approach when many unnecessary polynomials are added in Gröbner basis computation of subsets. On the other hand, for a large set of polynomials where Gröbner basis of many subsets can be independently computed and combined very fast, one can expect the hybrid approach to outperform the relaxation approach.

Mora's example provides some insights on the effect of scaling the problem size on the pseudo-speedups achieved by both the relaxation and the hybrid approaches. From Table 2 one can observe that the relaxation and hybrid approaches provide better pseudo-speedups for large problem sizes than for the small ones. The fixed-size speedup increases up to a certain number of processors and then degrades due to the overheads discussed in section 4. As the problem size increases, the degradation in speedup is observed at higher number of processors. In Table 2, the quantities in the parenthesis represent the execution time in seconds. Large problems in here represent the problems which are time-consuming or memory-demanding or both.

The examples considered in this paper are relatively small. Thus, the speedups achieved here can mislead to a conclusion that only moderate parallelism is exploited using both approaches for large number of processors (≥ 16). The problems considered in this paper are not large enough to efficiently utilize a large number of processors. The load imbalance due to limited parallelism, interprocessor communication, extra computation and additional interprocessor communication due to generation of unnecessary polynomials severely degrade performance for a large number of processors. Thus, the use of large number of processors for small problems is not justified. Empirical observations suggest that speedup improves as the problem size increases.

7 CONCLUSIONS AND FUTURE WORK

In this paper, two new approaches to Gröbner basis computation are introduced. The speedups achieved for the various examples shown in Table 1, suggest that reasonable parallelism is exploited by both the approaches in most cases. Depending on the problem, one of the two approaches gives better performance. On the other hand, poor speedups in Tables 3 and 4 indicate the need for algorithm improvements and for an understanding of how input and computation size are related to effective parallelism. For Arnborg4 example, both approaches did slightly better than Chakrabarti and Yelick's approach. However, a more complete performance comparison of both algorithms with Chakrabarti and Yelick's approach has yet to be done. This entails implementing all approaches on the same machine using comparable software (e.g. for infinite precision computation) and optimizations such as communication/computation overlap, multithreading and communication pipelining. Improvements to both approaches, concrete analysis of their scalability, and alternative techniques are also the subject of future research.

REFERENCES

[1] W. Boege, R. Gebauer, and H. Kredel, "Some Examples for Solving Systems of Algebraic Equations by Calculating Groebner Bases", *J. Symb. Comp.*, Vol.1, 1986, pp. 83–98.

[2] B. Buchberger, "A Criterion for Detecting Unnecessary Reductions in the Construction of Gröbner-bases", *EUROSAM*, 1979, pp. 1–21.

[3] B. Buchberger, "Gröbner Bases: An Algorithmic Method in Polynomial Ideal Theory", *Recent Trends in Multidimensional Systems*, 1985, pp. 184–232.

[4] B. Buchberger, "The Parallel L-Machine for Symbolic Computation", *EUROCAL*, 1985, pp. 541–542.

[5] S. Chakrabarti and K. Yelick, "Implementing an Irregular Application on a Distributed Memory Multiprocessor", *4th ACM Symp. on Principles & Practice of Parallel Programming*, 1993, pp. 169–178.

[6] D. Cox, J. Little, and D. O'Shea, *Ideals, Varieties, and Algorithms*, Springer-Verlag, 513 pp.

[7] J. Davenport, Y. Siret, and E. Tournier, *Computer Algebra Systems and Algorithms for Algebraic Computation*, Academic Press, San Diego, 1993, 298 pp.

[8] D. Hawley, "A Buchberger Algorithm for Distributed Memory Multi-Processors", *Proc. of 1st International ACPC Conf.*, 1991, pp. 385–390.

[9] C. Ponder, "Evaluation of "Performance Enhancements" in Algebraic Manipulation Systems", *Computer Algebra and Parallelism*, 1989, pp. 51–73.

[10] W. Schreiner and H. Hong, "A New Library for Parallel Algebraic Computation", *6th SIAM Conf. on Parallel Processing for Scientific Computing*, 1993, pp. 776–783.

[11] W. Schreiner and H. Hong, "PACLIB – A System for Parallel Algebraic Computation on Shared Memory Multiprocessors", *Parallel Systems Fair at the 7th International Parallel Processing Symp.*, 1993, pp. 56–61.

[12] S. A. Schwab, "Extended Parallelism in the Gröbner Basis Algorithm", *International J. Parallel Programming*, Vol. 21, No. 1, 1992, pp. 39–66.

[13] P. Senechaud, "Boolean Gröbner Basis and Their MIMD Implementation", *Computer Algebra and Parallelism*, 1990, pp. 90–99.

[14] P. Senechaud, "A MIMD Implementation of the Buchberger Algorithm for Boolean Polynomials", *Parallel Computing*, Vol. 27, 1991, pp. 29–37.

[15] H. V. Shah and J. A. B. Fortes, *Tree-structured Gröbner Basis Computation on Parallel Machines*, School of Electrical Engineering, Purdue University, Technical Report, TR-EE 94-30, October 1994, 18 pp.

[16] K. Siegl, "||MAPLE|| – A System for Parallel Symbolic Computation", *Parallel Systems Fair at the 7th International Parallel Processing Symp.*, 1993, pp. 62-67.

[17] K. Siegl, "Parallelizing Algorithms for Symbolic Computation using ||MAPLE||", *4th ACM SIGPLAN Symp. on Principles & Practice of Parallel Programming*, 1993, pp. 179–186.

[18] J. Vidal, *The Computation of Gröbner Bases on a Shared Memory Multiprocessor*, Computer Science Department, Carnegie Mellon University, Technical Report, CMU-CS-90-163, August 1990, 56 pp.

Table 1: Maximum Speedup for Various Examples

Problem	Approach used	Max. Speedup achieved	Number of processors	Ordering used
Arnborg4	Hybrid	6.96	16	Lexicographical
Arnborg4	Relaxation	5.35	13	Lexicographical
Arnborg4	Hybrid	4.65	8	Total degree lexicographical
Arnborg4	Relaxation	3.65	13	Total degree lexicographical
Mora(d=64)	Hybrid	.79	64	Lexicographical
Mora(d=64)	Relaxation	16.4	25	Lexicographical
Mora(d=64)	Hybrid	.87	64	Total degree lexicographical
Mora(d=64)	Relaxation	20.1	32	Total degree lexicographical
Apollonios	Hybrid	2.5	32	Lexicographical
Apollonios	Relaxation	3.42	21	Lexicographical
Petri net model	Hybrid	7.97	16	Lexicographical
Petri net model	Relaxation	5.63	16	Lexicographical

Table 2: Pseudo-speedup Table for Mora's Example

Relaxation Algorithm							
Problem Size	Pseudo-speedup achieved with p processors						
	$p=1$	$p=2$	$p=4$	$p=8$	$p=16$	$p=32$	$p=64$
$d=8$	1(.08)	.89(.09)	.8(.1)	.19(.43)	.17(.46)	.075(1.07)	.038(2.13)
$d=16$	1(.3)	1.19(.253)	1.43(.21)	.58(.52)	.52(.58)	.34(.89)	.19(1.57)
$d=32$	1(1.71)	1.47(1.16)	2.31(.74)	1.94(.88)	2.11(.81)	1.64(1.04)	.91(1.87)
$d=64$	1(30.3)	4.31(7.03)	7.71(3.93)	11.2(2.7)	14.9(2.03)	15.6(1.94)	11.9(2.54)
Hybrid Algorithm							
Problem Size	Pseudo-speedup achieved with p processors						
	$p=1$	$p=2$	$p=4$	$p=8$	$p=16$	$p=32$	$p=64$
$d=8$	1(.24)	.89(.27)	.75(.32)	.41(.58)	.43(.56)	.27(.9)	.19(1.29)
$d=16$	1(1.09)	1.16(.94)	1.24(.88)	1.16(.94)	1.43(.76)	1.08(1.01)	.63(1.74)
$d=32$	1(6.74)	1.36(4.95)	1.64(4.1)	1.91(3.52)	2.22(3.04)	2.46(2.74)	2.32(2.9)
$d=64$	1(237.4)	1.35(176.3)	2.51(94.4)	4.1(57.9)	4.85(48.9)	5.75(41.3)	6.17(38.5)

Table 3: Pseudo-speedup Table for Some Examples

Example	Problem Size	Approach	Pseudo-speedup (Execution Time)			
			$p=1$	$p=4$	$p=16$	$p=64$
Cylindrical equations	50 equations in 50 variables	Relaxation	1(163.1)	4.29(38)	8.91(18.3)	10.73(15.2)
		Hybrid	1(34.62)	1.31(26.32)	1.3(26.67)	1.16(29.77)
Apollonios' circle theorem	8 equations in 10 variables	Relaxation	1(5.83)	3.51(1.66)	5.11(1.14)	2.83(2.06)
		Hybrid	1(3.595)	1.9(1.897)	2.32(1.549)	1.73(2.084)
Petri net model of two communicating processes	8 equations in 11 variables	Relaxation	1(9.6)	.425(22.57)	9.95(.965)	3.42(2.809)
		Hybrid	1(5.434)	3.24(1.679)	7.97(.682)	2.65(2.41)

Table 4: Speedup Table for Some Examples

Example	Problem Size	Approach	Speedup (Execution Time)			
			$p=1$	$p=4$	$p=16$	$p=64$
Cylindrical equations	50 equations in 50 variables	Relaxation	.21(163.1)	.911(38)	1.89(18.3)	2.28(15.2)
		Hybrid	1(34.62)	1.31(26.32)	1.3(26.67)	1.16(29.77)
Apollonios' circle theorem	8 equations in 10 variables	Relaxation	.62(5.83)	2.17(1.66)	3.15(1.14)	1.75(2.06)
		Hybrid	1(3.595)	1.9(1.897)	2.32(1.549)	1.73(2.084)
Petri net model of two communicating processes	8 equations in 11 variables	Relaxation	.566(9.6)	.25(22.57)	5.63(.965)	1.93(2.809)
		Hybrid	1(5.434)	3.24(1.679)	7.97(.682)	2.65(2.41)

The Distributed Real-Time Control System of a Tokamak Fusion Device

H. Richter [1&2], R. Cole [3], M. Fitzek [3], K. Lüddecke [3],
G. Neu [2], G. Raupp [2], W. Woyke [4], D. Zasche [2], T. Zehetbauer [2]

[1]Lehrstuhl für Rechnertechnik und -organisation, Technische Universität München
Arcisstr. 21 D-80290 München, Germany
Tel. +49 89 2105 3256, email: richterh@informatik.tu-muenchen.de
[2]Max-Planck-Institut für Plasmaphysik, EURATOM Association
Boltzmannstr. 2, D-85748 Garching
[3]UCS GmbH, Gärtnerweg 2, D-82061 Neuried
[4]Bayernwerk AG, Nymphenburgerstr. 39, D-80335 München

Abstract

In high-energy and plasma physics large experimental devices have to be controlled by means of dozens of computer systems. This paper presents an overview of the distributed real-time control system of a tokamak fusion device. The computer setup and its general concepts are explained as well as the features of the underlying MULTITOP architecture. Based on transputers, parallel computers are used to match the real-time constraints as well as overcome the complexity of the control problem by using application-specific computer architectures. The most important real-time tasks are performed In less than 2 ms. A global clock is established with a precision of 2 µs between six parallel computers of different architectures.

Keywords: parallel, real time, control, transputers, tokamak.

I. Introduction

In the following, the environment of the distributed control system is described since plasma fusion experiments are something extraordinary. It will be seen that it is important to know about the environment because the set-up of the control system is reflected in the structure of the plasma experiment. Generally, in a tokamak, which is a Russian invention, magnetic fields that keep the plasma (i.e. a gas composed of charged atoms) confined in a toroidal space have to be controlled by means of computers in real time. In Europe's second largest fusion experiment, called ASDEX Upgrade [1], the timescale on which these magnetic fields are controlled is less than 2 ms. ASDEX Upgrade is a $ 200 Mio. device for investigating controlled nuclear fusion for energy production. It is based on plasma confinement by means of magnetic fields according to the tokamak principle. The first step in such devices is to pump hydrogen, deuterium and sometimes tritium gases into a ring-shaped vacuum vessel. In the second step the gases are ionized by a high-voltage electric discharge, and from then on the protons and electrons can be controlled by magnetic fields. In the third step the ionized gases - which are now called plasma - are heated to temperatures of up to 30 Mio. degrees. This heating is performed by inducing an electric current of up to 2 MA in the ring shaped plasma and applying external micro-wave heatings with approximately 15 MW of power. (All numbers given are for ASDEX Upgrade). If a dense plasma is kept long enough at such high temperatures, there is a significant probability that two deuterons will approach each other closely enough to start a nuclear fusion process. The resulting neutrons carry away kinetic energy, which in turn can be used to heat a cooling medium in a primary cooling circulation.

Unfortunately, the plasma behaves in an unstable manner as it is heated more and more. To overcome this problem, the protons and electrons can be confined by appropriate magnetic fields. The necessary strength of the fields produced by large external coils mounted around the vacuum vessel exceeds the earth´s magnetic field 10^4 times. Since the power to confine the plasma is very high, each so-called plasma "discharge" is limited to a maximum of 10 seconds; it then takes another 10 minutes to collect the required electric energy from public power stations. Then the whole experimental process is repeated with different plasma parameters to investigate the physical behavior of the plasma. Each cycle is called a "plasma shot" and the whole apparatus is intended to deliver at least 30,000 shots in a ten-year period. The construction and erection of ASDEX Upgrade took nearly another decade - it started its first operation in 1991 [2] - and so the life cycle of a tokamak experiment is roughly 20 years. Another two generations of tokamak devices have to be built in the future to gain the necessary databases of knowledge required for energy production for mankind.

The described steps for igniting the plasma and keeping it in a ring shape for several seconds is the concern of a tokamak control system [3]. Fortunately, despite the extraordinary technical efforts to produce, heat and maintain the plasma, the potential destructive power of such a burning plasma ring is surprisingly low. The worst case that can happen if the control system fails is that the coils and parts of the steel vacuum vessel, which have masses of 1,800 and 800 tons resp. will be damaged. But this is nothing compared with the potential danger inherent in a uranium and plutonium fission reactor. Nevertheless, the control system of a fusion device has to be very reliable. In fig. 1 the overall structure of ASDEX Upgrade and its computer systems is shown. At the bottom of fig.1 the toroidal vacuum vessel, which contains about $10 \, m^3$ of plasma, is symbolically depicted. Above the vessel two types of sensors continuously collect signals from the plasma and tokamak to inform the control system and the experimenting physicists what is going on. Counting all data from all sensors, about 100 Mbytes of data are collected in each 10-second experimental period. During the following 10 minutes all data have to be processed by the computer systems.

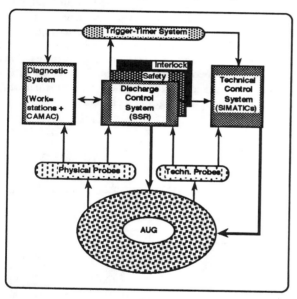

Fig. 1: Overview of the ASDEX Upgrade (AUG) computer systems.

There are three distinct classes of computers, the so-called "Diagnostic System", the "Fast Control System" and the "Technical Control System", as shown in fig. 1. The Diagnostic System is the measuring and computing equipment for the physicists. It consists of about 50 SUN workstations and as many CAMAC crates. On the other side, the Technical Control System is responsible for feed-forward and feed-back control of slow technical processes such as gas pumping and energy collecting for plasma discharges. It incorporates 17 SIMATIC programmable logical controllers (PLCs) from the Siemens company. The Fast Control System described here performs all feed-forward and feed-back control tasks such as plasma ignition and position control on a

short time-scale of several milliseconds. About 500 measured values from the tokamak and plasma have to be processed by the latter system in every millisecond cycle.

The tasks of the Fast Control System and its underlying trigger-timer, safety and interlock systems are now explained in detail. They have to be explained because each computer of the Fast Control System of ASDEX Upgrade is responsible for a specific task, since the system was built from scratch to fit optimally to that specific problem. Therefore, the structure of the tokamak control problem is not directly reflected in the programs of the computer but in the computers themselves. In other words, for this control system a hardware-oriented solution was chosen by means of special computers.

II. The Fast Control, Trigger-Timer, Safety and Interlock Systems

The Fast Control System has six main tasks [4]: 1) it initiates the sequences of events required for operating the technical machinery of the tokamak; 2) it outputs and protocols the command values for the electric coils, whose magnetic fields confine the plasma; 3) it detects, disseminates and protocols significant events in the values measured by the physical and technical probes; 4) it cyclically determines the current status of the tokamak and plasma and branches from this state to one of various scenarios previously provided by the physicists as an experimental schedule; 5) it detects and protocols alarming situations which could cause harm to the machine; 6) it provides the whole experiment, including Diagnostics and Technical Controls, with an obligatory synchronization in time. To perform these tasks, the Fast Control System uses three sources of information: As a prime source, subsets of the physical and technical probes are periodically scanned within a timescale of 1 to 50 ms to collect real-time data. As a second source, previously prepared lists called "shot programs" [5] are fed into the system before a new shot begins, to let the system know what the desired experimental goals of the physicists are. As a third source, built-in procedures, viz. the roughly 200,000 lines of code stored in the computers, are used to cope with standard situations.

To impose a global clock together with the sequence of events necessary to initiate, control and protocol the fusion processes, the Fast Control System is assisted by a Trigger-Timer System [6]. The task of this subsystem is to reliably bridge the distance of several hundred meters between the core computers of the Fast Control System, the tokamak hall, the Diagnostics, the Technical Controls, the RF heating facilities, the power supplies and the flywheel generator, which produces the energy for the whole plasma discharge. The Trigger-Timer System consists of a central transmitter, about 100 glass fiber transmission lines and as many receivers in the field.

Each receiver is an intelligent box [7] in itself, translating the received data into local signals, which in turn control all the said subsystems. Therefore, the Trigger-Timer System can be seen as one of the basic networks of ASDEX Upgrade. Of course, the transmission of these data has to be performed very reliably and in real time. To enhance reliability, an encoding scheme similar to FM modulation (i.e. "Manchester modulation") is used, and, additionally, data is assembled into packets which are guarded by error-detecting and correcting circuitry in the central transmitter and all receivers. Since its first operation in 1991, not one unrecoverable transmitting error has been detected. To match the close real-time constraints to which the central server is subjected to by the many clients distributed in the field, a data packet consisting of 32 bits of payload is assembled and transmitted every 10 μs by the Trigger-Timer System. Additionally, every 100 ns a clock pulse is transmitted through the glass fiber lines to be received at most 2 μs later (due to the speed of light). By means of these pulses and a global clock-reset signal, local clocks are guaranteed to be phase-coupled with the originating transmitter clock. All participants of the system are enabled to have the same notion of time with a precision of less than 2 μs. This in turn is a prerequisite for comparing measured values that are gathered on remote sites. To be able to compare each single value, the maximum data sampling frequency implemented in the Diagnostics is 500 KHz, the inverse of 2 μs. The imposing of a global time inside each parallel computer and between parallel computers, even if they are of different architectures, is an important factor. If the method shown here to be possible was be applied in other research fields, e.g. performance measurements of computers, new and very accurate results could be obtained.

To maintain personnel safety and machine integrity after detecting an alarming situation, the Fast Control System enlists the help of the so-called "Safety System". This is a very simple hardware device (a coax cable ring carrying a constant current) which signals all subsystems already connected together by the Trigger-Timer System that they should locally take emergency shut-down action. The difference between the Trigger-Timer star and the Safety ring is that each participant can open the ring so that current flow stops, which in turn is detected by the other participants. On the same low level the Interlock System performs basic Boolean operations to exclude contradictory machine settings. An example of such a bad setting would be trying to heat the plasma without having previously filled the vacuum vessel with gas with the result that there is no plasma in it. A closer look is now taken by blowing up fig. 1 into fig. 2.

III. Workstation and PLC Setup

In fig. 2 a more detailed view is given of the said three categories of computers, the Diagnostic computers (top),

the Fast Control System computers (middle) and the PLCs of the Technical Control System (bottom right). The approx. 50 Diagnostic Computers together with their Camac crates are coupled by a proprietary glass fiber net based on exclusive point-to-point links. These links are hierarchically concentrated in nodes forming a star. The transmission speed is guaranteed to be 2 Mbits/s between any of the Diagnostic Computers and their cluster nodes for all computers simultaneously. One computer of this category plays a special role: The "Mirnov shot number server" provides the participants in the system with an obligatory number for each plasma discharge. This is important to allow coherent collection and afterwards retrieval of the measured data. Each data set from any computer system can be retrieved by its shot number from a 3 TB mass-storage archiving system. The Technical Control System consists of 16 subordinate PLCs and one master PLC called SLS. The PLCs are connected together by the SIEMENS SINEC H1 net, which is a derivative of the standard ethernet, especially suited to high-reliability applications in an industrial environment. The subordinate PLCs take care of gas refuelling, power supplies, the flywheel generator, the coil currents, the RF heating, vessel heating and coil cooling facilities. Their timescale is 0.1 second to 10 minutes. The master PLC is equipped with several color displays with visualization software to present the ongoing technical processes.

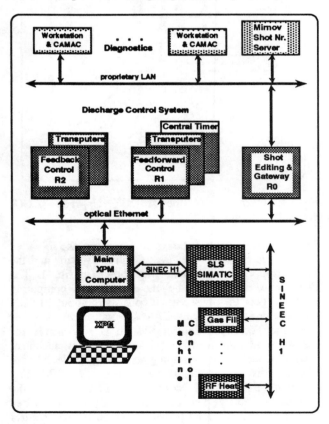

Fig. 2: Workstation and PLC setup.

Virtually and literally, the Fast Control System is situated between the Diagnostics and the Technical Controls. The System is connected to the former by a gateway to separate the proprietary net from a glass-fiber-based ethernet. It is coupled to the latter by a bus bridge which is a SUN Sbus interface and a SINEC H1 member. This optical ethernet is necessary for the Control System computers to allow reliable connection even in the presence of high electric and magnetic fields, radioactivity and RF radiation. Fig. 2 shows the core of the computers of the Fast Control System, which have to perform both real-time and non-real-time duties. For the non real-time tasks there is the "R0 computer", which allows lists of experimental schedules (i.e. the shot programs) to be user friendly entered by means of graphical widgets. Additionally, the R0 acts as the gateway between the Diagnostics and the Fast Control System. A standard SUN workstation is sufficient for that purpose. We also have a so-called "Experiment Management Computer (XPM)" [8] with high-resolution graphics. This computer is the master of all workstations of the Fast Control System. Its software automates handling and operation of the complex tokamak device. Without such an "autopilot" it would be very hard to operate the machine. For this purpose, the XMP machine is equipped with a sophisticated graphical user interface to operate the tokamak on a mouseclick.

The core of the real-time part of the Fast Control System of ASDEX Upgrade consists of the "R1" and "R2" computers, which each have two major components. The first component is a SUN workstation responsible for downloading real-time code and shot programs and for uploading debug information and protocol data. The second component is a proprietary transputer-based parallel computer. The parallel computer parts inside R1 and R2 are responsible for the most important real-time tasks of ASDEX Upgrade. These tasks are plasma position and shape control and generation of the system wide relevant status information, i.e. event generation. For the latter purpose, the R1 is equipped with a transputer-based "Central Timer" to provide the experiment with a global time, a fixed list of events to be scheduled at preprogrammed times, and with non-fixed but online detected events that occur during execution of the shot program. All three types of information are transmitted by means of the Trigger-Timer System mentioned before. Additionally, the R1 generates real-time output values like a function generator does, i.e. it outputs preprogrammed values at preprogrammed times on its output channels.

To summarize, the R1 parallel computer is responsible for generating time and prescheduled events, for detecting and protocolling online events, for determining the status of the tokamak and plasma, for branching to alternative scenarious in the shot program, for detecting and protocolling alarm situations, and for outputting values in real time. Additionally, R1 was chosen as the master for

all parallel computers of the Fast Control System. This means that the transputer component of R1 broadcasts the real-time code it has received from the workstation part to the other transputer computers by means of high-speed transputer links. After shot completion it collects their local protocol buffers in the reverse direction.

The most important task of the Fast Control System of ASDEX Upgrade is performed by the R2 parallel computer. It measures the actual position of the surface of the plasma ring located inside the vacuum vessel and computes the command values for the coils which produce the magnetic fields. Figure 3 helps to illustrate the problem.

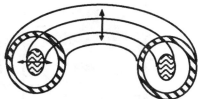

Fig. 3: Cross-section through the vaccum vessel showing the plasma ring.

Here one sees part of the ring-shaped vacuum vessel and the plasma inside. Due to its extraordinarily high temperatures of 30 Mio. degrees it has to be kept away from the inner vessel wall at a distance of several tens of centimeters. Then, the vacuum between the plasma and wall can act as temperature insulation. Unfortunately, at such high temperures the plasma is unstable in the sense that it "wants" to move with a speed of about 10 centimeters per millisecond away from the center of the vessel, finally touching the vessel wall. To avoid this, its position and shape have to be maintained by magnetic fields on a timescale of at most 2 ms. Therefore, R2 incorporates a very sophisticated algorithm to · be computed periodically in less than 2 milliseconds. The algorithm consists of a principal components analysis and PID controllers. The principal components analysis is a mathematical method to extract information in condensed form from raw measured data. The PID controllers try to maintain the desired plasma position according to proportional, integral and differential rules. To keep the plasma stable is a critical task. During the experiments it sometimes happens that R2 does not know how to do this. In these cases the 2 MA contained in the plasma ring are deposited on the inner carbon walls of the vacuum vessel, thereby reducing the life of the device. Next, we go into more detail to exhibit the internal structure of the Fast Control System and the architecture of the transputer-based parallel computers.

IV. Transputer Systems Setup

In fig. 4 we focus on the Fast Control System showing the tokamak, the Diagnostics and SIMATIC systems as peripherals on the left, upper and right sides of the picture, respectively.

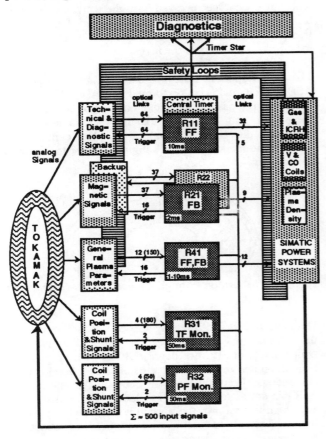

Fig. 4: Transputer Systems Setup (FF = feed-forward, FB = feed-back).

At this magnification step it becomes apparent, that the R2 computer has a shadow companion R22. Although not installed yet, it is intended to be used later for on-the-fly comparison of the command values of R21 and R22. Since the two parallel computers have the same input from magnetic field sensors, they should produce the same outputs. If this is not the case, an emergency shutdown should be started, since it is very likely that an essential part has failed. Fig. 4 shows three computers R31, R32 and R41 [9], which are each constructed of two parts: workstation + transputer-based parallel computer. The R31 and R32 parallel computers scan input signals from position probes indicating the precise mechanical position and stress a coil has. They continuously monitor these signals and initiate shutdown by interrupting the current in the Safety System loop if mechanical movements exceed an upper limit. So, their task is to ensure the physical integrity of the tokamak.

The R41 parallel computer performs various feed-forward and feed-back tasks on timescales between approx. 1 and 10 ms. It controls, for example, the distance between the RF antennae and the plasma, or the amount of RF power to be injected into the plasma to heat it up to a certain temperature. The timescales of the other computers R1-R3 vary between 2 and 50 ms. As the tasks of the computers are different, the number of transputers differs from 5 for the R31 and R32 to 14 for the R2s and 29 for the R1 computer. The number of input signals that have to be processed in the individual cycle times range from 12 (in future 150) signals for the R41 to 37 for the R2s and 64 for the R1 computer. The R3s have 4 physical inputs carrying up to 50 or 150 signals respectively by multiplexing them onto the input lines.

V. Architecture of the Parallel Computers

All parallel computers, viz. the R1, R2s, R3s and R4 have many features in common. Firstly, they are application-specific computers. This means that each machine was built to optimally perform a specific task. Therefore, their internal structure reflects the job they have to do. To understand their architecture requires knowing what a tokamak control system has to do. For this reason the tasks of the Fast Control System have already been explained. Our experience during the past few years has been that, if you have a machine ideally suited to solving a special task, the problem of controlling a tokamak can be better managed if the complexity of this task can be forgotten as soon as the machine is in operation and meets its requirements. Without using application-specific, transputer-based parallel computers we would not have been able to build the Fast Control System of ADEX-Upgrade.

The second common feature of all our computers is that they are based on the "MULTITOP" architecture [10], which says: use message passing as programming paradigm, take processor-memory modules with transputers as processors and connect them via two fixed and one variable interconnection network. The topology of the first fixed net is a star with many processor-memory modules at the rays of the star and one module in the center, which acts as a star master. The topology of the second fixed net is a chain which lines up all modules like a string of pearls. The variable interconnection network as the third connection scheme is reserved for user applications. The topology of the variable interconnection network is switched over during user program execution to adapt the interconnection scheme to the needs of the program. This means, if two remote processors want to communicate during algorithm execution that they are connected together dynamically. In a later phase of the program they can be disconnected to open the way for other communication partners.

The most important fact of the MULTITOP concept, besides the triplex interconnection network, is that there are only two basic building blocks [11] that can be grouped together in arbitrary numbers to form the desired machine. Each parallel computer in the Fast Control System of ASDEX Upgrade is an example of this property. The two basic building blocks are called CPU0 and MCPU. In fig. 5 we see the block diagram of the first of the two, the CPU0 board.

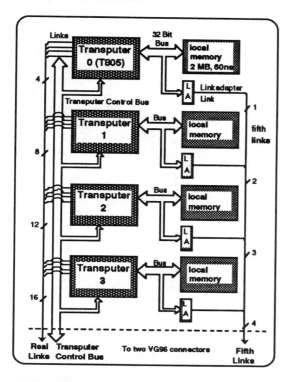

Fig. 5: The CPU0 Transputer Board.

This board contains four identical transputer-memory modules, which are connected by a transputer control bus (TSB). Additionally, each module has a so-called link adapter, which is a peripheral chip from the INMOS company that fits onto the transputer bus on one side and that emulates a transputer link on the other side. Normally, each transputer has only four links to communicate with other transputers, but with the link adapter chip a fifth link can be formed. And with many fifth links, a star can be made the first of the two fixed connection schemes. The second fixed connection scheme is implemented by the transputer control bus to form the chain of processors.

Additionally, the four real links of each module, making 16 per board, are fed into the third means of interconnection, the dynamic network. Each of the three coupling schemes has its specific task to perform. The fifth links are for synchronization, debugging and operating system purposes, the transputer control bus is for resetting, booting and single stepping the processors and the real links are for high-speed communication in the user application. The purpose of one or many CPU0s

is to be the "computing workhorse" in a MULTITOP machine. Each module is therefore equipped with a fast static RAM matching the speed of the processor.

If you have followed this far, you may ask who is the master of the bus and of the fifth-link star. The second building block, the MCPU now comes into the play, as depicted in fig. 6.

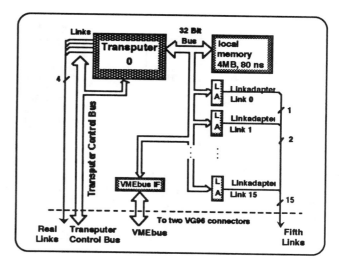

Fig. 6: The MCPU Transputer Board.

The MCPU contains one processor-memory module and 16 link adapters. It can be the star master of up to 4 CPU0 boards since each board contains four modules. Additionally, the MCPU is the master of an arbitrary number of CPU0 boards. Because they are connected in a chain, no limit in the number of processors is given exept that it may take too long to transmit data from the master to the last one in the chain. Beside its task as a master for the fifth-link star and/or for the transputer control bus, the MCPU is the gateway between the real-time transputer world and the non-real-time UNIX workstation, which is used as a server for accessing the keyboard and screen and for disk and ethernet I/O.

The gateway function is performed via a VMEbus Interface on the MCPU which translates the signals of the VMEbus-based workstation into transputer link signals. Both MCPU and CPU0 are VMEbus-compatible boards using the P2 connector for special purposes. To see how nicely the CPU0s and MCPUs fit together by using the reported features, let us analyze the architecture of the R2 feed-back control computer [13].

VI. The R2 Feed-back Control Computer

Fig. 7 shows the block diagram of the R2. In this computer 3 CPU0s, 2 MCPUs, one variable interconnection network and a couple of converter boards are connected together by their real links, fifth links and the transputer control bus.

Since the TSB is not visible in normal operation, it is not shown here. The converter boards have glass fiber inputs because all data to and from the computers of the Fast Control System of ASDEX Upgrade are optically transmitted for reasons of reliability. The task of the is to transform the electric signals converters into light and vice versa. The MCPU at top of fig. 7 is the TSB- and fifth-link master for the I/O processor and for the 12 processor-memory-modu-les located on three CPU0 boards. The 12 processor-memory modules compute the 12

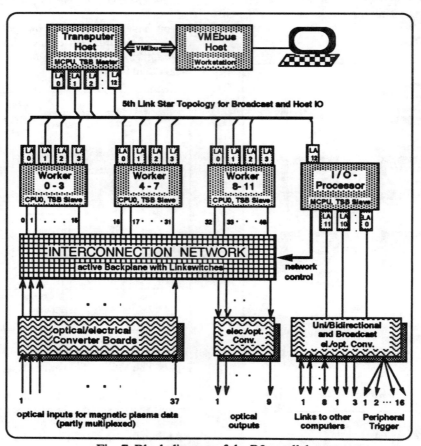

Fig. 7: Block diagram of the R2 parallel computer.

dynamic switching of the transputer links was required to be faster than 2 ms, which is the maximum allowed cycle time to keep the plasma stable. Because of the fact that the algorithm could not be divided into more than 12 parts, the only way to meet the cycle time was to minimize the communication, thereby maximizing the efficiency of the parallel program. This goal was reached by switching over the interconnection topology three times each cycle. The tasks assigned to the R2 computer could be performed successfully by this kind of architecture.

A completely different structure was realized with the same building blocks to create the R1 parallel computer. A block diagram is shown in fig. 8. R1 was scaled up to 29 processors, which are divided into three different groups. The first group of 16 transputers located on 4 CPU0 boards form a input funnel through which data has to be passed to extract high level information out of low level measured values. After that, the condensed and abstracted data is handed over to the second group of processors. There, 5 transputers cooperatively determine the current plasma and tokomak status. A decision is made where to branch next in the shot program. Additionally, the system-wide significant plasma events are created here. Because of the heavy memory requirements of these tasks, 5 MCPUs with 4 MB RAM each were chosen to meet the needs. In the third group of processors the output values of the preprogrammed functions are produced in real time. Here, speed is mandatory, and so the faster CPU0s were selected. The requirements for the number of input and output channels were 64 inputs and 16 outputs. This could be easily accomplished by installing 4 CPU0s on the input- and 2 CPU0s on the output side of R1.

To summarize, the R1 parallel real-time computer is a example for the close relationship between the problem

principal components of the input values, and in the second phase of the cycle 9 of the processors are used to form 9 PID controllers.

The input values are delivered from the converter boards through the variable interconnection network, which is configured by the I/O processor. In the beginning of the computing cycle all transputer links are switched to the data input sources to perform a massively parallel input. Then the network is reconfigured to a 3D torus topology which best fits to the communication needs of the principal components algorithm. After completion of the PID calculation the net is switched over a third time, so that the 9 results can be delivered at the output ports. Beside switching over the network, the I/O processor, which is implemented physically by a second MCPU, is responsible for communicating with the corresponding I/O processors of the R1, R3 and R4 computers. Additionally, it sends the data acquisition trigger to the peripheral probes, which in turn start a computing cycle by delivering new measured values. All data transferred between the periphery and the computers are optically transmitted. The that has to be solved and the computer's architecture. Especially the triplex interconnection networks of the MULTITOP architecture has proven to facilitate greatly the system integration of the various parts of the R1 computer since they allow a lot of flexibility. To be able to choose computing power, number of I/O channels and processor interconnection topology as needed, a very high degree of freedom is achieved that can be used to tackle a problem with maximum efficiency.

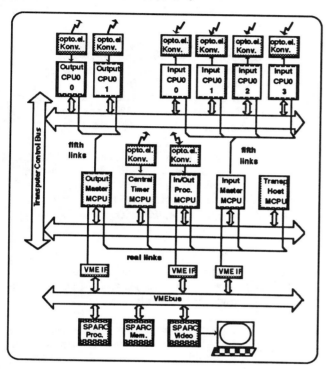

Fig. 8: Block diagram of the R1 parallel computer.

To complete the collection of application-specific computers used in the Fast Control System of ASDEX Upgrade, we finally present the R31 and R32 machines. It is recalled that these computers were responsible for monitoring a lot of sensor signals that determine the mechanical position of the toroidal (TF) and poroidal (PF) coils and their thermal and mechanical stresses. This is a task, which can be performed on a relatively relaxed time scale. The required cycle time was 50 ms. This means that not many processor-memory modules are necessary, especially, since the calculations did not consist in much more than summing up input values over a longer period of time. Therefore, we decided to use multiplexed input channels to save money because each optical transmission line costs about 5 TDM. Two variants, as depicted in fig. 9 and 10, emerged as being suited to do the job. The difference between these two architectures is that in the first case the input channels are formed by fifth links, while real links are used to build a link star. In the second case it is vice versa (fig. 10). The advantage of using fifth

links on the input side is that they are more robust against probe failures since the sensors are not directly connected to the processor.

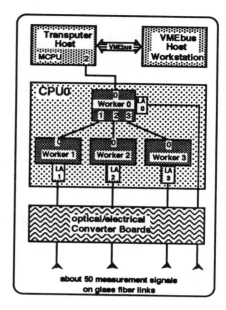

Fig. 9: The R31 TF coil monitor.

The disadvantage of using fifth links is that they are much slower than real links because the processor has to be interrupted for each input value. Since the R32 has to read 3 times as many signals as the R31, real links are used here as input channels, while the R31 was chosen to be more independent of sensor failures.

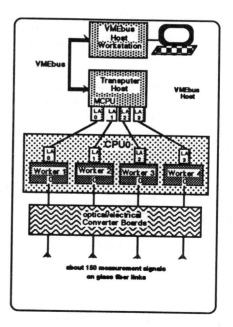

Fig. 10: The R32 PF coil monitor.

VII. Summary and Conclusions

This paper describes the plasma physical environment of the computer systems of the ASDEX Upgrade tokamak as well as the computer systems themselves. The tasks of one of those, the Fast Control System, were presented from a problem oriented point of view. The setup of this system, its architectural components and its real-time parts were analyzed in detail. The close relationship between the real-time requirements, the basic building blocks and the MULTITOP architecture was demonstrated. Our experience was that the use of application-specific, parallel computers with selectable structure, speed, and I/O capacity has greately simplified the task of building a control system for such a big machine as ASDEX Upgrade. Additionally, it was shown that it is possible to establish a common clock inside a parallel computer as well as between parallel computers with a precision of 2 μs. This global time proved to be very valuable for testing the Control System and operating the tokamak. The timescale of less than 2 ms to keep the plasma stable was very short compared with the possibilities of contemporary real-time controls. We met the requirements by applying parallel processing techniques based on a variable interconnection network, thus minimizing interprocessor communication. The methods and concepts employed here are considered to be of general significance and applicable wholely or in part in many similar cases.

VIII. References

[1]   W. Köppendörfer et al. "Completion of Assembly and Start of Technical Operation of ASDEX Upgrade," *Proc. 16th Symp. on Fusion Technology*, Vol. 1, London 1990, Eds. B.E. Keen, M. Huguet, R. Hemsworth, North-Holland Publ. Amsterdam 1991, p. 208-212.

[2]   B. Streibl et al. "ASDEX UPGRADE: The first period of Operation," *Proc. 17th Symp. on Fusion Technology*, Rome, Elsevier Amsterdam 1992.

[3]   G. Raupp, H. Bruhns, K. Förster, F. Hertweck, R. Huber, A. Jülich, G. Neu, H. Richter, U. Schneider, B. Streibl, W. Woyke, D. Zasche, T. Zehetbauer, "ASDEX Upgrade Discharge Control and Shot Management," *Proc. 17th Symp. on Fusion Technology*, Rome, Elsevier Amsterdam 1992, p. 1072-76.

[4]   G. Raupp, C. Aubanel, V. Mertens, , G.Neu, H.Richter, D. Zasche, "Real Time Processes for ASDEX Upgrade Control," *15th Symposium On Fusion Engineering*, Princeton University, 1993.

[5]   D. Zasche, H.Bruhns, R.Cole, K.Förster, R.Huber, K.Lüddecke, A.Jülich, G.Neu, G.Raupp, H.Richter, U.Schneider, "Tokamak discharge description at ASDEX Upgrade," *Proc. 8th Conf. REAL TIME '93 on Computer Applications in Nuclear, Particle and Plasma Physics*, Vancouver (CAN) 1993.

[6]   G. Raupp, H. Richter, "The Timing System for the ASDEX Upgrade Experiment Control," Proc. 7th Conf. REAL TIME '91 on Computer Applications in Nuclear, Particle and Plasma Physics, Juelich (DE) 1991, *IEEE Transactions on Nuclear Science*, Apr. 1992, Vol. 3, Nr.2, p. 198-203.

[7]   H. Richter, G.Neu, G.Raupp, D.Zasche, "System Integration of the ASDEX Upgrade Timing System," Proc. 8th Conf. REAL TIME '93 on Computer Applications in Nuclear, Particle and Plasma Physics, Vancouver, Kanada, 1993, *IEEE Transactions on Nuclear Science*, Vol. 41, Nr.1, Feb. 1994.

[8]   H. Richter, R. Cole, M. Fitzek, K. Förster, K. Lüddecke, G. Neu, G. Raupp, W. Woyke, D. Zasche, T. Zehetbauer, "Overview of the ASDEX Upgrade Experiment Management Software," *Proc. 17th Symp. on Fusion Technology*, Rome, Elsevier Amsterdam 1992, p.1077-81.

[9]   G.Neu, C. Aubanel, V. Mertens, G.Raupp, H.Richter, D. Zasche, "An Enhanced Plasma Control for ASDEX Upgrade," *15th Symposium On Fusion Engineering*, Princeton University, 1993.

[10]   H. Richter, "Multiprocessor with Dynamically Variable Topology," *Computer Systems Science and Engineering 5, 1,*Butterworth London 1990, p. 29-35.

[11]   G. Raupp, H. Richter, "The MULTITOP Parallel Computers for ASDEX Upgrade," *Proc. Int. Conf. on Parallel Processing*, St. Charles (US) CRC Press Boca Raton USA 1991, I, p. 656-657.

[12]   H. Richter, G. Raupp, "Control of a Tokamak Fusion Experiment by a Set of MULTITOP Parallel Computers," Proc. 7th Conf. REAL TIME '91 on Comp. App. in Nuclear, Particle and Plasma Physics, Juelich (DE) 1991, *IEEE Transactions on Nuclear Science*, Apr. 1992, Vol. 3, Nr.2, p. 192-197.

Increasing Production System Parallelism via Synchronization Minimization and Check-Ahead Conflict Resolution *

Chang-Yu Wang and Albert Mo Kim Cheng
Department of Computer Science
University of Houston
Houston, TX 77204-3475
E-mail:cheng@cs.uh.edu

Abstract – *This paper explores new techniques to improve the performance of production or rule-based systems. In particular, a rule-splitting mechanism using constraint domain analysis and a check-rule mechanism are proposed to extract internal parallelism among rules. This rule-splitting mechanism can also be used to solve the load balancing problem caused by culprit rules which require comparisons with many more working memory elements than those of other rules. In addition, a new conflict resolution strategy, called Check-Ahead Conflict Resolution, based on the last-in first-out scheduling discipline is presented to select multiple instantiations to fire simultaneously. Simulation results show that the speed-up of production systems obtained by applying the Check-Ahead Conflict Resolution strategy with the rule-splitting and check-rule mechanisms is superior to other existing techniques.*

1 Introduction

Production expert systems are increasingly used in a variety of financial, transportation, and industrial applications. Those used to monitor and control the operations of complex safety-critical real-time systems must meet certain stringent timing requirements imposed by the environment. Therefore, making a real-time expert system run as fast as possible to ensure that it will meet these timing requirements is a fundamental problem. One approach to speed up the execution of production expert systems is to introduce parallelism in both the pattern-matching phase and the rule firing phase of the recognize-act cycle.

Since the match procedure of a production system usually consumes around 90 percent of the total execution time [2], Stolfo [13] and Gupta [4] proposed several parallel pattern-matching algorithms for implementation on a tree-structured multiprocessor machine called DADO [14]. To further reduce the overall execution time, several parallel rule firing methods have been proposed ([1], [5], [6], [7], [11]). Yet another approach to reduce the execution time of production systems is to extract more internal parallelism from existing systems. Ishida and Stolfo [5] developed a data dependency graph approach for checking if a set

of rules can be fired in parallel. Based on a data dependency graph of production systems, Ishida [6],[7] proposed a parallel firing model which fires multiple rule instantiations simultaneously on a multiprocessor system. In Ishida's model, the instantiations of two rules cannot be parallelly fired if these two rules have disabling and/or conflicting relations. However, since not all of the parallelism of production systems can be determined at compilation time [8], a large amount of unnecessary synchronization is imposed by static analysis models. Therefore, on-line techniques for detecting rule interference must be developed to exploit more internal parallelism from production systems at run-time.

To reduce needless rule synchronization, Schmolze [11] proposed a refined parallel firing model that analyzes the relations among rules at compilation time and among rule instantiations in the select phase of the recognize-act cycle at run time. The parallel firing models proposed by Ishida and Schmolze, however, can work only with a conflict resolution strategy which has no constraint on the firing order of instantiations, but not with the sequential conflict resolution strategies such as LEX and MEA for the OPS5 [3] language and CLIPS [12] language in which the instantiations have to be fired in some order.

In [9] and [10], more production system parallelism is extracted by creating constrained copies of culprit rules and distributing them to different processors for parallel processing. Because the condition elements of culprit rules require comparisons with many more working memory elements than those of other rules, these culprit rules are copied, with each of the copies constrained to match only a subset of the working memory elements matched by the original rule. Each copy of a culprit rule needs less computation time and the load balancing among the rules becomes better. This in turn improves the performance of production systems. However, this copy-and-constraint method can accelerate the execution of production systems only when there is a large number of culprit rules. Also, it requires a hashing mechanism to implicitly confine the elements of the constraint domains of the attribute terms in each constrained copy of a culprit rule.

In this paper, a rule-splitting mechanism based on constraint domain analysis is proposed. It divides the rules which are involved in potentially parallelism-inhibiting relations into their corresponding sets of sub-rules to make the potential relations more re-

---

* This material is based in part upon work supported by the National Science Foundation under Award No. CCR-9111563, the Texas Advanced Research Program under Grant No. 3652270, and the University of Houston Institute of Space Systems Operations.

stricted, making it possible to exploit more internal parallelism in production system programs. This rule-splitting mechanism can also solve the load balancing problem caused by culprit rules which require comparisons with many more working memory elements than those of other rules. Furthermore, a dynamic check-rule mechanism is used to determine if potential interference relations among instantiations will actually occur at run time and thus eliminate unnecessary synchronization among these instantiations. To guarantee that parallel firing will produce the same result as sequential firing with a traditional conflict resolution strategy such as LEX which uses the last-in-first-out (LIFO) scheduling discipline, we propose a new conflict resolution strategy, called Check-Ahead Conflict Resolution (CACR). It selects multiple instantiations to fire at the same time. The result of firing rule instantiations in parallel with the CACR strategy is proved to be serializable.

The remaining sections of this paper are organized as follows. Section 2 briefly describes production systems and formally defines data dependency and conflict relations among rules. Section 3 addresses the serialization problems with parallel firing and provides two approaches to exploit more internal parallelism of production system programs and eliminate unnecessary synchronization among rule instantiations. The parallel firing model with the new conflict resolution strategy is described in section 4. Conclusions are in section 5.

2 Data Dependency and Conflict Relations

Before defining the terms and concepts needed to analyze the relations among rules, we briefly describe production systems. A production system is a program consisting of a set of conditional rules called productions that are stored in the *production memory* (PM) and work on assertions called elements in the *working memory* (WM). The left hand side (LHS) of a production consists of a conjunction of condition elements (CEs) and the right hand side (RHS) of a production consists of a set of actions that can add, delete, or modify *working memory elements* (WMEs), call external procedures to update WMEs or perform input/output (I/O) operations when the *instantiations* of the rules are fired. Since a modify action is defined to be equivalent to a delete action followed by an add action, we do not consider it in this paper. In addition, if the I/O operations of a set of rules have implementation-dependent sequential constraints, we can impose certain conflict relations among these rules to enforce them to fire in the original order. To simulate the actions, such as adding or deleting WMEs, taken by external procedures, we assume these external actions are known in advance and can be replaced by some production rule(s).

The general format of a production is as follows:

```
(<indicator of rule> <rule name>
<<sign>> <<CE>>
  ...
```

```
->
<<action>>
... )
```

where a CE makes up the attribute-expression tests that may contain variables and the sign of a CE can be positive or negative. A variable in the LHS can appear either once or several times. In the latter case, all occurrences of the variable have to match the same value, i.e., the variable is *bound* to the value it matches. Thus these variables are called bound variables. However, there is no such limit on variables that occur only once in the LHS. Therefore, they are called *free* variables. All of the variables appearing in the RHS are bound variables because their values are determined by their corresponding occurrence(s) in the LHS.

The template of patterns is [+/-] (class-identifier [<attribute-1> <constraint-1> ... <attribute-i> <constraint-i> ... <attribute-n> <constraint-n>]). If a pattern is positively referred in LHS or added in RHS, it has a positive sign which can be omitted. Otherwise, if a pattern is negatively referred in LHS or deleted in RHS, it has a negative sign.

Basically, there are two scalar data types, number and symbolic atom, in production systems. The entire constraint domain (CD) of an attribute is the set of real or integer numbers, R, if its data type is number or symbolic string, or S, if its data type is symbolic atom. When an action pattern adds an element into the WM, a production system automatically assigns a special value, *nil*, to each of these class attributes which do not appear in the action pattern. We treat *nil* as an element of both data types in this paper. The following is an example.

Example 1.

Suppose a class, say class1, has more than five attributes and a pattern of class1 is: (class1 ^A_1 5 ^A_2 <<10 20>> ^A_3 {<> 10 = <X>} ^A_4 {<> bird = <Y>} ^A_5 {> 0 < 10}). Then the CDs of these five attributes are $\{a_1 | a_1 \in R, a_1 = 5\}$, $\{a_2 | a_2 \in R, a_2 = 10 \text{ or } 20\}$, $\{a_3 | a_3 \in R, a_3 \neq 10\}$, $\{a_4 | a_4 \in S, a_4 \neq "bird"\}$, and $\{a_5 | a_5 \in R, 0 < a_5 < 10\}$, respectively, whether the pattern is in LHS or RHS. For the attributes that do not appear in the pattern, there is no constraint on them if the pattern represents a CE or a delete action. In this case, their CDs are either $\{a_i | a_i \in R\}$ or $\{a_i | a_i \in S\}$ which depends on their data types. However, if the pattern represents an add action, the CDs are $\{nil\}$ for the attributes that do not appear in the pattern. ◇

We are now ready to define the interactions between rules. A term is an attribute-constraint pair <attribute> <constraint>. A pattern P_i *potentially* matches another pattern P_j if for each term T_{jk} in P_i and its corresponding term T_{jk} which has the same attribute as T_{jk} in P_j, the CD of T_{jk} includes the CD of T_{jk}, and the term T_{jk} has bound variable(s); or the CDs of T_{jk} and T_{jk} partially overlap each other. A pattern P_i *absolutely* matches another pattern P_j if for each term T_{jk} in P_i and its corresponding term T_{jk} in P_j, the CD of T_{jk} includes the CD of T_{jk} and the term T_{jk} has no bound variable(s).

In those definitions, we do not take into account the signs of the patterns. When there is a *bound* variable in a pattern and the CD of every term in this pattern includes the CD of its corresponding term in another pattern, whether the former pattern absolutely matches the latter pattern is determined by the WMEs at run time and cannot be detected by static analysis. This case is thus classified as a potential category. Based on these pattern-matching definitions, we define the enabling and disabling relations among rules.

A rule R_i *potentially* enables another rule R_j if there exists a RHS action pattern, say $P_m(i,r)$, of R_i and a LHS condition element pattern, say $P_n(j,l)$, of R_j such that $P_n(j,l)$ potentially matches $P_m(i,r)$ and both patterns have the same sign. The symbols $P_m(i,r)$ and $P_n(i,l)$ stand for the m-th RHS action pattern and the n-th LHS condition element pattern of the rule R_i, respectively. A rule R_i *absolutely* enables another rule R_j if for each CE pattern $P_n(j,l)$ of R_j, there exists an action pattern $P_m(i,r)$ in R_i such that $P_n(j,l)$ *absolutely* matches $P_m(i,r)$ and both patterns have the same sign.

A rule R_i *potentially* or *absolutely* disables another rule R_j if there exists an action pattern $P_m(i,r)$ in R_i and a CE pattern $P_n(j,l)$ in R_j such that $P_n(j,l)$ *potentially* or *absolutely* matches $P_m(i,r)$ and they have opposite signs. A rule R_i *potentially* or *absolutely* conflicts with another rule R_j if there exists an action pattern $P_m(i,r)$ in R_i and an action pattern $P_n(j,r)$ in R_j such that $P_m(i,r)$ and $P_n(j,r)$ *potentially* or *absolutely* matches with each other and have opposite signs.

In forward-chaining production systems, firing a rule may enable other rules which in turn will probably disable and/or conflict with some other rules. This means that a rule may indirectly disable and/or conflict with other rules. To cover these direct and indirect relations, the following derivative relations among rules are defined to complement the above definitions.

The enabling rule (ER) graph G = (V, E, W) of a set of rules is a weighted directed graph. V is a nonempty set of vertices such that there is one vertex for each rule; E is a possibly empty set of arcs (directed edges) and W is a set of binary values {0, 1} such that there is an arc weighted 1 (or 0) from the vertex corresponding to the rule R_i to the vertex corresponding to the rule R_j if and only if R_i potentially (or absolutely) enables R_j.

A rule R_i has *potentially* or *absolutely derivative* enabling (PDE or ADE) relation on another rule R_j if and only if there exists at least one path from the vertex corresponding to R_i to the vertex corresponding to R_j in the ER graph. Note that the length of the path (the sum of the weights of all arcs along the path) is at least zero.

A self-looping ER (SLER) graph is constructed as an ER graph except that each vertex has an extra self-looping arc whose weight is zero. A rule R_i has *loose* or *tight mastership* on another rule R_j if there exists

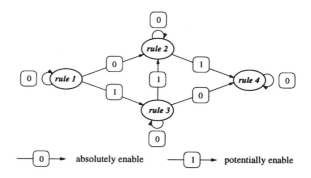

Figure 1: Example of SLER graph

at least one path from the vertex corresponding to R_i to the vertex corresponding to R_j in the SLER graph and the length of the path is at least zero.

A rule R_i has *potentially derivative* disabling (PDD) relation on another rule R_j if and only if there exists a rule R_k such that (1) R_i has loose mastership on R_k, and R_k absolutely or potentially disables R_j; or (2) R_i has tight mastership on R_k, and R_k potentially disables R_j. A rule R_i has *absolutely derivative* disabling (ADD) relation on another rule R_j if and only if there exists a rule R_k such that R_i has tight mastership on R_k, and R_k absolutely disables R_j.

Two rules R_i and R_j have *potentially derivative* conflict (PDC) relation with each other if and only if there exists a rule R_k such that

(1) R_i (or R_j) has loose mastership on R_k, and R_k potentially or absolutely conflicts with R_j (or R_i); or

(2) R_i (or R_j) has tight mastership on R_k, and R_k potentially conflicts with R_j (or R_i); or

(3) R_i (or R_j) has loose mastership on R_k, and R_k is potentially or absolutely disabled by R_j (or R_i); or

(4) R_i (or R_j) has tight mastership on R_k, and R_k is potentially disabled by R_j (or R_i).

Two rules R_i and R_j have *absolutely derivative* conflict (ADC) relation with each other if and only if there exists a rule R_k such that R_i (or R_j) has tight mastership on R_k, and (1) R_k absolutely conflicts with R_j (or R_i) or (2) R_k is absolutely disabled by R_j (or R_i).

Example 2.

Assume a set of rules has a SLER graph and direct interference relations as shown in Figure 1 and Figure 2. Then the derivative interference relations among these rules are:

(1) each rule in the rule set {rule1, rule2, rule3} has PDD relation on every rule in the rule set {rule5, rule6};

(2) rule3 has ADD relation on rule6;

(3) each rule in the rule set {rule1, rule2, rule3} has PDC relation with every rule in the rule set {rule7, rule8, rule9, rule10}; and

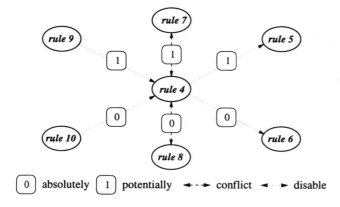

Figure 2: Example of interference relations

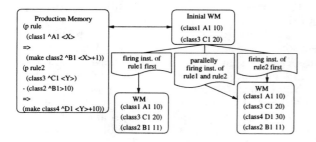

Figure 3: Sequential and parallel firing results of two instantiations having disabling relation.

(4) rule3 has ADC relation with rule8 and rule10.

Rule3 simultaneously has both PDD and ADD relations on rule6, and both PDC and ADC relations with rule8 and rule10. In this case, only the absolute relations are taken into account when synchronizing the parallel firing sequences of rule instantiations. ◇

3 Analysis of Synchronization

3.1 Serializability Problems with Parallel Firing

Ensuring that the parallel execution of a production system can produce a serializable result is the major concern when firing multiple instantiations simultaneously. Basically, sequential and parallel executions will generate two different results under two conditions. First, the firing of a higher priority instantiation will cause the lower priority instantiations to be disabled. Consider the example shown in Figure 3. Both rule1 and rule2 are satisfied by the elements in the initial WM. If the instantiation of rule1 has higher priority and fires first, then the instantiation of rule2 will be disabled and the final WME set is { (class1 A_1 10), (class3 C_1 20), (class2 B_1 11)}. In contrast, when the instantiation of rule2 has higher priority and fires first, the instantiation of rule1 still remains in the conflict set and will fire in the next cycle. Thus, the final WME set will be { (class1 A_1 10), (class3 C_1 20), (class4 D1 30), (class2 B_1 11)} which is

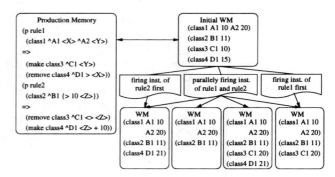

Figure 4: Sequential and parallel firing results of two instantiations having conflict relation.

the same as the result of firing the instantiations of rule1 and rule2 simultaneously. Obviously, the parallel execution of set of instantiations is serializable for the second case in which the higher priority instantiation will not disable the lower priority instantiation. Second, there exist conflict relations among the rule instantiations. Figure 4 shows the result of sequential execution and the possible results of parallel execution of two instantiations which have conflict relation with each other. In this case, parallel firing may yield four different results since the actions of one instantiation may interleave with the actions of the other. Only one of the possible results of the parallel firing is the same as that of the sequential firing sequence.

To produce a serializable result, the instantiations of rules need to be synchronized if these rules have derivative disabling or conflict relations. The following two subsections introduce two methods to eliminate the unnecessary synchronization among rules or rule instantiations at compilation time and at run time.

3.2 Elimination of Synchronization at Compilation Time

Since the potentially derivative relations among rules are evoked only under certain WM environment, the lower the overlap the CDs resulting in the potential relation, the lower the evoked probability of the potential relation. To eliminate unnecessary synchronization and exploit more internal parallelism at compilation time, each rule pattern involved in the potential relation can be split into several sub-patterns according to the overlap situations of the CDs. This means that a rule will also be divided into a corresponding set of sub-rules if one or more of its patterns are split and only one of these sub-rules will keep the potential relation (which sometimes may become the absolute relation) of the original rule. If n patterns of a rule are split and each of these patterns is divided into m sub-patterns, then the rule will be split into m_n sub-rules.

Consider the example shown in Figure 3. Since the intersection of the CDs of B_1's terms in rule1 and rule2 is $\{a|a \in R, a > 10\}$, the add action pattern of rule1 can be divided into two subpatterns, (make class2 $^{\wedge}B_1$ {> 10 <X> + 1}) and (make class2 $^{\wedge}B_1$ {<= 10 <X> + 1}), by splitting B_1's term into two sub-terms, one's CD is $\{a|a \in R, a > 10\}$ and the other's CD is $\{a|a \in R, a \le 10\}$. Then the CE pattern of rule1 also needs to be split because it is tied up with the add action pattern by variable X and X's possible value domains of these two sub-terms are $\{a|a \in R, a > 9\}$ and $\{a|a \in R, a \le 9\}$, respectively. Thus, the two corresponding sub-rules of rule1 are:

```
(p rule1-1
   (class1 ^A1 {> 9 <X>})
   ->
   (make class2 ^B1 {> 10 <X> +1}))
(p rule1-2
   (class1 ^A1 {<= 9 <X>} ^A2 <Y>)
   ->
   (make class2 ^B1 {<= 10 <X> +1}))
```

Now only rule1-1 has a disabling relation on rule2 and the relation becomes the absolute disabling relation. Thus by static analysis, we can exactly conclude that the instantiations of rule1-1 and rule2 can never be fired in parallel if the instantiation of rule1-1 has higher priority, but the instantiations of rule1-2 and rule2 can always be executed simultaneously. If rule1 is not split, a closer conclusion which can be made to ensure the serializability of the parallel execution is that the instantiations of rule1 and rule2 should not be fired in parallel if the instantiation of rule1 has higher priority. However, the latter conclusion may cause unnecessary synchronization under some WM environments. For instance, assume that the initial WM shown in Figure 3 is changed to $\{$(class1 A_1 9), (class3 C_1 20)$\}$ and the instantiation of rule1 has higher priority, then the instantiations of rule1 and rule2 can be fired in parallel because there are actually no conflict relation between them. However, by the latter conclusion, these two instantiations cannot be executed in parallel. When rule1 is divided into rule1-1 and rule1-2, the instantiations of rule1-2 and rule2 will be created and fired simultaneously.

Obviously, more specific relations among rules can be obtained by splitting a rule's pattern(s) based on the intersection of the constraint domains of each pair of the corresponding attribute terms. Therefore, a more precise synchronization analysis can be made at compilation time and more parallelism can be exploited by processing these split sub-rules in parallel. Moreover, the intra-node parallelism can be converted to rule-level parallelism.

Besides, the rule-splitting method can also be applied to divide a culprit rule into its corresponding set of sub-rules. Each of these sub-rules matches only part of the WMEs which satisfies the original culprit rule and has fewer intermediate tokens and inter-condition tests. Thus, the overall joint tests can be reduced in order to solve the bottleneck problem of parallel processing caused by the culprit rule and speed-up the execution of the production system.

When compared to the copy-and-constraint method proposed by Pasik, our rule-splitting approach has two additional advantages. First, the rule-splitting approach needs no hashing mechanism which implicitly confines the elements of the CDs of the attribute terms in each constrained copy of a culprit rule. Second, the rule-splitting approach can make the relations among rules more specific and extract more parallelism from production system programs at compilation time because it explicitly indicates the CDs of the attributes in each split copy of a culprit rule and each of these CDs is restricted to a certain range.

3.3 Elimination of Synchronization at Run Time

When there are potential relations among rules, static analysis alone cannot determine whether these potential relations will actually happen at run time. The simplest way to guarantee the serializability of the parallel execution is to assume that these potential relations always occur under any WM environment. However, this assumption would impose unnecessary synchronization on the rules involved in these potential relations. Dynamic checking is one of the approaches which can be used to reduce these unnecessary synchronization.

In this paper, we assume that the parallel computer for executing the production system has sufficient processors to perform the parallel pattern-matching tests regardless of the number of rules. Hence, check rules are added to the original rule set to decide if the potentially derivative relations may occur or not. Since the check rules are not allowed to take any action to affect the whole system environment, they have the CE patterns only. Actually, for each pair of rules which have a potential relation, the CEs of their check rules are obtained by modifying only part of their CEs. Therefore, if the data flow network used by RETE, TREAT, or other similar match algorithms is properly constructed, the check rules will only create a few number of extra two-input nodes and memory nodes [2]. Furthermore, since these check rules are obtained by using static analysis at compilation time, they do not increase the overhead at run time.

First, consider an instance of a rule, called an enabled rule, which is potentially enabled by a set of rules, called enabling rules. There exists a case where these enabling rules luckily add (or delete) some elements to (from) the WM such that all of the enabled rule's positive (negative) CEs matching a part of their action patterns are satisfied. This leaves the fewest number of enabled rule's CEs that need to match part of the current WMEs in order to create an instantiation. These leftover CEs of the enabled rule form the LHS of the correspondent enabling-check rule. Thus, the correspondent enabling-check rule does not create any two-input node and memory node in the data flow network. Moreover, if the instantiation of an enabled rule can be created after firing part or all of instantiations of these correspondent enabling rules, then an instantiation of the correspondent enabling-check rule is also created. This means that if no instantiation of the correspondent enabling-check rule is created, then no instantiation of the enabled rule is created either, i.e., no enabling rule can evoke the enabled rule under the current WM environment.

From the above observation, we have developed algorithms for obtaining the CEs of an enabling-check rule, for obtaining the CEs of disabling-check rule, and fro obtaining the CEs of a conflict-check rule. They are omitted there due to space limitations.

4 Parallel Rule-Firing Model

The parallel execution model we propose can be implemented on parallel computers with shared memory or distributed memory. It needs synchronization at the select phase to guarantee that the result of the parallel execution is serializable. The recognize-act cycle of this execution model is as follows.

(a) **Match:** The processors simultaneously perform pattern-matching tests.

(b) **Select:** There are two steps in this phase. In the first step, processors simultaneously determine the interference relations among the instantiations of each pair of rules which are in the synchronization list created at compilation time. If a pair of instantiations do have interference relation(s), then an unfirable or synchronized mark is placed on the lower priority instantiation if it will be disabled by or conflicts with the higher priority instantiation. In the second step, processors repeatedly perform the following check for each pair of instantiations until no new marked instantiation is reported to a designated processor: an instantiation is marked to synchronize if it has interference relation(s) with another instantiation which has a synchronized mark and a higher priority.

(c) **Act:** All unmarked instantiations are fired in parallel.

To further speed up the parallel execution of production systems, the act phase can overlap with the match phase of the next cycle to effectively reduce the execution time of each cycle if the actions do not create, delete, or modify WMEs accessed by the next match. In the select phase, the parallel firing of the unmarked instantiations after completing the first step is not serializable because some of these instantiations may disable other instantiations which have higher priorities and synchronized marks (these instantiations cannot be fired in this cycle, but will be executed eventually).

Based on the LIFO scheduling discipline, a new selection mechanism called the check-ahead conflict resolution (CACR) strategy is proposed to select multiple instantiations to fire at each recognize-act cycle. The final result of parallel execution is also serializable. The CACR strategy consists of two algorithms to determine the interference relations among rules or instantiations at compilation and run time, respectively.

Compilation time algorithm of the CACR strategy:

This algorithm generates a corresponding program OP for each input production system program IP and constructs a related synchronization list SL containing related check paths for all rules in OP. It consists of the following steps:

(1) copy each rule of the input program IP into its corresponding rule set RS_i, $i = 1$ to n, where n is the number of rules in the program;

(2) set splitting flag SF to false;

(3) for $i = 1$ to $n - 1$
 for $j = i + 1$ to n
 for each rule R_{ik} in RS_i and each rule R_{jl} in RS_j, if they have any potential relation and can be further split, split them into their corresponding sets of sub-rules and replace them by their corresponding sets of sub-rules and set SF to true;

(4) if SF is true, go to step (2);

(5) put all rules in each individual rule set RS_i into the output program OP;

(6) identify the direct relation between each pair of rules in OP and construct the SLER graph;

(7) identify the derivative interference relation(s) between each pair of rules in OP. For each interference relation identified, make and attach a node with the related relation-chaining path(s) to SL. (If the length of one of these paths is zero, then set all paths to NULL;)

(8) make a check-rule for each direct relation between two rules and add it to OP.

Suppose that a rule set has k sub-rules, a rule has l patterns, and a pattern has m terms, then in the worst case, it needs $O(n_2 \cdot k_2 \cdot l)$ pattern comparisons to detect relations among rules and $O(n_2 \cdot k_2 \cdot m \cdot l)$ rule-splitting operations in step (3) assuming there are at least $l \cdot max(l, 2^{m+1})$ processors for performing parallel pattern-comparison and parallel rule-splitting. Thus the time complexity of the algorithm for obtaining check-rules is also polynomial.

Run-time algorithm of the CACR strategy:

This algorithm uses the synchronization list SL generated at compilation time together with the conflict set of instantiations to determine the interference relations among these instantiations and select a subset of instantiations in the conflict set to fire at run time. Its steps are:

(1) for each pair of instantiations
 determine their firing priorities according to the LIFO strategy such as LEX;

(2) for each pair of instantiations I_i and I_j of rules R_i and R_j, where I_i has higher priority
 if R_i and R_j are a pair of rules listed in SL, then
 for each interference relation between I_i and I_j
 if the check path is NULL or all check rules along a path have instantiations in the conflict set, then
 if I_i disables I_j, then place an unfirable mark on I_j;

else if I_i conflicts with I_j, then place a synchronized mark on I_j;
else set the synchronization flag between I_i and I_j to true;

(3) repeat
for each pair of instantiations I_i and I_j, where I_i has higher priority
if the synchronization flag between them is set and there is a synchronized mark on I_i and I_j is not marked, then place a synchronized mark on I_j;
until no new marked instantiation;

(4) select these unmarked instantiations to fire.

In this algorithm, a pair of instantiations are assigned to a group of processors in which every interference relation between the pair of instantiations is checked by a dedicated processor. Every processor has its own memory for storing information related to the pair of instantiations it is processing. Thus, all pairs of instantiations and all relations between a pair of instantiations can be processed in parallel.

Suppose that there are n non-check-rule instantiations in the conflict set and each instantiation has m interference relations whose average length of the check path is l. Then in the worst case, the time complexities of selecting instantiations in steps (2) and (3) are $O(l)$ and $O(n)$, respectively, if there are at least $m \cdot n \cdot (n - 1)/2$ processors. In addition, if an instantiation has k WMEs, the time complexity for determining the firing priorities of a pair of instantiations is $O(k \cdot logk + k)$. Therefore, the time complexity of the run-time algorithm of the CACR strategy is $O(k \cdot logk + k + l + n)$. However, since the average number of CEs per rule of production system programs is usually less than 10 and it is around 4 in most cases [4], the run-time algorithm actually runs in linear time.

Since the run-time algorithm requires each processor computer to independently check the interference relations between each pair of instantiations, synchronization in the select phase is needed to avoid firing the lower priority instantiations earlier than the higher priority instantiations when the former have the interference relations with the latter. For example, assume the conflict set contains three instantiations I_1, I_2, and I_3 in which I_1 conflicts with I_2, I_3 disables I_2, and their firing priorities are $I_1 > I_2 > I_3$. After completing the second step of the run-time algorithm, only I_2 is marked not to fire at this execution cycle. However, the firing of I_3 will disable I_2, so I_2 finally will not be fired. Thus, the parallel execution cannot produce a serializable result because all three instantiations must be fired according to the sequential execution. To prevent nonserializable result like the case just mentioned, the third step of the run-time algorithm repeatedly synchronizes the select phase until no new marked (unselected) instantiation is reported. Only I_1 is selected to fire in the first cycle after the third step. I_2 and I_3 will be selected to fire simultaneously in the second cycle. Therefore, the parallel

execution requires only two cycles to fire these three instantiations but the serial execution requires three cycles.

To prove that the parallel execution with the CACR strategy produces a serializable result, we define several terms and derive some results which are stated as theorems. Owing to space limitations, the proofs of these theorems are omitted but can be found in [15].

An instantiation I_i is a successor of another instantiation I_j if I_i is directly or indirectly enabled by firing I_j. Two instantiations are *compatible* if the higher priority instantiation has no disabling relation on or conflict relation with the lower priority instantiation.

We derive the following theorems whose proofs are omitted here due to space limitations.

Theorem 1.

If an instantiation I_i has no interference relation with another instantiation I_j, then I_i and I_i's successors have no interference relation with I_j and I_j's successors.

Theorem 2.

All firing sequences of two instantiations I_i and I_j produce the same result if there is no interference relation between them.

Theorem 3.

The parallel execution and the sequential execution of a set of instantiations produce the same result if every pair of these instantiations are compatible.

Theorem 4.

The result of parallel execution with the CACR strategy is serializable.

A software simulation environment is developed and implemented to simulate the parallel firing of production systems with the CACR strategy. Due to space limitations, the complete simulation results are reported in an upcoming paper.

When compared to the parallel rule firing models proposed by Ishida and Stolfo or by Schmolze, our parallel rule firing model can gain more speed-up since the rule-splitting and the check-rule mechanisms can eliminate more unnecessary synchronizations among rules and rule instantiations at both compilation time and run time. Unlike the copy-and-constraint method proposed by Pasik, the rule-splitting mechanism needs no extra hashing function to implicitly confine a rule to match a certain part of WMEs and can make the relations among rules more specific. Thus, the speed-up gained by the copy-and-constraint method is also lower.

5 Conclusions

We have introduced a rule-splitting and a check-rule mechanisms to exploit more internal parallelism in production system programs and a new parallel conflict resolution strategy called Check-Ahead Conflict Resolution to select multiple instantiations to fire in parallel. Both the rule-splitting and the check-rule mechanisms can eliminate unnecessary synchronization among rule instantiations and thus increase the parallelism in production systems. The new parallel conflict resolution strategy includes two algorithms for checking the relations among rules or instantiations. The compilation time algorithm has polynomial time complexity whereas the on-line algorithm has linear time complexity. The simulations show that the execution speed of production systems is improved when applying the CACR strategy with the rule-splitting and the check-rule mechanisms.

In addition to the initial WM environment, the amount of speed-up depends on the degree of internal parallelism which is characterized by the enabling, disabling, conflict, and splitting ratios of the rules as well as by the number of independent rule clusters. The degree of internal parallelism of a program is always reduced when the disabling, conflict, or splitting ratio is increased, but it may be increased or reduced when the enabling ratio increases if there exist interference relations among rules. If the rules of a program can be tailored into more independent rule clusters, its degree of internal parallelism can be increased. The more the rule instantiations which can be created during a run of a program, the higher the degree of internal parallelism which can be actually exploited, and a higher speed-up ratio thus can be obtained by parallel firing. Our simulations show that the magnitude of the speed-up ratio can be over 200. The memory size for storing synchronization information is increased when the disabling, conflict, or enabling (if there exist interference relations among rules) ratio increases. It ranges from several kilo-bytes up to tens of giga-bytes. Therefore, the data structure, distribution manner, and access method of synchronization information have to be properly designed and implemented to reduce the overhead of the contention and memory access. However, when a program's enabling ratio is at most 0.1 and the total number of rules is at most 200, the memory needed for synchronization is less than 10 mega-bytes. If a program's enabling ratio is 0.2, the maximum needed memory is around 80, 160, or 400 mega-bytes when the number of rules of the program is 50, 100, or 200, respectively.

In the future, a translator will be constructed to convert a production program into a corresponding data set used as input to the current simulator. In addition, precise cost-estimation methods can be incorporated to study the overhead caused by dynamic checking and to compare the execution times of parallel and sequential firings. Furthermore, we also need to investigate the performance improvement obtained when the Check-Ahead Conflict Resolution strategy with the rule-splitting and the check-rule mechanisms are implemented by different numbers of processors and interconnection topologies. In particular, a compiler and an execution environment are being implemented on a Kendall Square Research (KSR) system to validate the results obtained by simulation.

References

[1] A. Cheng, "Parallel Execution of Real-Time Rule-Based Systems," *Proc. of the 7th Int'l. Parallel Processing Symposium*, Newport Beach, Calif., Apr., 1993, pp. 779–786.

[2] C. Forgy, *On the Efficient Implementation of Production Systems*, Dept. of Computer Science, Carnegie-Mellon Univ., Ph.D. Dissertation, 1979.

[3] C. Forgy, *OPS5 User's Manual, Dept. of Computer Science*, Carnegie-Mellon Univ., Tech. Report CMU-CS-81-135, Jul., 1981.

[4] A. Gupta, *Parallelism in Production Systems*, Dept. of Computer Science, Carnegie-Mellon Univ., Ph.D. Dissertation CMU-CS-86-122, Mar., 1986.

[5] T. and S. Stolfo, "Towards the Parallel Execution of Rules in Production System Programs," *Proc. of the 1985 IEEE Int'l. Conf. on Parallel Processing*, Aug., 1985, pp. 568–575.

[6] T. Ishida, "Methods and Effectiveness of Parallel Rule Firing," *Proc. of the Sixth Conf. on Artificial Intelligence Applications*, Los Alamitos, Calif., Apr., 1990, pp. 116–122.

[7] T. Ishida, "Parallel Rule Firing in Production Systems," *IEEE Trans. on Knowledge and Data Eng.*, 1991, Vol. 3, No. 1, pp. 11–17.

[8] A. Oshisanwo and P. Dasiewicz, "A Parallel Model and Architecture for Production Systems," *Proc. of the Int'l. Conf. on Parallel Processing*, 1987, pp. 147–153.

[9] A. Pasik, "A Source-to-Source Transformation for Increasing Rule-Based System Parallelism," *IEEE Trans. on Knowledge and Data Eng.*, 1992, Vol. 4, No. 4, pp. 336–342.

[10] A. Pasik and S. Stolfo, *Improving Production System Performance on Parallel Architectures by Creating Constrained Copies of Rules*, Tech. Report, Computer Science Dept., Columbia Univ., 1987.

[11] J. Schmolze, "Guaranteeing Serializable Results in Synchronous Parallel Production Systems," *Journal of Parallel and Distributed Computing*, 1991, Vol. 13, No. 13, pp. 348-365.

[12] Software Technology Branch, *CLIPS Reference Manual*, Software Technology Branch, Information Systems Directorate, Lyndon B. Johnson Space Center, NASA, Houston, Texas, 1991.

[13] S. Stolfo, "Five Parallel Algorithms for Production System Execution on the DADO Machine," *Proc. of the National Conf. on Artificial Intelligence*, AAAI, Austin, Texas, Aug., 1984, pp. 300–307.

[14] S. Stolfo and D. Shaw, "DADO: a Tree-Structured Machine Architecture for Production Systems," *Proceedings of the National Conf. on Artificial Intelligence*, AAAI, Aug., 1982, pp. 242–246.

[15] C.-Y. Wang, *New Algorithms for Parallel Execution of OPS5 Rule-Based Systems and Their Performance Analysis*, Master's Thesis, Dept. of Computer Science, Univ. of Houston - Univ. Park, Aug., 1994.

FEEDBACK BASED ADAPTIVE RISK CONTROL PROTOCOLS IN PARALLEL DISCRETE EVENT SIMULATION

Donald O. Hamnes and Anand Tripathi
Department of Computer Science
University of Minnesota
Minneapolis, Minnesota 55455
E-mail: hamnes@mail.cs.umn.edu, tripathi@cs.umn.edu

Abstract -- Several adaptive risk control protocols for parallel and distributed discrete event simulation are proposed and evaluated. Three approaches that we refer to as optimization, time equalization and rate equalization are investigated as techniques to achieve a stable algorithm. Feedback of information to immediate predecessor nodes is used. Actual implementations of the algorithms on a CM-5 are used to evaluate the protocols. It is seen that the algorithms are stable and consistently perform better than an optimistic algorithm.

1.0   INTRODUCTION

Recently, significant research in parallel discrete event simulation has been done on adaptive synchronization protocols that are able to span the complete range of protocols from the conservative [6] to the optimistic [4] and perform better than either one alone [3].

Adaptability [7] is the ability of a process to change the values of its own design variables based on information about the state of the system. Two design variables [7] identified were aggressiveness and risk. *Aggressiveness* measures the tendency of a process to optimistically process events before it is known whether another event with a lower timestamp will arrive. *Risk* is the degree to which a process will release messages based on optimistic processing.

The investigation in [2] provided a general framework in which to understand adaptive risk control protocols; however, the risk control protocols with complete aggressiveness did not appear to be stable. That observation and the idea that the use of additional feedback might be used to produce adaptive risk control protocols which are stable are the motivation for the present investigation.

The contributions of this work are stable adaptive risk control protocols based on feedback which perform better than a corresponding optimistic protocol. These protocols use complete aggressiveness in input processing. An optimizing protocol is stabilized, and another protocol which was expected to yield good performance is examined. The best protocol is identified.

A brief survey of related work follows. However, the reader is assumed to be familiar with basic concepts such as those found in [4] and [6].

Several protocols have involved risk control. At one extreme no risk is taken. This is true of the conservative protocol as noted in [7]. Several additional protocols allow events to be processed optimistically, but the effect of that processing may not be propagated to other nodes until it is certain that the messages will never need to be rolled back. Examples of this approach are the use of local rollback [1] and breathing time buckets [8]. The breathing time warp approach [8] results in an intermediate value of risk when averaged over time since the approach alternates between phases, one of which takes no risk and one which takes complete risk.

The unified distributed simulation algorithm [5] attempts to combine an adaptive risk control protocol with an adaptive aggressiveness control protocol. The term *send window* is defined to be a period in simulated time during which a process may send a message. The window is based relative to GVT. Using the framework in [2], the definition of a send window is extended to be a window in simulated time relative to the time of the last send. A *holding window* is defined to be a period in real time during which output messages are held.

Feedback is used in a conservative adaptive demand driven protocol in [9]. Requests from successor nodes are used to reduce a message queueing threshold by half.

2.0   ADAPTIVE RISK CONTROL

The class of risk control protocols to be investigated have the following characteristics. Each LP examines the timestamps of its output messages. If it falls within the send window, then it will be sent immediately. If that is not the case, then a real time holding window is computed for the message and it is inserted into a pending output queue to be sent at the estimated release time (present time plus holding time). The expectation is that many of the messages which are queued will be canceled locally before the release time is reached. The pending output queue is a FIFO queue maintained in both increasing simulated timestamp order and increasing real release time order. The equations to be used to compute the adaptive send window (SW') and adaptive holding window (HW') are those derived in [2]:

$$HW'_i = \frac{k_i}{\alpha_i}\left[snow_i - (slast_i + SW'_i)\right]$$

$$SW'_i = \frac{\alpha_i}{k_i}(rnow_i - rlast_i)$$

the subscript i identifies the values as pertaining to output channel i. The variables are defined as follows:

$rnow_i$ = current value of real time.

$rlast_i$ = real time at which a message was last sent on channel i.

$snow_i$ = timestamp of current message to be sent or queued.

$slast_i$ = simulated time at which a message was last sent on channel i.

α_i = average amount of increase in simulated time per unit real time on output channel i.

k_i = a constant to be adjusted to maximize α_i.

The quantity $\alpha'_i = \alpha_i / k_i$ was seen to be the effective rate at which we are trying to drive the system [2]. Thus, $1/k_i$ can be thought of as a forcing factor; or the ratio by which we are trying to increase the performance of the system.

The objective of the algorithms is to maximize α_i; that is, to maximize the average simulation rate. The intent is to minimize the finish time of the over all simulation. The algorithms considered have the additional characteristic that the values of α_i and k_i are measured at the immediately downstream LP and sent back in some transformed manner to the preceding LP as required.

Three different methods to achieve a stable algorithm are examined:

- **optimization**: The optimization procedure is investigated.
- **time equalization**: This method attempts to minimize stragglers by equalizing the simulation time on the channel which is presently behind.
- **rate equalization**: This method tries to equalize the rate of simulation time increase on all channels.

These methods will be examined by specifying when to send feedback; what information to send back; and what actions are to be taken by the receiver. The term *trigger event* designates a special event which causes feedback to be sent.

Optimization Procedure

The optimization procedure is executed periodically by the downstream node for each of its input channels. The objective is to maximize α_i as a function of k_i. Initially, three fixed values of k_i are used. Each value is used during one observation period and the resulting α_i is noted. In all cases the first value of k_i is 0 since this is the smallest possible value and corresponds to a policy of complete risk; messages are not held. At the end of each period, the observed α_i and the next k_i are sent to the upstream nodes to be used in computing SW' and HW' during the next period of time. The goal is to find three successive (k_i, α_i) pairs such that a quadratic can be fit to the values to predict which k_i value should be considered next to maximize α_i. If the first three fixed values of k_i

do not produce appropriate α_i values, then a search has to be conducted to find such appropriate points.

The original search procedure used 0, 0.5 and 1.0 as the initial fixed search values for k_i. In addition, the search procedure had a climb mode in which it doubled the previous value of k_i as the next value of k_i to consider. As will be noted in the rate equalization investigation (section 3), it appears that values of k_i greater than one are to be avoided and hence a climb mode which asymptotically approaches one is more desirable. This would be coupled with lower initial values for the fixed portion of the search. The reason for this is suggested by recalling that $1/k_i$ can be thought of as a forcing factor. When k_i is greater than one, then we are trying to force α_i to be less than it was on the previous iteration. Repeated instances of this could lead to poor performance. Thus, the revised climb mode of the search procedure sets k_i to the previous k_i + (1 - previous k_i)/2. A final issue to be investigated is the sensitivity of the optimization to the size of the real time interval, ΔT, over which each α_i is computed.

Time Equalization

The second method is to equalize the time of the farthest behind channel to that of the second farthest behind channel. This is to be done in addition to the optimization described in the preceding section. Suppose that the greatest timestamp in the input queue of channel j is t_j; if the input queue is empty, then the timestamp of the last processed event from that channel is used. When a trigger event is encountered, such as two consecutive stragglers on one channel, m, then the time value $T = \min_{j \ne m}\{t_j\}$ is sent to the predecessor LP on the channel

on which the trigger event occurred. The predecessor LP then uses the T value to empty its output queue of events with timestamps up to and including simulation time T. Release times for any remaining items in the output queue are recalculated in order to maintain consistency with future insertions. This flushing is expected to help alleviate the straggler problem on channel m and yet not put it too far ahead since one other channel has events with timestamps up to time T. Time equalization is to be viewed as a small perturbation on the whole process of optimization.

Rate Equalization

The final method to be considered is rate equalization. The basic premise is that by equalizing the output α_i values at the predecessor LPs on the input channels to each LP, the LP will experience fewer straggler events in the future and hence achieve better performance. The expectation is that LPs which were sending many rollback messages because they had been taking too much risk will have to increase the holding time for messages which are presently in the queue as well as for future messages inserted into the queue. On the other hand, LPs which were sending many stragglers because they were advancing at a low rate, will be forced to empty their pending output queue and retain fewer messages in the queue in the future.

When a trigger event occurs, such as two consecutive straggler events on one channel, the rate equalization procedure is executed. This procedure sends a mid range α value and a k value to all predecessor LPs. These LPs use the new α and k values to recompute the holding times for any messages in their pending output queues. This approach computes one input α value per LP. Two variants of this strategy will be considered. The first uses no optimization. Rather it assumes that equalization alone will optimize performance. The second variant performs optimization over periods of time which typically span many instances of rate equalization. For both approaches a mid range α, which is denoted by $\hat{\alpha}$, is computed as follows:

$$\hat{\alpha} = \left(\min_{j}\{\alpha_j\} + \max_{j}\{\alpha_j\} \right)\Big/ 2.$$

Both approaches use a rate equalization procedure (RATE_EQ) which has one input parameter which will be called: specified_k. RATE_EQ performs the following steps:

i. If insufficient new events have occurred since the last computation of $\hat{\alpha}$, then return.

ii. Compute the $\hat{\alpha}$ value based on the current state of the system.

iii. Reset trigger counters.

iv. Send the $\hat{\alpha}$ value and specified_k value to all predecessor nodes; these nodes will use the values as their α_i and k_i values, respectively, for their associated output channel.

The first, basic, rate equalization method calls RATE_EQ with specified_k = 1 whenever a trigger event occurs. Note that a k value of 1 implies that there is no forcing factor. In addition, if a sufficient interval of real time passes without the occurrence of a trigger event, then RATE_EQ is executed anyway with a specified_k value of 0.9. This value of k was chosen so that the predecessor LPs will be forced to take a little more risk in releasing messages with a resulting higher α value.

The second, optimizing, rate equalization method calls RATE_EQ in the same instances as the basic rate equalization method; however, in all cases specified_k = k_{opt}. k_{opt} is an optimized k value as described below. Since $0 \leq k_{opt} < 1$, k_{opt} corresponds to a forcing factor greater than one and will be used to attempt to drive the LP at a faster rate than that at which it has been running.

The optimization procedure for determining k_{opt} is executed periodically, every ΔT seconds, and will usually span multiple executions of the rate equalization procedure, RATE_EQ. When the optimization procedure executes, an $\hat{\alpha}$ is determined which represents an average over the whole ΔT units of real time. $\hat{\alpha}$ is now optimized by a procedure similar to that in the preceding

section; the currently used k value is referred to as k_{opt}. However, $\hat{\alpha}$ and k_{opt} are never sent to the predecessor nodes by the optimizer. Rather k_{opt} is sent by the rate equalizing procedure, RATE_EQ. Thus, the optimizing rate equalization method sends to predecessor LPs a long term average optimal k value with mid range α values.

3.0   PROTOCOL EVALUATION

The amount of elapsed time required to simulate a given closed queueing system to a fixed simulation time limit is used for comparing the different protocols. The protocols will be evaluated for various numbers of jobs in the system. Stability of the protocols is also considered to be very important.

The particular closed queueing system chosen for this investigation is an 8×8 torus. Each node represents a server with two input and two output channels. Equal branching probabilities are used on the output channels. A biased exponential service time distribution is used for the server. The fixed bias is set to one percent of the average service time. A non-uniform distribution of average service times is used since this has been found to be an interesting configuration in other investigations [3]. Alternating rows have servers with average service times of 10 and 20 units. The simulation time limit was fixed at 250000 units of simulated time.

The hardware used for this study is a 64-node partition of a CM-5 Connection Machine. A host/node program was developed to test these algorithms in this environment. A controller program to handle administrative tasks runs in the host while server-simulators run in the nodes.

The graph in this section is presented as a function of *message density* which is defined to be the number of jobs in the closed queueing system divided by the number of nodes in the system.

Optimization

A climb function which doubles the previous value as the next one to consider produced instability. However, a function which is asymptotoically bounded by 1 was stable. A couple of different choices for the initial fixed three search points were also considered (all less than 1). Though they had relatively little effect on the performance, the values of 0, 0.25 and 0.5 produced slightly better results.

The optimization interval ΔT is the period of real time between executions of the optimization function. It was found that the interval choice is not too critical. Longer intervals produce better results, probably because they average the behavior of the system over a longer period of time and so adapt in a more correct manner.

When appropriate the simulations in the following sections make use of the optimum results found in this section by using $\Delta T = 2$ seconds and a bounded optimization function.

Time Equalization

Results of the time equalization protocol appear in Figure 1 below. Two consecutive stragglers on a channel was used as the trigger event. The protocol was stable and yielded acceptable results; however, they were not better than the optimization protocol. A more extensive investigation of this protocol was not performed since it is directly based on the optimization protocol.

Rate Equalization

When the basic rate equalization procedure was tried, it was realized that the assumption that a minimum elapsed time would result solely because the rates were being equalized was incorrect. For example, it was estimated that the final elapsed time at a message density of 128 would have been over 800 seconds. As a result, an investigation was performed to test the sensitivity of the procedure to using different fixed values of k in place of the values of 1.0 or 0.9 as stated in section 2. These results for message density 128 showed that best performance was obtained for a k value between 0.25 and 0.5. Very poor performance was obtained for k values of 0.9 or higher. The high sensitivity to k caused us to design the rate equalization method with optimization.

An investigation into what is an acceptable trigger event was then conducted All of the trigger events were based on the number of stragglers observed: two consecutive on a channel, two or four on a channel or two or four at the node. One observation noted is that using 2 or 4 stragglers at a node does not appear to be a consistent predictor of the need to equalize the rates. This conclusion is reached because the elapsed time using these triggers increases sharply around a message density of 8. A somewhat more selective trigger such as four stragglers on a channel yields better performance. This trigger event was used for rate equalization results appearing in Figure 1.

Comparison of the Protocols

A comparison of the protocols appears in Figure 1. This shows that all protocols are stable and perform better than the optimistic protocol on the average. The optimization only protocol performed best of all.

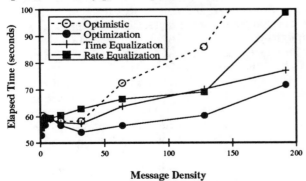

Figure 1: Comparison of the adaptive risk protocols with the optimistic protocol.

4.0   CONCLUSION

Several feedback based adaptive risk control protocols have been presented. Their performance has been evaluated through implementation and execution on a CM-5. Stable versions of all have been found which perform better than a non-adaptive optimistic protocol. The most important observation from this study is that care must be taken in the design of the optimization function. In particular, the function should concentrate on k values in the interval [0, 1). An alternative using feedback has been considered to the optimization method. The idea that basic rate equalization would optimize performance by minimizing stragglers has been seen to be incorrect; the performance is unacceptable because of the lack of a forcing factor greater than 1. This helps validate the necessity of explicit optimization.

Acknowledgment

Sponsored in part by the Army HPC Research Center under the auspices of the Department of the Army, Army Research Laboratory. The content does not necessarily reflect the position or the policy of the government, and no official endorsement should be inferred.

5.0   REFERENCES

[1]   P. M. Dickens and P. F. Reynolds Jr., "SRADS with Local Rollback", *Proc. of the SCS Multiconf. on Distrib. Simulation*, 22 (1), (1990), pp. 161 - 164.

[2]   D. O. Hamnes and A. Tripathi, *A Comparative Study of Adaptive Risk vs. Adaptive Aggressiveness Control in Parallel and Distributed Simulation*, Department of Computer Science, University of Minnesota, TR 94-66, (Dec. 1994).

[3]   D. O. Hamnes and A. Tripathi, "Evaluation of a Local Adaptive Protocol for Distributed Discrete Event Simulation", *1994 International Conf. on Parallel Processing*, Vol. III, pp. 127 - 134.

[4]   D. R. Jefferson, "Virtual Time", *ACM Transactions on Programming Languages and Systems*, 7 (3), (July 1985), pp. 404 - 425.

[5]   J. McAffer, "A Unified Distributed Simulation System", *Proc. of the 1990 Winter Simulation Conference*, pp. 415 - 422.

[6]   J. Misra, "Distributed Discrete-Event Simulation", *Computing Surv.*, 18 (1), (Mar. 1986), pp. 39 - 65.

[7]   P. F. Reynolds, Jr., "A spectrum of options for parallel simulation", *Proc. of the 1988 Winter Simulation Conference*, pp. 325 - 332.

[8]   J. S. Steinman, "Breathing Time Warp", *7th Workshop on Parallel and Distributed Simulation*, (1993), pp. 109 - 118.

[9]   W. Su and C. L. Seitz, "Variants of the Chandy-Misra-Bryant distributed discrete-event simulation algorithm", *Proc. of the SCS Multiconf. on Distributed Simulation*, 21 (2), (1989), pp. 38 - 43.

An Optimal Ear Decomposition Algorithm with Applications on Fixed-Size Linear Arrays* (Extended Abstract)

Ying-Min Huang[†]and Joseph JáJá[‡]
Department of Electrical Engineering
University of Maryland at College Park
College Park, MD 20742
yingmin@src.umd.edu, joseph@umiacs.umd.edu

ABSTRACT

The decomposition of a graph into simpler pieces, called ears, provides an important technique for designing efficient graph algorithms. In particular, it can be used to determine the biconnected components of a graph. Determining the ear decomposition and the biconnected components of a connected, undirected graph $G(V, E)$ can each be solved sequentially in $O(m + n)$ time, where $|V| = n$ and $|E| = m$. We develop parallel algorithms that solve these problems in $\Theta(\frac{m}{p} + n + p)$ time on a p-processor linear array. Our algorithm for ear decomposition generates in fact an open ear decomposition for each biconnected component; we refer to such decomposition as a pseudo-open ear decomposition. Parallel algorithms to identify the bridges and the articulation points of $G(V, E)$ are also described. The input is assumed to consist of a set of m edges distributed arbitrarily among the p processors. These algorithms are the best possible on our model.

Keywords: parallel algorithms, connected components, minimum spanning forest, ear decomposition, biconnected components, bridges, articulation points, and linear array.

1 INTRODUCTION

The decomposition of a graph into simpler pieces,

called *ears*, provides an important technique for designing efficient graph algorithms. In particular, it can be used to determine the biconnected components of a graph. The ear decomposition, pseudo-open ear decomposition, and biconnected components of connected graph $G(V, E)$ can each be determined sequentially in $O(m + n)$ time, where $|V| = n$ and $|E| = m$. Parallel PRAM-type algorithms for determining open ear decompositions for biconnected graphs have appeared in [6, 7]. In this paper, we assume a parallel model based on a p-processor linear array. We develop parallel algorithms that solve these problems in $\Theta(\frac{m}{p} + n + p)$, where the input consists of m edges distributed equally among the p processors. In fact, our ear decomposition algorithm generates an open ear decomposition for each biconnected component, which we call *pseudo-open ear decomposition*.

We consider a computation model based on a fixed-size linear array. The simplicity of the communication network of a linear array makes it easy to build, maintain and extend. However the task of designing fast algorithms becomes more difficult due to the network diameter and the limited communication bandwidth. Our algorithms are carefully tailored to avoid these potential bottlenecks.

We next review some terminology and a few well-known facts.

2 PRELIMINARIES

Given an undirected **connected** graph $G(V, E)$ with n vertices and m edges, G is *2-edge connected* if it contains no bridges. An *ear decomposition* $D = [\mathcal{P}_0, \mathcal{P}_1, \cdots, \mathcal{P}_r]$ of an undirected 2-edge connected graph $\mathcal{G}(\mathcal{V}, \mathcal{E})$ is an ordered partition of $\mathcal{E}$ into disjoint simple paths $\mathcal{P}_0, \mathcal{P}_1, \cdots, \mathcal{P}_r$, where $\mathcal{E} = \mathcal{P}_0 \cup \mathcal{P}_1 \cup$

---

*Partially supported by NSF Grant No. CCR-9103135, NSF Engineering Research Center Program NSFD CD 8803012.

[†]Is with Cadence Design Systems, Inc. 555 River Oaks Parkway, MS 2B1, San Jose, CA 95134.

[‡]Also, Institute for Advanced Computer Studies, Institute for Systems Research, University of Maryland, College Park, MD 20742.

$\mathcal{P}_2 \cup \cdots \cup \mathcal{P}_r$. $\mathcal{P}_0$ is an edge, $\mathcal{P}_0 \cup \mathcal{P}_1$ is a simple cycle, and $\mathcal{P}_i$ is a simple path and only its two endpoints belong to $\mathcal{P}_0 \cup \mathcal{P}_1 \cup \mathcal{P}_2 \cup \cdots \cup \mathcal{P}_{i-1}$, $1 < i \leq r$. If each $\mathcal{P}_i$ is a simple path, $1 \leq i \leq r$, then D is an *open ear decomposition*. The *pseudo-open ear decomposition* determines an open ear decomposition for each biconnected component of G. For any undirected connected graph, its pseudo-open ear decomposition is not unique. An undirected connected graph has an ear decomposition if and only if it is 2-edge connected, i.e. *bridgeless*.

Given a pseudo-open ear decomposition, let $label(e)$ be the open ear of edge e of G. Let $b(v)$ be the biconnected component of non-articulation vertex $v \in V$, and $Bicc(e)$ be the biconnected component of edge $e \in E$. Before going further, let us state the following well known result.

Theorem 1. A graph is biconnected if and only if it has an open ear decomposition.

Let T be a spanning tree of a given graph $G(V, E)$. Consider any two vertices u and v from a given tree T. The *lowest common ancestor* (LCA) of u and v, represented by $lca(u, v)$, is the vertex that is an ancestor of both u and v and farthest from the root. Any nontree edge e of G induces a *fundamental cycle* $\mathcal{C}_e$ of e respect to T in the graph $T \cup \{e\}$. $\mathcal{C}_e$ consists of the path from $lca(e)$ to one end point of e, followed by the edge e, followed by the path from the other end point of e to $lca(e)$. The preorder traversal of a tree consists of a traversal of the root r, followed by the preorder traversal of the subtrees of r from left to right. The sequential ordering of a vertex v in this traversal starting from r is the preorder numbering of v. Let $preorder(v)$ be the preorder numbering of v in T, $level(v)$ be the distance of v from the root of T, and $parent(v)$ be the parent of v in T. For any vertex v in T, let T_v be the subtree of T rooted at v including v. Let $f(v)$ be the smallest vertex (in the smallest level) any fundamental cycle including edge $(v, parent(v))$ can reach in T. $f(v) = \infty$ if no such edge exists.

The *Euler tour* of a tree is a simple directed circuit which traverses each tree edge exactly once. The Euler tour of a rooted tree traverses the tree starting from the root, followed by the inorder traversals of the subtrees from left to right. Let $\mathcal{EA}$ be the Euler tour array of T by replacing each directed edge (u, v) of the Euler tour of T by the vertex v. Let $l(v)$ and $r(v)$ be the indices of the leftmost and rightmost appearances of v in $\mathcal{EA}$.

We show how to compute an open ear decomposition

of each biconnected component of an undirected connected graph $G(V, E)$ in Section 3. Then, by using the information of the pseudo-open ear decomposition of G we show how to compute the biconnected components of G in Section 4.

3 PSEUDO-OPEN EAR DECOMPOSITION

Before we describe our pseudo-open ear decomposition algorithm, we review the strategy of the PRAM ear decomposition algorithm. The basic idea of the algorithm is as follows. Given an undirected 2-edge connected graph $\mathcal{G}(\mathcal{V}, \mathcal{E})$, an arbitrary spanning tree $\mathcal{T}(\mathcal{V}, \mathcal{E}_\mathcal{T})$ is first computed, and then the non-tree edges $e \in \mathcal{G} - \mathcal{T}$ are given labels according to their $lca(e)$ values. These $lca(e)$ values are sorted in nondecreasing order and the labels of corresponding non-tree edges are renumbered as $1, 2, 3, \ldots$. Finally, each tree edge e is assigned the minimum label of any non-tree edge whose fundamental cycle contains e. With such assignment on the edges of $\mathcal{G}$, an ear decomposition D is generated. Each ear contains a non-tree edge and part of the edges on the fundamental cycle of this non-tree edge. If there exists an ear containing the whole fundamental cycle of a non-tree edge, then D is not an open ear decomposition.

Let $T = (V, E_T)$ be a spanning tree of G. Consider two non-tree edges (u, v) and (x, y) in $G - T$, such that $lca(u, v) = lca(x, y) = k$. Let (k, a) be the first edge on the path from k to u and (k, b) be the first edge on the path from k to v in T. The edges (k, a) and (k, b) are called $lca(u, v)\_edges$. The edges (k, a), (k, b), and (u, v) are on a common cycle. Similarly, let (k, c) and (k, d) be the $lca(x, y)\_edges$ of (x, y). The edges (k, c), (k, d), and (x, y) are on a common cycle. Note that (u, v) connects subtree T_a to T_b, and (x, y) connects subtree T_c to T_d. If $b = c$, T_a, T_b, and T_d form a biconnected component.

In the ear decomposition algorithm, every non-tree edge (u, v) is given a label according to its $lca(u, v)$ value, which may cause two $lca(u, v)\_edges$ being given the same label and hence will form a closed ear. We want to relabel all non-tree edges $e \in E - E_T$ with the same $lca(e)$ such that no two $lca(e)\_edges$ have the same labels except the first ear in each biconnected component of G. This can be achieved as follows.

Given G and the spanning tree T of G, T is rooted at an arbitrary vertex r and vertices are numbered in preorder traversal. From now on, vertices are represented by their preorder numberings. We now introduce the following function:

$$f(v) = \min\{lca(e) \,|\, e \in E - E_T \text{ and } e \text{ is incident on an vertex in } T_v\}, \forall v \in V - \{0\},$$

that is, $f(v)$ is the smallest vertex any fundamental cycle involving edge $(v, parent(v))$ can reach in T, and $f(v) = \infty$ if no such e exists. We mark v if $f(v) < parent(v), v \in V - \{0\}$. In order to have different labels on every pair of $lca()\_edges$ we introduce a lca_edges_graph $H(V', E')$ to relabel the two distinct vertices of each pair of $lca()\_edges$. For any non-tree edge (u, v) in $G - T$, let $k = lca(u, v)$ and $(k, a), (k, b)$ be the $lca(u, v)\_edges$. Then an edge (a, b) is constructed. If $a = b$, it is a self-loop. The set E' of $m - n + 1$ such edges, induced by the non-tree edges in $G - T$ and $V' = V - \{0\}$ vertices, constitute a lca_edges_graph $H(V', E')$. Next, the connected components of H are computed.

Consider each connected component C of H. Let $label(C) = parent(a)$ be the *label of connected component* C, for any vertex $a \in C$. All non-tree edges (u, v) inducing edges in the same connected component C of H have the same $lca()$ value. In fact, we have that $lca(u, v) = label(C)$. Note that many connected components may have the same labels. In order to identify every connected component, we set $l(C) = label(C)$, sort these $l(C)$'s, and renumber them.

Lemma 1. All the edges of the fundamental cycles that induce the edges in a connected component $C \in H$ belong to the same biconnected component.
Proof: Let $label(C) = k$. Consider two non-tree edges (x, y) and (u, v) such that the $lca(u, v)\_edges$ are (k, a) and (k, b), and the $lca(x, y)\_edges$ are (k, b) and (k, c). The edges (a, b) and (b, c) are in C. The intersection of fundamental cycles of (u, v) and (x, y) includes $\{ (k, b) \}$. Then, the union of these two cycles belongs to a biconnected component. Therefore, for any two adjacent edges (a, b) and (b, c) of C, the union of fundamental cycles containing (k, a) and (k, b) and (k, b) and (k, c) is in a biconnected component. By induction, all fundamental cycles, which induce all edges in a connected component C in H, are in a biconnected component. Notice that the converse is not true. $\square$

For every C of H, we find a spanning tree $\mathcal{T}_C$, root $\mathcal{T}_C$ at a marked vertex if there is one in C otherwise root $\mathcal{T}_C$ at an arbitrary vertex, and renumber vertices of $\mathcal{T}_C$ in a preorder traversal as $0, 1, 2, 3, \cdots, k - 1$. Let $preorder(a)$ be the preorder of a in $\mathcal{T}_C$, and $par(preorder(a))$ be parent of $preorder(a)$ in $\mathcal{T}_C$. Note that each $par(a)$ in $\mathcal{T}_C$ may not be the same as the $parent(a)$ in T.

Now, we describe how to label edges of C. Every tree edge $(par(preorder(a)), preorder(a))$ of $\mathcal{T}_C$ is labeled

as $(l(C), preorder(a))$. Every non-tree edge (a, b) of $\mathcal{T}_C$, including multiple copies and self-loops, is labeled as $(l(C), j)$, where $j = k, k + 1, \cdots, \#$ of edges in C. Here, each non-tree edge is given a different label by the following parallel steps. We first count the number of vertices in each C as $count(C)$. In each processor P_i, we count the number of non-tree edges of C in $P_1, P_2, \cdots, P_{i-1}$, $1 \leq i \leq p$, as $sum_i(C)$. Non-tree edges of C are given the labels $(l(C), count(C) + sum_i(C))$, $(l(C), count(C) + sum_i(C) + 1)$, $\cdots$. Together with the edge labeling of H, we assign labels of non-tree edges e of T, $e \in E - E_T$, as follows:

$$label(e) = \text{label of the edge in } H \text{ which was induced by } e.$$

Let a set of numbers n_l, $l = 1, 2, \cdots$, be given, and let $f_1(n_l), f_2(n_l), \cdots, f_k(n_l)$ be functions on the n_l's. We define a lexicographical ordering of n_l's as follows. $n_i \leq n_j$ if and only if there exists an integer c, $1 \leq c \leq k$, such that $f_1(n_i) = f_1(n_j), f_2(n_i) = f_2(n_j), \cdots, f_{c-1}(n_j) = f_{c-1}(n_j)$, and $f_c(n_i) \leq f_c(n_j)$.

Let $label(v)$ be the lexicographically minimum label of any non-tree edge whose fundamental cycle contains v, $\forall v \in V$. Finally, each tree edge $(parent(v), v)$ of T, $\forall v \in V$, is labeled with the lexicographically minimum label of any non-tree edge whose fundamental cycle contains it, i.e. $label(parent(v), v) = label(v)$. The labels of edges in C are the open ears of G. Note that we need to take care of bridges and single-edge biconnected components ear labeling. All such edges are labeled ∞.

We now describe the overall algorithm for determining an pseudo-open ear decomposition of a connected graph.

Pseudo-Open Ear Decomposition Algorithm

Input: An undirected connected graph $G(V, E)$ whose m edges are distributed evenly among the p processors $\{P_i\}$, $1 \leq i \leq p$, of a linear array.
Output: An ear decomposition such that the labeling induces an open ear decomposition for each biconnected component of G.

Step 1: Find a spanning tree T of G, root T at a root r, and number the vertices of T in preorder as $0, 1, 2, \cdots, n - 1$.
Step 2: Compute $parent(v)$, $level(v)$, and the Euler path of T, $\forall v \in V$, in P_p. Broadcast these to all processors P_i's.
Step 3: Assign ear numbers to non-tree edges $e \in E - E_T$.

3.1. For all $v \in V - \{0\}$, compute
$f(v) = \min\{lca(e) \,|e \in E - E_T$ and e is
incident on an vertex in $T_v\}$.
$f(v) = \infty$ if no such e exists.
Mark v if $f(v) < parent(v)$.

3.2. Construct the *lca_edges_graph* H of G.

3.3. Find the connected components $\{C\}$ of H.

3.4. Assign labels to the edges of H as follows:

3.4.1. In P_p, for each C of H, let
$label(C) = parent(a), \exists a \in C$, and
$l(C) = label(C)$. Sort the $l(C)$'s and renumber
them as 1, 2, 3, $\cdots$. Every $v \in V - \{0\}$ and
$label(v)$ and $l(v)$. Broadcast these to all P_i's.

3.4.2. Find a spanning tree $\mathcal{T}_C$ for each connected component C of H.

3.4.3. In P_p, root each $\mathcal{T}_C$ at a marked vertex
if one exists, and number vertices of $\mathcal{T}_C$ in
preorder traversal as 0, 1, 2, $\cdots$, $k - 1$.
Let $preorder(a)$ be the preorder of a in $\mathcal{T}_C$.
Compute $count(C) = $ # of vertices in C,
$par(preorder(a)), \forall a \in H$.
Broadcast $preorder(a)$ and $par(preorder(a))$
to all P_i.

3.4.4. Compute $sum_i(C) = $ total number of
non-tree edges of C in $P_1, P_2, P_3, \cdots, P_{i-1}$,
$1 \le i \le p$.

3.4.5. For all P_i, $1 \le i \le p$, label edges of H
as follows:
For each tree edge $(par(preorder(a))$,
$preorder(a)) \in \mathcal{T}_{label(preorder(a))}$,
$label(par(preorder(a)), preorder(a)) =$
$(l(preorder(a)), preorder(a))$.
$count(C) = count(C) + sum_i(C), \forall C \in H$.
For each non-tree edge
$(preorder(u), preorder(v)) \in C_{label(preorder(a))}$,
$label(preorder(u), preorder(v)) =$
$(l(preorder(u)), count(C_{label(preorder(a))}))$,
$count(C_{label(preorder(a))}) =$
$count(C_{label(preorder(a))}) + 1$.

3.5. For all non-tree edges $e \in E - E_T$, let
$label(e) = $ label of the edge in H which was
induced by e.

Step 4: Assign ear numbers to tree edges in T.

4.1. For all $v \in V$, compute
$w(v) = $ lex_min$\{label(e) \mid e \in G - T$ and e is
incident on $v\}$.
$w(v) = \infty$ if no such e exists.
These $w(v)$'s are moved to P_p.

4.2. In P_p, using depth first search, compute
$label(v) = $ lex_min$\{w(u) \mid u \in T_v\}$.
Broadcast $label(v)$'s to all P_i.

4.3. For all P_i, $1 \le i \le p$, label tree edges $e \in T$
in P_i, $label(parent(v), v) = label(v)$.

4.4. For all P_i, $1 \le i \le p$, edges $(a, b) \in H$,
if $parent(a) \ne 0$ and
$label(a) = label(b) = label(parent(a))$, then
$label(parent(parent(a)), parent(a)) = \infty$.

Step 5: For each closed ear in a biconnected
component, relabel the non-tree edge (u, v) with
$label(l(preorder(u)), 1)$ to $label(l(preorder(u)), 0)$.

Note that each e with $label(e) = \infty$ is a single-
edge biconnected component or a single-edge open ear.
After T is computed, the preorder numbering of the
vertices, $parent(v)$, and $level(v)$ can be computed in
$O(n)$ sequential time in P_p. These values and T are
broadcasted to all P_i in $O(n + p)$ time. With these
values, $lca(u, v)$ and $lca(u, v)$_edges of edges $(u, v) \in$
$E - E_T$ in each P_i can be computed in $O(\frac{m}{p} + n)$
time. $f(v)$ is computed by two substeps: (1) compute $w(v) = min\{lca(v, u) \mid (v, u) \in E - E_T\}$. If
no such (v, u) exists then $w(v) = \infty$, (2) compute
$f(v) = min\{w(x) \mid x \in T_v\}$. Let $w_i(v)$ be local w value
in P_i. $w_i(v)$ can be computed in $O(\frac{m}{p})$ time in P_i,
$\forall v \in V$. Then, $w(v)$ can be computed in $O(n + p)$ time
using pipelining and the results are stored in P_p. With
$w(v)$, $f(v)$ cab be computed in $O(n)$ sequential time
in P_p. Broadcasting $f(v)$ to all P_i's takes $O(n + p)$
time. Steps 3.4.3. and 3.4.4. compute $count(C)$ and
$sum_i(C)$. $count(C) + sum_i(C)$ in P_i indicates the beginning label index of non-tree edges and self-loops of
C.

Complexity Analysis:
Both Steps 1 and 3.4.2 take $O(\frac{m}{p} + n + p)$ time using
Minimum Spanning Tree algorithm [3] while Step 3.3
can be computed using Connected Component algo-
rithm in $O(\frac{m}{p} + n + p)$ time [3]. Step 2 takes $O(n + p)$
sequential time in processor P_p. Step 3.1 is computed
in $O(\frac{m}{p} + n + p)$ time. Also, $lca(e)$ and $lca(e)$_edges,
$\forall e \in E - E_T$, are computed in this step. In $O(\frac{m}{p})$ time,
every processor P_i, $1 \le i \le p$, executes Step 3.2. To
assign labels to the edges of H in Step 3.4, Steps 3.4.1
and 3.4.3 take $O(n)$ computation time and $O(n + p)$
broadcasting time, Step 3.4.5 takes $O(\frac{m}{p})$ sequential
time, and Step 3.4.4 is done in two substeps:
(1)$g_i(C) = $ total number of non-tree edges of C in
P_i, $\forall C \in H$.
(2)$sum_i(C) = \sum_{j=1}^{i-1} g_j(C)$.
Summing up these substeps, Step 3.4 takes $O(\frac{m}{p}) +$
$O(n + p) = O(\frac{m}{p} + n + p)$ time. Thus, the non-tree
edges of G can be simultaneously labeled in all processors using Step 3.5 in $O(\frac{m}{p})$ time.
The assignment of ear labels to tree edges in T is

accomplished in four steps. Step 4.1 is done in two substeps:

(1) $w_i(v) = \text{lex\_min}\{label(e) \mid e \in E - E_T \text{ and } e \text{ is incident on } v \text{ and } e \in P_i\}$.

(2) $w(v) = \text{lex\_min}\{w_i(v)\}_{i=1}^{p}$.

The $w(v)$'s are moved to processor P_p. All these steps take $O(\frac{m}{p} + n + p)$ time. Step 4.2 takes $O(n)$ computation time and $O(n + p)$ broadcasting time and Steps 4.3 and 4.4 are computed in $O(\min\{\frac{m}{p}, n\})$ sequential time. Finally, the smallest ear, which is a closed ear, in each biconnected component is relabeled using at most $O(\min\{\frac{m}{p}, n\})$ sequential time in Step 5. Therefore, Open Ear Decomposition can be computed in $O(\frac{m}{p} + n + p)$ time. $\square$

Theorem 2. The Pseudo-Open Ear Decomposition of a n-vertex and m-edge graph on a p-processor linear array takes $O(\frac{m}{p} + n + p)$ time.

Consider a connected component $C \in H$ such that C contains no marked vertex. According to the definition of $f(v)$, $label(C)$ turns out to be the smallest vertex on all the fundamental cycle of non-tree edges which induce edges in C. If $label(C) \neq 0$, then $label(C)$ is an articulation point, which separates biconnected component involving all $(label(C), a)$, $\forall a \in C$, from the biconnected component containing $(parent(label(C)), label(C))$.

Lemma 2. For each $C \in H$ such that C contains no marked vertex, there is one closed ear of G with the smallest label of C. Further, it is the smallest ear in its biconnected component.

Proof: Let $label(C) = k$ and $a \in C$. Since there is no marked vertex in C, no edge $(k, a) \in T$ belongs to any fundamental cycle of non-tree edge with $lca < k$. The fundamental cycle of $e \in E - E_T$, which has the smallest label of C, contains $lca(e)\_edges$ and causes $lca(e)\_edges$ both labeled by this smallest index. Hence, the fundamental cycle of e forms a closed ear.

Note that k is an articulation point. From Lemma 1 we observe the following property:
Any other fundamental cycle in the same biconnected component of edges (k, a), $\forall a \in C$, which induced another $C' \in H$, has $lca > k$, and its edges have labels larger than any label of C. Therefore, the fundamental cycle of e with smallest label of C is the smallest ear

in its biconnected component. $\square$

Lemma 3. For each biconnected component, every ear $\mathcal{P}_i$ is an open ear except the first one.
Proof: Consider a non-tree edge $(u, v) \in E - E_T$ with $ear(u, v) = (l, s)$. Let $lca(u, v) = k$ and $lca(u, v)\_edges$ be (k, a) and (k, b). Then, a and b belong to the same C in H. Without loss of generality, let $preorder(a) \leq preorder(b)$ in $\mathcal{T}_C$. We will prove that (k, a) belongs to an ear labeled less than (l, s).

Consider the edge (k, a). If a is marked, then (k, a) is on a fundamental cycle with $lca < k$. Hence, $ear(k, a) < (l, s)$. If a is not marked and a has a parent $par(preorder(a)) = d$ in $\mathcal{T}_C$, then (d, a) has $label(d, a) = (l, s - 1)$, which is less than $label(a, b) = (l, s)$. Let the non-tree edge inducing (d, a) in C be (x, y). Then $ear(x, y) < ear(u, v) = (l, s)$. Obviously, (k, a) is on the fundamental cycle of (x, y) and has $ear(k, a) \leq ear(x, y)$. Thus, $ear(k, a) < (l, s)$. If a is not marked and a is a root of $\mathcal{T}_C$, then (a, b) has the lexicographically minimal label in C. From Lemma 2, (k, a) and (k, b) are on the fundamental cycle of (u, v), which is a closed ear and the smallest one in its biconnected component. $\square$

Note that for each $(u, v) \in C$ of H, $parent(u) = parent(v) = k$ in T and $k = label(C)$ in H. According to the labeling of tree edges and vertices, $label(parent(v)) \leq label(v)$, $\forall v \in V$. Further, $ear(parent(k), k) \leq ear(k, v)$.

In Open Ear Decomposition algorithm, we are able to find all bridges, articulation points, and single-edge biconnected components while the edges of each biconnected component of G are labeled with ear numbers.

Lemma 4. Using the Open Ear Decomposition algorithm, all bridges, articulation points, and single-edge biconnected components of G can be identified as follows:

(1) Vertex 0 is an articulation point iff there exist C and C' in H such that $label(C) = label(C') = 0$. Further, C and C' belong to different biconnected components.

(2) Vertex k is an articulation point iff there exists an edge (u, v) in H such that $label(u) = label(v)$ in T and $parent(u)(= parent(v)) = k \neq 0$. Further, the connected component of (u, v) has no marked

vertex and the fundamental cycle of non-tree edge inducing (u, v) forms a closed ear, which is the smallest in its biconnected component.

(3) If there exists an edge $(u, v) \in H$ such that $label(u) = label(v) = label(k)$ in T and $k \neq 0$, where $k = parent(u)$, then $(parent(k), k)$ is a bridge. k and $parent(k)$ are articulation points.

(4) If there exists a vertex v such that $label(v) = \infty$ in T, then $(parent(v), v)$ is a single-edge biconnected component and $parent(v)$ is an articulation point.

Proof: (1) let the children of k be $u_1, u_2, \cdots, u_r, u_{r+1}, u_{r+2}, \cdots, u_{r+s}$, where $u_1, u_2, \cdots, u_r \in C$ and $u_{r+1}, u_{r+2}, \cdots, u_{r+s} \in C'$. There is no non-tree edge of G connecting a vertex in subtrees of $T_{u_1}, T_{u_2}, \cdots, T_{u_r}$ and a vertex in subtrees of $T_{u_{r+1}}, T_{u_{r+2}}, \cdots, T_{u_{r+s}}$. If there is one, then C and C' would be connected. In order to merge C and C' into a biconnected component, a path connecting subtrees $\{T_{u_1}, T_{u_2}, \cdots, T_{u_r}\}$ and subtrees $\{T_{u_{r+1}}, T_{u_{r+2}}, \cdots, T_{u_{r+s}}\}$ not going through k is needed. This path will go through sibling(s) or ancestor(s) of k. If $k = 0$, then such path does not exist. Therefore, $k = 0$ is an articulation point. C and C' belong to different biconnected components.

(2) Recall that $label(parent(u)) \leq label(u)$, $\forall u \in V$. So, the edges $(parent(parent(u)), parent(u))$ are in some ears whose number are $\leq label(u)$. Since $label(u) = label(v)$, $ear(k, u) = ear(k, v) =$ the closed and smallest ear of its biconnected component. Any fundamental cycle in this biconnected component contains no vertex less than k, according to Lemma 2. Hence $(parent(k), k)$ will not be in the same biconnected component of (k, u) or (k, v). Therefore, k is an articulation point, if $k \neq 0$. The second part is proved in Lemma 2.

(3) Since $label(u) = label(v) = label(k)$ and k is an articulation point according to Lemma 4.(2), $(parent(k), k)$ is not on any fundamental cycle. The $label(parent(k), k)$ was given the $label(k, u)$ in Step 4.3, and reassigned to ∞ in Step 4.4. Therefore, $parent(k)$ and k are articulation points, and $(parent(k), k)$ is a bridge.

(4) The fact that $label(v) = \infty$, implies that $(parent(v), v)$ is not on any fundamental cycle. Therefore, $(parent(v), v)$ is a single-edge biconnected component. Moreover, $parent(v)$ is an articulation point separating $(parent(v), v)$ from the biconnected component containing $(parent(parent(v)), parent(v))$. $\square$

4 BICONNECTED COMPONENTS

Before we discuss how to compute biconnected components of an undirected connected graph G, let us make some observations concerning the Pseudo-Open Ear Decomposition algorithm and its implications on the biconnected components of the input graph.

Recall that $l(C)$ is the sorted sequence of $label(C)$. In order to keep track of the inverse relation of $l(C)$ to $label(C)$ for each $C \in H$, we define a function B such that $B(l(C)) = label(C)$. We add this step to Step 3.4.1 of the algorithm. We use $C_{label(C)}$ to identify a connected component with $label(C)$. The articulation points are computed according to Lemma 4. Let A be the set of articulation points of G.

From Lemma 1, ears are grouped into at most $n - 1$ sets, each set has labels from a connected component of H, and ears in each set belong to the same biconnected component. Since H has $n-1$ vertices, the total number of connected components is at most $n - 1$. To identify each set of ears, it is easy to notice that ears with the same $l()$-value, which is the first component of $label$, are in the same biconnected component.

To merge groups of ears further, we construct another auxiliary graph $\tilde{H}(\tilde{V}, \tilde{E})$, where $\tilde{V} = \{C_i \mid C_i \in H\}$ and $(C_i, C_j) \in \tilde{E}$ if and only if C_i and C_j are in the same biconnected component.

Lemma 5. Consider an arbitrary $C_k \in H$ and the set of articulation points, A, of G. For all $a \in C_k$, if $a \notin A$ and C_a exists, then $(C_k, C_a) \in \tilde{E}$.
Proof: There exists a path from every $x \in C_a$ to every $y \in C_k$ avoiding a. Therefore, C_k and C_a are in the same biconnected component. $\square$

Again, let us consider the connected components $C_{parent(v)}$ of H, $\forall v \in H$. For each $v \in C_{parent(v)}$, $f(v) = k =$ smallest vertex that the fundamental cycle containing $(parent(v), v)$ can reach. The C_k exists. If $k \neq parent(v)$, then:

1. $(parent(v), v)$ is on the fundamental cycle of $e \in E - E_T$, where $label(e) = label(v)$.

2. Let $label(v) = (s, t) = label(e)$. Then $B(s) = k = f(v)$, where s is the sorted index of C_k. Let (k, a) and (k, b) be the $lca(e)\_edges$. Then $a, b \in C_k$, and $(parent(v), v)$ having $label(parent(v), v) = (s, t)$ is in the same biconnected component of (k, a). Further, (k, a), (k, b), and $(parent(v), v)$ are on a common cycle. Let the path from $parent(v)$ to a and the path from v to b on this cycle be Q_1 and Q_2, respectively.

Lemma 6. $C_{f(v)}$ and $C_{parent(v)}$ are in the same biconnected component.

Proof: $C_{parent(v)}$ is biconnected. All vertices $x \in C_{parent(v)}$ are such that $parent(v)$, v, and x are on a common cycle. There exists two vertex-disjoint paths from x to $parent(v)$ and from x to v, say q_1 and q_2, respectively. Similarly, there exist two vertex-disjoint paths from y to b and from y to a, say q_3 and q_4, respectively, $\forall y \in C_k$. So, paths from x to y are: $\text{path}(x, parent(v)) + \text{path}(parent(v), a) + \text{path}(a, y) = q_1 + Q_1 + q_4$, or $\text{path}(x, v) + \text{path}(v, b) + \text{path}(b, y) = q_2 + Q_2 + q_3$. We conclude that $(C_{f(v)}, C_{parent(v)}) \in \tilde{E}$. □

If there are more than one connected component with $label = f(v) = k$ in H, we connect $C_{parent(v)}$ to the C'_k in $\tilde{H}$, where $label(v) = (s, t)$, $l(C'_k) = s$, $B(s) = C'_k$. Next, we find the connected components of $\tilde{H}$. Each connected component of $\tilde{H}$ is a biconnected component of G. Note that $|\tilde{V}| = n$ and $|\tilde{E}| \leq 2n$. Let $Bicc(e)$ be the biconnected component of $e \in E$. Hence e and g are in the same biconnected component if and only if $Bicc(e) = Bicc(g)$.

Biconnected Components Algorithm

Input: A graph $G(V, E)$ whose m edges are distributed evenly among p processors $\{P_i\}$, $1 \leq i \leq p$, of a linear array.

Output: Label the edges e with $Bicc(e)$ such that $Bicc(e) = Bicc(g)$ iff e and g are in the same biconnected component.

Step 1. Apply the Open Ear Decomposition Algorithm Steps 1-4.

Step 2. In all P_i, for each $e \in E$ such that $label(e) = \infty$ is a single-edge biconnected component. Let $Bicc(e) = \infty$.

Step 3. Identify the articulation points of G in P_p.
  3.1. For all $C \in H$,
    If $(a, b) \in H$ such that $label(a) = label(b)$ in T and $parent(a) \neq 0$, then $parent(a) \in A$.
    If there exist C and C' such that $label(C) = label(C') = k$ and $k = 0$, then $0 \in A$.
    If $(a, b) \in H$, such that $label(a) = label(b) = label(k)$, where $k = parent(a)$, then k, $parent(k) \in A$.
  3.2. For all $v \in V$, if $label(v) = \infty$, then $parnet(v) \in A$.

Step 4. Construct the auxiliary graph $\tilde{H}(\tilde{V}, \tilde{E})$,

$\tilde{V} = \{C_i \mid C_i \in H\}$ and
  4.1. In P_p, for each C_k, $\forall a \in C_k$, if $a \notin A$ and C_a exists, then $(C_k, C_a) \in \tilde{E}$.
  4.2. In P_p, for each $C_{parent(v)}$, $\forall v \in C_{parent(v)}$, let $label(v) = (s, t)$, and $B(s) = C'_{f(v)}$.
    If $f(v) \neq parent(v)$ then $(C_{f(v)}, C_{parent(v)}) \in \tilde{E}$.

Step 5. Compute connected components of $\tilde{H}$ in P_p, and broadcast this information to all P_i.

Step 6. For all P_i, For each $(u, v) \in E$ such that $label(u, v) = (s, t)$, let $Bicc(u, v) = $ connected component of $B(s)$.

Complexity Analysis:

Step 1 takes $O(\frac{m}{p} + n + p)$ time as described in the Open Ear algorithm complexity analysis. Each of Steps 2 and 6 can be computed simultaneously in all processors in $O(\frac{m}{p})$ time while Steps 3 and 4 can be computed in processor P_p in (n) sequential time. Step 5 takes $O(n)$ computation time and $O(n + p)$ communication time. Therefore, biconnected components can be computed in $O(\frac{m}{p} + n + p)$ time. □

Theorem 3. The biconnected components of a n-vertex and m-edge graph on a p-processor linear array can be determined in $O(\frac{m}{p} + n + p)$ time.

REFERENCES

[1] P. S. Gopalakrishnan, I. V. Ramafrishnan, and L. N. Kaanal, "An Efficient Connected Components Algorithm on a Mesh-Connected Computer," *Proc. 1985 International Conference on Parallel Processing*, pp. 711-714, 1985.

[2] T. Hagerup, "Towards Optimal Parallel Bucket Sorting," *Information and Computation 75*, pp. 39-51, 1987.

[3] Ying-Min Huang and Joseph F. JáJá, "Optimal Algorithms on Fixed-Size Linear Arrays," *Proc. 1993 Conference on Information Sience and Systems*, Johns Hopkins, pp. , March 1993.

[4] Ying-Min Huang, "Optimal Graph Algorithms on Linear Arrays," *ISR Thesis Report Ph.D.*, Institute for Systems Research, University of Maryland at College Park, 1994.

[5] Joseph F. JáJá, *An Introduction to Parallel Algorithms*, Addison-Wesley Press, 1992.

[6] Yael Maon, Baruch Schieber, and U. Vishkin, "Parallel Ear Decomposition Search(EDS) and st-Numbering in Graphs," *Theoretical Computer Science*, Vol. 47, pp. 277-298, 1986.

[7] Vijaya Ramachandran, "Parallel Open Ear Decomposition with Applications to Graph Biconnectivity and Triconnectivity," *Synthesis of Parallel Algorithms*, Morgan Kaufman, San Mateo, CA, 1991.

[8] C. Savage and J. F. JáJá , "Fast, Efficient Parallel Algorithms for Some Graph Problems," *SIAM Journal of Computing*, Vol. 10, No. 4, pp. 682-690, November, 1981.

[9] Y. Shiloach and U. Vishkin, "An $O(\log n)$ Parallel Connectivity Algorithm," *Journal of Algorithms*, Vol. 3, No. 1, pp. 57-67, 1982.

[10] Y. H. Tsin and F. Y. Chin, "Efficient Parallel Algorithms for a Class of Graph Theoretic Problems," *SIAM J. COMPUT*, Vol. 13, No. 3, pp. 580-599, July 1984.

[11] R. E. Tarjan and U. Vishkin, "An Efficient Parallel Biconnectivity Algorithm," *SIAM J. COMPUT*, Vol. 14, No. 4, pp. 864-874, Nov. 1985.

Embedding Two-Dimensional Grids into Hypercubes with Dilation 2 and Congestion 5

Yuan Cai, Shou-Hsuan Stephen Huang and Chien-Chi Lin

Department of Computer Science

University of Houston

Houston, TX 77204-3475

E-Mail: s_huang@cs.uh.edu

Abstract — Two-dimensional grids can be embedded into optimal hypercubes with load one and dilation two. The congestion, i.e. the maximal number of paths passing through an edge in the host graph corresponding to edges in the guest graph, of grid-to-hypercube embedding is known to be bounded by a constant. Straightforward analysis of M.Y.Chan's dilation 2 algorithm did not yield satisfactory upper bound on the congestion. We present a modified version of that algorithm with dilation 2 and congestion 5. An example is also found to show that the upper bound of 5 is tight.

Keywords: Embedding, Dilation, Congestion, Hypercubes, Simulation.

1 Introduction

Given two undirected connected graphs $G = < V^G, E^G >$ and $H = < V^H, E^H >$, an embedding from G to H is a mapping from V^G to V^H, $\Phi : V^G \rightarrow V^H$ in such a way that for any edge $e^G = < v_1^G, v_2^G >$, where $e^G \in E^G$ and $v_1^G, v_2^G \in V^G$, we can find in H at least one path which starts at v_1^H and ends at v_2^H with $v_1^H = \Phi(v_1^G)$ and $v_2^H = \Phi(v_2^G)$. By doing so, we can use H to simulate G [7][8][9]. Four parameters, namely, expansion, dilation, congestion and load are commonly used for measuring the quality of an embedding. The definitions of these four parameters can be found in [1][11].

In this paper, G is a two-dimensional grid of size $\alpha \times \beta$, H is the smallest (optimal) hypercube with at least as many nodes as G, (we will use **point** to denote vertex in G and **node** to denote vertex in H), and Φ is a one-to-one mapping. Such embedding will have optimal expansion and unit load. The problem is how to reduce the dilation and the congestion under such restrictions.

Let $\tilde{\alpha} = 2^{\lfloor \log \alpha \rfloor}$ and $\tilde{\beta} = 2^{\lfloor \log \beta \rfloor}$. We know that G is a subgraph of H, if and only if $\tilde{\alpha} = \alpha$ or $\tilde{\beta} = \beta$ or $\alpha\beta > 2\tilde{\alpha}\tilde{\beta}$. M. Y. Chan presented a dilation 2 algorithm in [3] for the case that $\alpha\beta \leq 2\tilde{\alpha}\tilde{\beta}$,

$\tilde{\alpha} < \alpha$ and $\tilde{\beta} < \beta$. However, Chan's algorithm did not take the congestion into consideration.

In [2], we studied the importance of congestion as well as introduced a congestion upper bound of 8 for any dilation 2 embedding [3][4][7][12] of two dimensional grids. In deriving this upper bound, we only make use of the topological characteristics of G and the condition of dilation 2.

Unfortunately, we are unable to show directly that the embedding algorithm [3] as presented by Chan yields congestion 5. In order to analysis the congestion, we need to "localize" the adjustments made in the last step of Chan's algorithm. In this paper, Section 2 presents the modified embedding algorithm and its dilation 2 property. Section 3 derives the congestion of this embedding. Section 4 concludes the paper.

2 Modified Embedding Algorithm

Since an n-dimensional hypercube (n-cube) is an undirected graph with 2^n nodes and each node has n neighbors, we can label every node with an n-bit binary number in such a way that two nodes are adjacent if and only if their labels differ in one bit. Therefore, any embedding problem that has an n-dimensional hypercube as the host graph can be transferred into the problem of labeling the vertices in guest graph with n-bit binary numbers.

In the algorithm of [3], the whole embedding process consists of two stages: the first stage is to partition all the grid points into $\tilde{\alpha}$ chains so as to decide the first $\lfloor \log \alpha \rfloor$ bits of labels. The partition scheme ensures that any two adjacent points in the two-dimensional grid belong to the same or adjacent chains, so that the first $\lfloor \log \alpha \rfloor$ bits of the labels assigned to grid points would differ in at most one bit after the chain numbers being encoded into binary-reflected Gray code. The $\tilde{\alpha}$ chains are stacked on top of each other. Each chain uses 1 or 2 points from each column of the grid. One of the critical results in this stage is a special partitioning matrix A of size $\tilde{\alpha} \times \tilde{\beta}$, which is an integer matrix comprising of 1s and 2s. The element $a_{i,j}$ in A indicates how many points from column j of grid G belong to chain i. The first column of matrix A is the vector:

$$
\begin{bmatrix} a_{1,1} \\ a_{2,1} \\ a_{3,1} \\ \vdots \\ a_{\tilde{\alpha},1} \end{bmatrix} = \begin{bmatrix} \lceil \alpha/\tilde{\alpha} \rceil \\ \lfloor \alpha/\tilde{\alpha} \rfloor \\ \lfloor 2\alpha/\tilde{\alpha} \rfloor - \lfloor \alpha/\tilde{\alpha} \rfloor \\ \vdots \\ \lfloor (\tilde{\alpha}-1)\alpha/\tilde{\alpha} \rfloor - \lfloor (\tilde{\alpha}-2)\alpha/\tilde{\alpha} \rfloor \end{bmatrix} .
$$

The entire matrix is produced by cyclic shifting of the first column, i.e. for all $1 \leq i < \tilde{\alpha}$ and $1 \leq j < \beta$, $a_{i+1,j+1} = a_{i,j}$ and $a_{1,j+1} = a_{\tilde{\alpha},j}$. For example, for $\alpha = 11$ and $\beta = 11$, the matrix A is shown below.

$$
A = \begin{bmatrix} 2 & 1 & 2 & 1 & 1 & 2 & 1 & 1 & 2 & 1 & 2 \\ 1 & 2 & 1 & 2 & 1 & 1 & 2 & 1 & 1 & 2 & 1 \\ 1 & 1 & 2 & 1 & 2 & 1 & 1 & 2 & 1 & 1 & 2 \\ 2 & 1 & 1 & 2 & 1 & 2 & 1 & 1 & 2 & 1 & 1 \\ 1 & 2 & 1 & 1 & 2 & 1 & 2 & 1 & 1 & 2 & 1 \\ 1 & 1 & 2 & 1 & 1 & 2 & 1 & 2 & 1 & 1 & 2 \\ 2 & 1 & 1 & 2 & 1 & 1 & 2 & 1 & 2 & 1 & 1 \\ 1 & 2 & 1 & 1 & 2 & 1 & 1 & 2 & 1 & 2 & 1 \end{bmatrix} .
$$

Based on A, two auxiliary functions $COLSUM$ and $ROWSUM$ are defined as

$$
COLSUM(y; x, z) = \sum_{i=x}^{z} a_{i,y}
$$

and

$$
ROWSUM(x; y, z) = \sum_{j=y}^{z} a_{x,j} .
$$

In the second stage, the last $\lfloor \log \beta \rfloor + 1$ bits are determined. This stage can be further divided into two substages. The first substage assigns $\lfloor \log \beta \rfloor + 1$ bits to the grid points. In this assignment, two properties are guaranteed: (1) in the same chain, every point should have unique last $\lfloor \log \beta \rfloor + 1$ bits; (2) for adjacent points, the last $\lfloor \log \beta \rfloor + 1$ bits differ in at most two bits. The second substage adjusts the last bit of labels to ensure that for adjacent points, their labels differ in at most two bits.

The adjustment is "global", i.e. changing the last bit of a specific grid point label might affect the label of another grid point which is far away. This global adjustment is undesirable for the study of congestion. Thus, we first modified this substage.

The original algorithm of [3] is outlined below:

(1) Assume that, without loss of generality, $\alpha \leq 3\tilde{\alpha}/2$. (Either $\alpha \leq 3\tilde{\alpha}/2$ or $\beta \leq 3\tilde{\beta}/2$; otherwise, $\alpha\beta > 9\tilde{\alpha}\tilde{\beta}/4 > 2\tilde{\alpha}\tilde{\beta}$.) Construct partitioning matrix $A(\tilde{\alpha}, \tilde{\beta})$.

(2) In guest grid G, for each point $[x, y]$, compute $CHAIN[x, y]$, $(1 \leq CHAIN[x, y] \leq \tilde{\alpha})$, where $CHAIN[x, y] = z$ if and only if $COLSUM(y; 1, z-1) < x \leq COLSUM(y; 1, z)$.

(3) For each point $[x, y]$ in G, compute $NUMBER[x, y]$, $(1 \leq NUMBER[x, y] \leq \lceil \frac{\alpha\beta}{\tilde{\alpha}} \rceil)$, where $NUMBER[x, y] = 1 + \delta[x, y] + ROWSUM(CHAIN[x, y]; 1, y-1)$ and

$$
\delta[x, y] = \begin{cases} 1, & CHAIN[x+1, y] = CHAIN[x, y] \\ 0, & otherwise. \end{cases}
$$

(4) For each $[x, y]$, compute $MARK[x, y]$, where

$$
\begin{aligned} MARK[x, y] = & CHAIN[x, y] + NUMBER[x, y] \\ & - COLSUM(1; 1, CHAIN[x, y]). \end{aligned}
$$

(5) Let $GRAY(t, p)$ denote the $((p-1) \bmod 2^t + 1)$th element of the t-bit binary reflected Gray code sequence [10]. Let the first $\lfloor \log \alpha \rfloor$ bits of the label given to point $[x, y]$ of G be $GRAY(\lfloor \log \alpha \rfloor, CHAIN[x, y])$ and the last $(\lfloor \log \beta \rfloor + 1)$ bits be $GRAY(\lfloor \log \beta \rfloor + 1, MARK[x, y])$.

(6) Modify the last bit of each node so that

(a) two points with the same first $(\lfloor \log \alpha \rfloor + \lfloor \log \beta \rfloor)$ bits differ in their last bit, and

(b) two grid neighbors with first $(\lfloor \log \alpha \rfloor + \lfloor \log \beta \rfloor)$ bits differing in exactly two bits have the same last bit.

For this algorithm, the solution in Step (6) is not unique. Implementationwise, the strategy is to find out all the pairs of "twins" and "critical pairs", then build "dependency graph" G' of G, where G' is a forest. By traversing all the trees in G' level by level, the last bit of every point is decided. An 11×11 grid G together with the "dependency graph" G' was given as an example in [3].

In our algorithm, we modify Step (6) in the following way.

(6*)-1 In matrix A, find 2×2 submatrix satisfying

$$
\begin{cases} A[i, j] = A[i+1, j+1] = 2 \\ A[i+1, j] = A[i, j+1] = 1 \\ COLSUM(j; 1, i-1) = \\ \qquad COLSUM(j+1; 1, i-1), \end{cases}
$$

where $1 \leq i < \tilde{\alpha}, 1 \leq j < \beta$. Each submatrix is corresponding to a rectangular block structure in the grid, which is named tile. A tile is identified by the index of its upperleft point.

(6*)-2 If a tile found in (1) intersects with another tile, i.e.

$$\begin{cases} A[i,j] = A[i+1,j+1] = 2 \\ A[i+2,j+2] = 2 \\ A[i+1,j] = A[i,j+1] = 1 \\ A[i+2,j+1] = A[i+1,j+2] = 1 \\ COLSUM(j;1,i-1) = \\ \quad COLSUM(j+1;1,i-1) \\ COLSUM(j+1;1,i) = \\ \quad COLSUM(j+2;1,i), \end{cases}$$

where $1 \leq i < \tilde{\alpha} - 1, 1 \leq j < \beta - 1$, assuming $z = COLSUM(j+1;1,i)+1$, exchange the labels of two points $[z, j+1]$ and $[z+1, j+1]$.

(6*)-3 If a tile $[z, j]$ does not intersect with other tiles, choose one of the following exchange model according to the last bit L of $MARK[z,j]$:

Case 1. If $L = 1$, exchange the labels for $[z+1, j]$ and $[z, j-1]$.

Case 2. If $L = 0$, exchange the labels for $[z+1, j+1]$ and $[z+2, j+2]$.

For the special case that if one point in the pair is absent, "exchange" will be replaced by "assign", i.e. we can assign the label corresponding to the absent point directly to that point. both $+$ and $-$ operations are modulo arithmetic.

Using the same 11×11 grid example, the tiles found in Step 6* and the corresponding exchanging pairs are shown in Fig. 1.

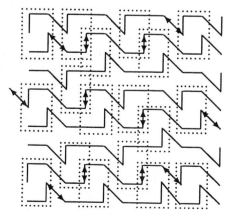

Fig. 1 Tiles and Exchanging Pairs Found in (6*)

Lemma 1. $2 \leq A[i,j] + A[i+1,j] \leq 3;$ $\forall 1 \leq i < \tilde{\alpha}$ and $1 \leq j \leq \beta$.

Proof: Since $A[i,j] \geq 1$ and $A[i+1,j] \geq 1$, it is clear that $2 \leq A[i,j]+A[i+1,j]$. To prove $A[i,j]+A[i+1,j] \leq 3$, two cases are to be considered.

Case 1. For j=1, if i=1,

$$\begin{cases} A[1,1] = \lceil \alpha/\tilde{\alpha} \rceil = 2 \\ A[2,1] = \lfloor \alpha/\tilde{\alpha} \rfloor = 1 \end{cases} \implies A[1,1]+A[2,1] = 3;$$

if $2 \leq i < \tilde{\alpha}$,

$$\begin{aligned} A[i,1] + A[i+1,1] &= \lfloor i\alpha/\tilde{\alpha} \rfloor - \lfloor (i-2)\alpha/\tilde{\alpha} \rfloor \\ &\leq \lceil 2\alpha/\tilde{\alpha} \rceil \\ &\leq 3. \end{aligned}$$

Thus $A[i,1] + A[i+1,1] \leq 3$.

Case 2. For any $2 \leq j \leq \beta$, column j can be obtained by j-1 times cyclic shifting of column 1, since

$$\begin{aligned} A[\tilde{\alpha},1] &= \lfloor (\tilde{\alpha}-1)\alpha/\tilde{\alpha} \rfloor - \lfloor (\tilde{\alpha}-2)\alpha/\tilde{\alpha} \rfloor \\ &= \alpha - 2 - (\alpha - 3) \\ &= 1, \end{aligned}$$

we have $A[i,j] + A[i+1,j] \leq 3$.

According to Case 1 and 2, $2 \leq A[i,j] + A[i+1,j] \leq 3$ for any $1 \leq i < \tilde{\alpha}$, $1 \leq j \leq \beta$. $\square$

The above lemma indicates that there is no vertically adjacent 2s in matrix A. Because of the way of constructing A, neither is there horizontally adjacent 2s in A.

According to [3], all adjacent pairs X and Y with their labels differing in 3 bits before Step (6) have the same 3×2 style shown in Fig. 2.

```
         Column j   j+1   Row
                          z
Chain i      X    Y       z+1
Chain i+1                 z+2
```

Fig. 2 A 3*2 Tile with a Distance 3 Pair

Let $X = [z+1, j]$, $Y = [z+1, j+1]$, $CHAIN(X) = i$ and $CHAIN(Y) = i+1$ as shown in Fig. 2. It is easy to see that the relationship expressed in terms of the matrix A is:

$$\begin{cases} A(i,j) = A(i+1,j+1) = 2 \\ A(i,j+1) = A(i+1,j) = 1 \\ COLSUM(j;1,i-1) = \\ \quad COLSUM(j+1;1,i-1). \end{cases}$$

Step (6*)-1 in our modified algorithm identifies all such tiles to prepare for the following label adjustments. For the intersecting tiles, the adjustments can be made in the following two ways.

Case 1: Single Intersection. This is the case when two tiles intersect with each other as shown in Fig. 3(a). After exchanging the labels for X and Y, it can be easily verified that the labels' Hamming distance[10] between any pair of adjacent points in the local area is at most 2.

Case 2: Multiple Intersection. A double intersecting case is given in Fig. 3(b). After exchanging the labels for X_1 and Y_1, X_2 and Y_2, the labels' Hamming distance between any pair of neighbors in this local area is less than or equal to 2. Since Y_1 and Z_1, Z_2 and X_2 have the same MARK values, X_1 and Y_2 have the same value of MARK. After exchanging the labels, the labels for Y_1 and X_2 differ by one bit. As to intersection with more than 3 tiles, the analysis is similar.

It is clear that Step (6*)-2 eliminate all the distance 3 neighbors (i.e. reduce the distance to 1 or 2) in the intersecting tiles without affecting the labels of other points in the structure. Thus we "localize" the adjustment within the intersecting tiles.

For non-intersecting tiles, there is only one configuration as shown in Fig. 3(c). The labels of any adjacent points in a non-intersecting tile differ in 2 bits or less except the pair of X_2 and Y_4, which differs in 3 bits. We cannot simply exchange X_2 and X_3, or Y_3 and Y_4, since so doing will increase the distance between other pair of neighbors to 3. According to the last two bits of $MARK[X_3] - 1$, we have the four cases shown in Fig. 4.

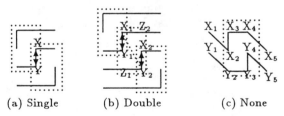

(a) Single   (b) Double   (c) None

Fig. 3 Three Intersection Cases among Tiles

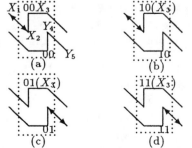

Fig. 4 Four Cases for Non-overlap Tiles

For (a) and (b) in Fig. 4, after exchanging labels of X_1 and X_2, the labels' Hamming distance

betwen X_2 and Y_4 is reduced to 2. The same distance exists for X_2 and Y_4 if we exchange the labels of Y_4 and Y_5 in (c) and (d). The reason lies in the following lemma.

Lemma 2. If a and b are n-bit binary numbers $(n > 1)$, $b - a = 3$ and $a \bmod 2 = 0$, the binary reflected Gray code corresponding to a and b differs in exactly one bit.

Proof: Let $a_{n-1}a_{n-2}...a_1a_0$ and $b_{n-1}b_{n-2}...b_1b_0$ be the n-bit binary representation of a and b, respectively. Let $a'_{n-1}a'_{n-2}...a'_1a'_0$ and $b'_{n-1}b'_{n-2}...b'_1b'_0$ be the corresponding Gray code representation. We then have $a'_i = (a_{i+1} + a_i) \bmod 2$ and $b'_i = (b_{i+1} + b_i) \bmod 2$, where $0 \leq i < n$ and $a_n = b_n = 0$.

Since $a \bmod 2 = 0$, $a_0 = 0$. Let a_i, $1 \leq i < n$, be the rightmost bit left to a_0 that is 0 in the binary representation of a. We then observe that $a_j = b_j$ for $i < j \leq n - 1$ and $\overline{a}_k = b_k$ for $0 \leq k \leq i$.

Since for any two binary bits e_i and e_j

$$(e_i + e_j) \bmod 2 = (\overline{e}_i + \overline{e}_j) \bmod 2,$$

we have

$$a'_i = \overline{b'}_i$$
$$a'_j = b'_j, \quad 0 \leq j \leq n - 1 \ and \ j \neq i.$$

Hence the lemma is proved. $\square$

In Step (6*)-2, all adjustments are "localized" if the structure being adjusted is a part of intersecting tiles; we only exchange the labels of "internal" pairs without affecting the border points. There is no "side effect", i.e. to cause additional distance 3 pairs after Step (6*). But for non-intersecting tiles, the adjustment is made on the border. If we combine this kind of adjustment with others, can the "side effect" occur?

To simplify our discussion and without loss of generality, consider how the pattern extends to the right. (Assuming the pattern can be extend to the right; otherwise, the extension stops at some column or can be cyclically extended to the left).

In Fig. 3(c), let X_3 be in Column j and Chain i. It is easy to get the column number and chain number for all other points in the same Figure. (See Fig. 5(a).)

By the cyclic shifting characteristic of the matrix A, we have $A(i+2, j+2) = 2$, $A(i+2, j+1) = 1$, $A(i+1, j+2) = 1$. We also have $COLSUM(j+1; 1, i) \neq COLSUM(j+2; 1, i)$. If $COLSUM(j+1; 1, i) = COLSUM(j+2; 1, i)$, there will be a intersecting structure. So $COLSUM(j+1; 1, i) \neq COLSUM(j+2; 1, i)$.

If $A(i, j + 2) = 1$, the neighboring points are shown in Fig. 5(b). According to our adjustment strategy, K and L can not be exchanged. As to M and N, if we exchenge P and Q, the last bit L of $MARK(R)$ is 1, we cannot exchange M and N at the same time.

If $A(i, j + 2) = 2$, there are two possibilities as shown in Fig. 6.

In Fig. 6(a), $MARK(V) - MARK(U) = 3$ and $MARK(W) - MARK(U) = 2$; if we exchange K and L, M and N should also be exchanged (the rightmost bit for $MARK(U)$ and $MARK(V)$ are different). If this happens, M and N will not exchange; but no distance 3 neighboring pair is produced. As to the case of Fig. 6(b), because the right part is an intersecting structure, change anything by our rule will not affect the right part. Thus, after Step (6*), we get the dilation 2 embedding.

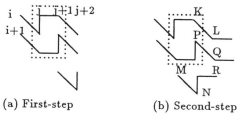

(a) First-step   (b) Second-step

Fig. 5 Extension of Fig. 3(c)

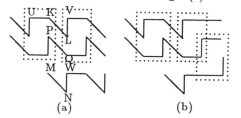

(a)   (b)

Fig. 6 Further Extension of Single Tile

3   Congestion of the Embedding

The above-mentioned modification to algorithm [3] is aimed at "localizing" the adjustments made of the last stage. Careful study of the adjustments in Step (6*), we can get a series of lemmas. Based on these lemmas, we prove the congestion in Theorem 1.

In the following discussion, a path in host graph H is an image of an edge in G if

- the length of the path is $\leq$ dilation 2,
- two end points of edge in G are respectively mapped into two end points of the path in H.

We use A, B, ... to denote points in grid and A', B' ... to denote corresponding nodes in hypercube.

Lemma 3. For any grid point A, there exists an adjacent point B such that A' and B' are adjacent in the hypercube.

Proof: In the adjustment of Step (6*), we only make exchanges as shown in Fig. 3 and Fig. 4. Extending Fig. 3(a), we get Fig. 7.

Fig. 7 Labels Exchange of X and Y

After exchanging the labels X and Y, X' and Y' are still adjacent. For X's other neighbor K, A and C, A' and H' are adjacent since they are in the same chain; if C does not exchange label with any other point, C' and D' are adjacent, if C and I exchange their labels, C' still have J' as its neighbor; K' and H' will be adjacent. Similar result can be gained for Y's neighbors B, E and D. It is straight-forward to generalize the proof to the multiple intersection case.

For non-intersecting cases, we shall prove the case shown in Fig. 4(a). Extend Fig. 4(a), we get Fig. (8).

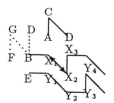

Fig. 8 Label Exchange of X_1 and X_2

After exchanging the labels of X_1 and X_2, X_2' and Y_1', X_3' and X_1', Y_2' and Y_3' will be adjacent in the hypercube. For X_1's neighbor A, A' will be adjacent to C' or D'; B' will be adjacent to at least one of E', F' or D'. Thus after our adjustment, every node has at least one neighbor whose image is still adjacent to the image of this point. □

If X and Y are adjacent in the grid, because the embedding algorithm has the property of dilation 2, one of the following cases must be true. (In the following discussion, we use $MARK'(X)$ to denote the number whose binary reflected Gray code representation is the same as the last $\lfloor \log \beta \rfloor + 1$ bits of X's label.)

$$(1) \quad \begin{cases} |MARK'(X) - MARK'(Y)| = 1 \\ CHAIN(X) = CHAIN(Y), \end{cases}$$

$$(2) \quad \begin{cases} MARK'(X) = MARK'(Y) \\ |CHAIN(X) - CHAIN(Y)| = 1, \end{cases}$$

$$(3) \quad \begin{cases} |MARK'(X) - MARK'(Y)| = 2 \\ CHAIN(X) = CHAIN(Y), \end{cases}$$

$$(4) \quad \begin{cases} |MARK'(X) - MARK'(Y)| = 1 \\ |CHAIN(X) - CHAIN(Y)| = 1. \end{cases}$$

Case (3) and (4) refer to X' and Y' at a distance of 2. To simplify our proof, we define Y as a **critical point** of X if X and Y satisfy the conditions in case (3) (**Type I critical point**) or case (4) (**Type II critical point**). X and Y are called **critical neighbors** in these two cases.

For any point X in grid, at most two of its neighbors are Type I critical points. If a point X has two Type I critical points, say A and B, the points distribution around X can only be those as shown in Fig. 9, where the number 1 or 0 represents the last bit of $MARK(X) - 1$. The reason is, if X has two Type I neighbors, at least two of its neighbors should be in the same chain as X. But X cannot have all of its four neighbors in the same chain. If X has only two neighbors in the same chain, as shown in Fig. 10, X_1 and X_2 cannot be X's Type I neighbors at the same time. If X has three neighbors in the same chain (Fig. 11), two critical points exist only if they have the nearby points distributed as Fig. 9.

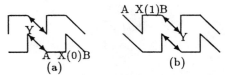

Fig. 9 Two Type I Critical Neighbors of X

Fig. 10 Two Neighbors in the Same Chain as X

Fig. 11 Three Neighbors in the Same Chain as X

Lemma 4. Among all neighbors of a particular point, at most two of them are its Type II critical points.

Proof: Assuming there are three neighbors of X satisfying condition (4), then there are at least

two of them in the same chain (say, be Y_1 and Y_2). Because the exchanges made in Step (6*) are only in the same chain, while two of X's vertical neighbors cannot be in the same chain, Y_1 or Y_2 should be X's horizontal neighbors.

Without loss of generality, let Y_1 be X's left neighbor. According to the strategy of making chain in [3], there is no edge connecting points [i,j] and [i-1,j+1], $(1 < i \leq \tilde{\alpha}, 1 \leq j < \beta)$. Thus the relationship of X, Y_1 and Y_2 can only appear as Fig. 12.

Even though X and Z_1 should exchange their labels in Step (6*), X's right neighbor Z_2 should be in the same chain as X. This contradicts with our assumption. □

Fig. 12 Three Neighbors Satisfying (4)

Lemma 5. For any edge $X'Y'$ in the hypercube, there are at most five paths going through this edge, with each path being the image of an edge in G.

Proof: We consider several cases depending on whether X' and Y' are images of points in G.

Case 1. If X' and Y' are not the images of any points in the grid, then there is no path of length 1 or 2 going through this edge (the two end points of each path are adjacent in the grid).

Case 2. If only one of X' and Y' is the image of some point in the grid, without loss of generality, assuming X' is the image of X. According to Lemma 3, there are at most three of X's critical points, whose images might be adjacent to Y' in the hypercube. It is obvious that the number of paths going through $X'Y'$ is at most 3.

Case 3. If X' and Y' are both images of X and Y, let the labels of X and Y be:

$$label(X) = X_{p+q+1}X_{p+q}\cdots X_{q+2}X_{q+1}X_q\cdots X_2X_1$$
$$label(Y) = Y_{p+q+1}Y_{p+q}\cdots Y_{q+2}Y_{q+1}Y_q\cdots Y_2Y_1$$

where, $p = \lfloor \log \alpha \rfloor$, $q = \lfloor \log \beta \rfloor$.

Case 3.1 If X and Y are in different chains, $X_{p+q+1}X_{p+q}\cdots X_{q+2}$ differs in one bit from $Y_{p+q+1}Y_{p+q}\cdots Y_{q+2}$; and

$$X_{q+1}X_q\cdots X_2X_1 = Y_{q+1}Y_q\cdots Y_2Y_1.$$

If A is a Type I critical point of X, and

$$label(A) = A_{p+q+1}A_{p+q}\cdots A_{q+2}A_{q+1}A_q\cdots A_2A_1,$$

$A_{q+1}A_q \cdots A_2 A_1$ and $X_{q+1}X_q \cdots X_2 X_1$ differ in two bits. Since

$$A_{p+q+1}A_{p+q} \cdots A_{q+2} = X_{p+q+1}X_{p+q} \cdots X_{q+2},$$

We will have $A_{p+q+1}A_{p+q} \cdots A_{q+2}$ differing from $Y_{p+q+1}Y_{p+q} \cdots Y_{q+2}$ in one bit. Since $label(A)$ and $label(Y)$ differ in three bits, A' cannot be adjacent to Y' in the hypercube. Therefore X has at most two neighbors whose images are adjacent to Y' in the hypercube.

Exchanging the roles of X and Y in our discussion, Y also has at most two neighbors whose images are adjacent to X' in the hypercube. Considering X and Y might be adjacent, there are at most five paths corresponding to edges in grid will go through $X'Y'$.

Case 3.2 If X and Y are in the same chain, i.e. $X_{p+q+1}X_{p+q} \cdots X_{q+2} = Y_{p+q+1}Y_{p+q} \cdots Y_{q+2}$, $X_{q+1}X_q \cdots X_2 X_1$ and $Y_{q+1}Y_q \cdots Y_2 Y_1$ differ in one bit.

Case 3.2.1: If $|MARK(X) - MARK(Y)| \neq 1$, for any X's Type II critical point A, we have $MARK(A) \neq MARK(Y)$; that is $A_{q+1} \cdots A_1$ and $Y_{q+1} \cdots Y_1$ differs in at least one bit. And because A and X are in different chains, A and Y are also in different chains, which means $A_{p+q+1}A_{p+q} \cdots A_{q+2}$ and $Y_{p+q+1}Y_{p+q} \cdots Y_{q+2}$ differs in one bit. A' and Y' cannot be adjacent in hypercube. Thus, there are at most two of X's neighbors, whose images will be adjacent to Y'.

Exchange the roles of X and Y, Y also has at most two neighbors whose image are adjacent to X'. Considering X and Y might be adjacent, there are at most five paths going through $X'Y'$, each of these paths is corresponding to an edge in the grid.

Case 3.2.2: If $|MARK(X) - MARK(Y)| = 1$, without loss of generality, assuming $MARK(X) + 1 = MARK(Y)$. If each of X and Y has at most two critical points, there are at most five paths going through $X'Y'$; otherwise, assuming X has three critical points A, B and C, furthermore, assuming their images A', B' and C' are adjacent to Y'. If A and B are Type II critical points of X, we have the following relation.

$$
\begin{aligned}
A_{q+1}A_q \cdots A_2 A_1 &= Y_{q+1}Y_q \cdots Y_2 Y_1 \\
&= B_{q+1}B_q \cdots B_2 B_1
\end{aligned}
$$

Thus, A and B lie in different chains. Assuming $CHAIN(A) < CHAIN(B)$,

$$
\begin{aligned}
CHAIN(A) + 1 &= CHAIN(Y) \\
&= CHAIN(B) - 1.
\end{aligned}
$$

As to C, we have

$$
\begin{cases}
MARK(C) - 1 = MARK(Y) \\
CHAIN(C) \quad = CHAIN(Y)
\end{cases}
$$

According to the conditions stated above, X, Y, A, B and C can only have the relationship shown in Fig. 13(a). Therefore, there will be at most the following five paths going through $X'Y'$ in hypercube:

$$
\begin{array}{ll}
A' \leftrightarrow Y' \leftrightarrow X' & (AX) \\
B' \leftrightarrow Y' \leftrightarrow X' & (BX) \\
C' \leftrightarrow Y' \leftrightarrow X' & (CX) \\
E' \leftrightarrow X' \leftrightarrow Y' & (EY) \\
F' \leftrightarrow X' \leftrightarrow Y' & (FY)
\end{array}
$$

If A and B are Type I critical points of X, since $MARK(X) + 1 = MARK(Y)$, the points near X are shown in Fig. 14. Thus there is only one path going through $X' \leftrightarrow Y' \leftrightarrow B'$ (XB). $\square$

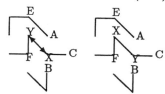

(a) Before (6*)   (b) After (6*)

Fig. 13 Exchange X and Y in Step (6*)

Fig. 14 Two Type I Neighbors of X

Lemma 6. The congestion upper bound proved in Lemma 5 is tight.

Proof: In the example of 11×11 grid, consider the intersecting grid with the upperleft corner in row 1 and column 3 (Fig. 15).

In the hypercube, there are five path going through $X'Y'$:

$$
\begin{array}{ll}
A' \leftrightarrow X' \leftrightarrow Y' & (AY) \\
B' \leftrightarrow X' \leftrightarrow Y' & (BY) \\
C' \leftrightarrow Y' \leftrightarrow X' & (CX) \\
D' \leftrightarrow Y' \leftrightarrow X' & (DX) \\
X' \leftrightarrow Y' & (XY). \quad \square
\end{array}
$$

Theorem 1. The congestion created by the modified algorithm is 5.
Proof: The proof follows from Lemma 5 and Lemma 6.

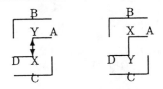

(a) Before step (6*) (b) After step (6*)

Fig. 15 Congestion 5 is tight.

4 Conclusion

In this paper, we presented a modifed algorithm for embedding two-dimensional grids into hypercubes with dilation 2 and congestion 5. Our algorithm is simpler than the original one. In a previous paper, we have deduced the upperbound of 8 in embedding any two-dimensional grids with dilation 2 [2]. The result in this paper is significant since two-dimensional grids and hypercubes are both important topologies [5][6][13]. According to our results, we can simulate two-dimensional grid by hypercube with the most communication delay of 10.

The congestion result obtained in this paper is based on the definition of embedding which only specifies how to map vertices of the guest graph into vertices of the host graph. Thus in counting congestion, we included all possible shortest paths. This is actually the worst case. If we can devise an algorithm to map any edge to a specific path, the congestion can be further reduced.

References

[1] S. N. Bhatt and C. F. Ipsen, "How to Embed Trees in Hypercubes," Yale University Research Report YALEU/DCS/RR-443, (Dec, 1985).

[2] Y. Cai and S. Huang, "Constant Congestions of Dilation 2 Embeddings," *University of Houston Research Report UH-CS-94-11,* (Jun, 1994).

[3] M. Y. Chan, "Embedding of Grids into Optimal Hypercubes," *SIAM J. Comput.* (Oct, 1991) Vol 20, No. 5, pp. 834-864.

[4] C. Y. Chen and Y. C. Chung, "Embedding Networks with Ring Connections in Hypercube Machines," *The International Conference on Parallel Processing,* (Aug, 1990) Vol III, pp. 327-334.

[5] C. T. Ho, "Efficient Submesh Permutation in Wormhole-Routed Meshes," *IBM Research Report RJ 9752(84638)* (Mar, 1994),

[6] C. T. Ho, "Embedding Meshes in Boolean Cubes by Graph Decomposition," *Journal of Parallel and Distributed Computing* (Aug. 1990), pp.325-339.

[7] S. Huang, H. Liu and R. M. Verma, "On Embedding Rectangular Meshes into Optimal Squares," *The International Conference on Parallel Processing,* (Aug, 1993), Vol III, pp. 73-76.

[8] S. L. Johnsson, "Binary Cube Emulation of Butterfly Networks Encoded by Gray Code," *Journal of Parallel and Distributed Computing* 20, (1994), pp.261-279.

[9] V. Pin Kumar, Ananth Grama, Anshul Gupta and George Kaypis, *"Introduction to Parallel Computing: Design and Analysis of Algorithm,"* Bejamin/Cummings, Redwood City, CA. (1994).

[10] G. Langholz, A. Kandel, J. L. Mott, *"Digital Logic Design,"* Wm. C. Brown Publishers, Dubuque, Iowa, (1988).

[11] F. T. Leighton, *"Introduction to Parallel Algorithm and Architecture: Arrays · Trees · Hypercubes,"* Morgan Kaufmann Publishers, Inc, San Matro, California, (1992).

[12] R. G. Melhem and G. Y. Huang, "Embedding Rectangular Grids into Square Grids with Dilation Two," *IEEE Trans. on Computers,* 39:12, (1990), pp. 1446-1455.

[13] A. S. Wagner, "Embedding the Computer Tree in the Hypercube," *Journal of Parallel and Distributed Computing* 20, (1994), pp.241-247.

Acknowledgement: This work has supported in part by Navy under grant N3039-93-C-0165 and by NASA under grant NAS 5-32672.

MULTILEVEL GRAPH PARTITIONING SCHEMES *

George Karypis and Vipin Kumar
Department of Computer Science, University of Minnesota, Minneapolis, MN 55455
{karypis, kumar}@cs.umn.edu

Abstract – In this paper we present experiments with a class of graph partitioning algorithms that reduce the size of the graph by collapsing vertices and edges, partition the smaller graph, and then uncoarsen it to construct a partition for the original graph. We investigate the effectiveness of many different choices for all three phases: coarsening, partition of the coarsest graph, and refinement. In particular, we present a new coarsening heuristic (called heavy-edge heuristic) for which the size of the partition of the coarse graph is within a small factor of the size of the final partition obtained after multilevel refinement. We also present a new scheme for refining during uncoarsening that is much faster than the Kernighan-Lin refinement. We test our scheme on a large number of graphs arising in various domains including finite element methods, linear programming, VLSI, and transportation. Our experiments show that our scheme consistently produces partitions that are better than those produced by spectral partitioning schemes in substantially smaller timer (10 to 35 times faster than multilevel spectral bisection). Also, when our scheme is used to compute fill reducing orderings for sparse matrices, it substantially outperforms the widely used multiple minimum degree algorithm.

1   Introduction

Graph partitioning is an important problem that has extensive applications in many areas, including scientific computing and VLSI design. The problem is to partition the vertices of a graph in p roughly equal parts, such that the number of edges connecting vertices in different parts is minimized. For example, the solution of a sparse system of linear equations $Ax = b$ via iterative methods on a parallel computer gives rise to a graph partitioning problem. A key step in each iteration of these methods is the multiplication of a sparse matrix and a (dense) vector. The problem of minimizing communication in this step is identical to the problem of partitioning the graph corresponding to the matrix A [26]. If parallel direct methods are used to solve a sparse system of equations, then a graph partitioning algorithm can be used to compute a fill reducing ordering that lead to high degree of concurrency in the factorization phase [26, 9]. The multiple minimum degree ordering used almost exclusively in serial direct methods is not suitable for parallel direct methods, as it provides very little concurrency in the parallel factorization phase.

The graph partitioning problem is NP-complete. However, many algorithms have been developed that find a reasonably good partition. Spectral partitioning methods are known to produce excellent partitions for a wide class of problems, and they are used quite extensively [33, 20]. However, these methods are very expensive since they require the computation of the eigenvector corresponding to the second smallest eigenvalue (Fiedler vector). Execution of the spectral methods can be speeded up if computation of the Fiedler vector is done by using a multilevel algorithm [2]. This multilevel spectral bisection algorithm (MSB) usually manages to speedup the spectral partitioning methods by an order of magnitude without any loss in the quality of the edge-cut. However, even MSB can take a large amount of time. In particular, in parallel direct solvers, the time for computing ordering using MSB can be several orders of magnitude higher than the time taken by the parallel factorization algorithm, and thus ordering time can dominate the overall time to solve the problem [14]. The execution time of MSB can be further speeded up by computing the Fiedler vector in parallel. The algorithm for computing the Fiedler vector, is iterative and in each iteration it performs a matrix-vector multiplication of a matrix whose graph is identical to the one we are trying to partition. These matrix-vector products can be performed efficiently on a parallel computer only if a good partition of the graph is available—a problem that MSB is trying to solve in the first place. As a result, parallel implementation of spectral methods exhibit poor efficiency since most of the time is spent in performing communication [21, 1].

Another class of graph partitioning techniques uses the geometric information of the graph to find a good partition. Geometric partitioning algorithms [17, 28, 29] tend to be fast but often yield partitions that are worse than those obtained by spectral methods. Among the most prominent of these scheme is the algorithm described in [28]. This algorithm produces partitions that are provably within the bounds that exist for some special classes of graphs. However, due to the randomized nature of these algorithms, multiple trials are often required to obtain solutions that are comparable in quality to spectral methods. Multiple trials do increase the time [13], but the overall runtime is still substantially lower than the time required by the spectral methods. However, geometric graph partitioning algorithms have limited applicability because often the geometric information is not available, and in certain problem areas (*e.g.*, linear programming), there is no geometry associated with the graph. Recently, an algorithm has been proposed to compute geometry information for graphs [4]. However this algorithm is based on computing spectral information, which is expensive and dominates the overall time taken by the graph partitioning algorithm.

---

*This work is sponsored by the AHPCRC under the auspices of the DoA, ARL cooperative agreement number DAAH04-95-2-0003/contract number DAAH04-95-C-0008, the content of which does not necessarily reflect the position or the policy of the government, and no official endorsement should be inferred. Access to computing facilities was provided by Cray Research Inc. Related papers are available via WWW at URL: http://www.cs.umn.edu/users/kumar/papers.html

Another class of graph partitioning algorithms reduce the size of the graph (*i.e.*, coarsen the graph) by collapsing vertices and edges, partition the smaller graph, and then uncoarsen it to construct a partition for the original graph. These are called multilevel graph partitioning schemes [3, 5, 15, 20, 7, 30]. Some researchers investigated multilevel schemes primarily to decrease the partitioning time, at the cost of somewhat worse partition quality [30]. Recently, a number of multilevel algorithms have been proposed [3, 20, 5, 15, 7] that further refine the partition during the uncoarsening phase. These schemes tend to give good partitions at reasonable cost. In particular, the work of Hendrickson and Leland [20] showed that multilevel schemes can provide better partitions than the spectral methods at lower cost for a variety of finite element problems. Their scheme uses random maximal matching to successively coarsen the graph until it has only a few hundred vertices. Then it partitions this small graph using the spectral methods. Now it uncoarsens the graph level by level, and applies Kernighan-Lin refinement periodically. However, even-though multilevel algorithms have been shown to be good alternatives to both spectral and geometric algorithms, there is no comprehensive study today on their effectiveness on a wide range of problems.

In this paper we experiment with various parameters of multilevel algorithms, and their effect on the quality of partition and ordering. We investigate the effectiveness of many different choices for all three phases: coarsening, partition of the coarsest graph, and refinement. In particular, we present a new coarsening heuristic (called heavy-edge heuristic) for which the size of the partition of the coarse graph is within a small factor of the size of the final partition obtained after multilevel refinement. We also present a new scheme for refining during uncoarsening that is much faster than the Kernighan-Lin refinement used in [20].

We test our scheme on a large number of graphs arising in various domains including finite element methods, linear programming, and VLSI. Our experiments show that our scheme consistently produces partitions that are better than those produced by spectral partitioning schemes in substantially smaller timer (10 to 35 times faster than multilevel spectral bisection). Compared with the scheme of [20], our scheme is about twice as fast, and is consistently better in terms of cut size. Much of the improvement in run time comes from our faster refinement heuristic. We also used our graph partitioning scheme to compute fill reducing orderings for sparse matrices. Surprisingly, our scheme substantially outperforms the multiple minimum degree algorithm [27], which is the most commonly used method for computing fill reducing orderings of a sparse matrix.

Even though multilevel algorithms are quite fast compared with spectral methods, they can still be the bottleneck if the sparse system of equations is being solved in parallel [26, 14]. The coarsening phase of these methods is easy to parallelize [23], but the Kernighan-Lin heuristic used in the refinement phase is very difficult to speedup in parallel computers [12]. Since both the coarsening phase and the refinement phase with Kernighan-Lin heuristic take roughly the same amount of time, the overall scheme cannot be speeded up significantly. Our new faster methods for refinement reduce this bottleneck substantially. In fact our parallel implementation [23] of this multilevel partitioning

is able to get a speedup of as much as 56 on a 128-processor Cray T3D for moderate size problems.

2 Graph Partitioning

The *k-way* graph partitioning problem is defined as follows: Given a graph $G = (V, E)$ with $|V| = n$, partition V into k subsets, $V_1, V_2, \ldots, V_k$ such that $V_i \cap V_j = \emptyset$ for $i \neq j$, $|V_i| = n/k$, and $\bigcup_i V_i = V$, and the number of edges of E whose incident vertices belong to different subsets is minimized. A k-way partition of V is commonly represented by a partition vector P of length n, such that for every vertex $v \in V$, $P[v]$ is an integer between 1 and k, indicating the partition at which vertex v belongs. Given a partition P, the number of edges whose incident vertices belong to different subsets is called the *edge-cut* of the partition.

The efficient implementation of many parallel algorithms usually requires the solution to a graph partitioning problem, where vertices represent computational tasks, and edges represent data exchanges. A k-way partition of the computation graph can be used to assign tasks to k processors. Because the partition assigns equal number of computational tasks to each processor the work is balanced among k processors, and because it minimizes the edge-cut, the communication overhead is also minimized.

Another important application of recursive bisection is to find a fill reducing ordering for sparse matrix factorization [9, 26, 16]. This type of algorithms are generally referred to as nested dissection ordering algorithms. Nested dissection recursively splits a graph into almost equal halves by selecting a vertex separator until the desired number of partitions are obtained. The vertex separator is determined by first bisecting the graph and then computing a vertex separator from the edge separator. The vertices of the graph are numbered such that at each level of recursion, the separator vertices are numbered after the vertices in the partitions. The effectiveness and the complexity of a nested dissection scheme depends on the separator computing algorithm. In general, small separators result in low fill-in.

The k-way partition problem is most frequently solved by recursive bisection. That is, we first obtain a 2-way partition of V, and then we further subdivide each part using 2-way partitions. After $\log k$ phases, graph G is partitioned into k parts. Thus, the problem of performing a k-way partition is reduced to that of performing a sequence of 2-way partitions or bisections. Even though this scheme does not necessarily lead to optimal partition, it is used extensively due to its simplicity [9, 16].

3 Multilevel Graph Bisection

The graph G can be bisected using a multilevel algorithm. The basic structure of a multilevel algorithm is very simple. The graph G is first coarsened down to a few hundred vertices, a bisection of this much smaller graph is computed, and then this partition is projected back towards the original graph (finer graph), by periodically refining the partition. Since the finer graph has more degrees of freedom, such refinements usually decrease the edge-cut.

Formally, a multilevel graph bisection algorithm works as follows: Consider a weighted graph $G_0 = (V_0, E_0)$, with weights both on vertices and edges. A multilevel graph bisection algorithm consists of the following three phases.

Coarsening Phase The graph G_0 is transformed into a sequence of smaller graphs $G_1, G_2, \ldots, G_m$ such that $|V_0| > |V_1| > |V_2| > \cdots > |V_m|$.

Partitioning Phase A 2-way partition P_m of the graph $G_m = (V_m, E_m)$ is computed that partitions V_m into two parts, each containing half the vertices of G_0.

Uncoarsening Phase The partition P_m of G_m is projected back to G_0 by going through intermediate partitions $P_{m-1}, P_{m-2}, \ldots, P_1, P_0$.

3.1   Coarsening Phase

During the coarsening phase, a sequence of smaller graphs, each with fewer vertices, is constructed. Graph coarsening can be achieved in various ways. In most coarsening schemes, a set of vertices of G_i is combined together to form a single vertex of the next level coarser graph G_{i+1}. Let V_i^v be the set of vertices of G_i combined to form vertex v of G_{i+1}. We will refer to vertex v as a ***multinode***. In order for a bisection of a coarser graph to be good with respect to the original graph, the weight of vertex v is set equal to the sum of the weights of the vertices in V_i^v. Also, in order to preserve the connectivity information in the coarser graph, the edges of v are the union of the edges of the vertices in V_i^v. In the case where more than one vertex of V_i^v, contain edges to the same vertex u, the weight of the edge of v is equal to the sum of the weights of these edges. This is useful when we evaluate the quality of a partition at a coarser graph. The edge-cut of the partition in a coarser graph will be equal to the edge-cut of the same partition in the finer graph.

Two main approaches have been proposed for obtaining coarser graphs. The first approach is based on finding a random matching and collapsing the matched vertices into a multinode [3, 20, 2], while the second approach is based on creating multinodes that are made of groups of vertices that are highly connected [5, 15, 7]. The later approach is suited for graphs arising in VLSI applications, since these graphs have highly connected components. However, for graphs arising in finite element applications, most vertices have similar connectivity patterns (*i.e.*, the degree of each vertex is fairly close to the average degree of the graph). In the rest of this section we describe the basic ideas behind coarsening using matchings.

Given a graph $G_i = (V_i, E_i)$, a coarser graph can be obtained by collapsing adjacent vertices. Thus, the edge between two vertices is collapsed and a multinode consisting of these two vertices is created. This edge collapsing idea can be formally defined in terms of matchings. A ***matching*** of a graph, is a set of edges, no two of which are incident on the same vertex. Thus, the next level coarser graph G_{i+1} is constructed from G_i by finding a matching of G_i and collapsing the vertices being matched into multinodes. The unmatched vertices are simply copied over to G_{i+1}. Since the goal of collapsing vertices using matchings is to decrease the size of the graph G_i, the matching should be of maximal size. That is, it should contain all possible edges, no two of which are incident on the same vertex. The matching of maximal size is called ***maximal matching***.

Note that depending on how matchings are computed, the size of the maximal matching may be different.

In the remaining sections we describe four ways that we used to select maximal matchings for coarsening. The complexity of all these schemes is $O(|E|)$.

Random Matching (RM)   A maximal matching can be generated efficiently using a randomized algorithm. In our experiments we used a randomized algorithm similar to that described in [3, 20]. The random maximal matching algorithm is the following. The vertices are visited in random order. If a vertex u has not been matched yet, then we randomly select one of its unmatched adjacent vertices. If such a vertex v exists, we include the edge (u, v) in the matching and mark vertices u and v as being matched. If there is no unmatched adjacent vertex v, then vertex u remains unmatched in the random matching.

Heavy Edge Matching (HEM)   While performing the coarsening using random matchings, we try to minimize the number of coarsening levels in a greedy fashion. However, our overall goal is to find a bisection that minimizes the edge-cut. Consider a graph $G_i = (V_i, E_i)$, a matching M_i that is used to coarsen G_i, and its coarser graph $G_{i+1} = (V_{i+1}, E_{i+1})$ induced by M_i. If A is a set of edges, define $W(A)$ to be the sum of the weights of the edges in A. It can be shown that $W(E_{i+1}) = W(E_i) - W(M_i)$. Thus, the total edge weight of the coarser graph is reduced by the weight of the matching. Hence, by selecting a matching M_i that has a maximal weight, we can maximize the decrease in the edge weight of the coarser graph. Now, since the coarser graph has smaller edge weight, it is more likely to have a smaller edge-cut.

Finding a matching with maximal weight is the idea behind the ***heavy-edge matching***. A maximal weight matching is computed using a randomized algorithm similar to that for computing a random matching described in Section 3.1. The vertices are again visited in random order. However, instead of randomly matching a vertex u with one of its adjacent unmatched vertices, we match u with the vertex v such that the weight of the edge (u, v) is maximum over all valid incident edges (heavier edge). Note that this algorithm does not guarantee that the matching obtained has maximum weight, but our experiments has shown that it works very well.

Light Edge Matching (LEM)   Instead of minimizing the total edge weight of the coarser graph, one might try to maximize it. This is achieved by finding a matching M_i that has the smallest weight, leading to a small reduction in the edge weight of G_{i+1}. This is the idea behind the ***light-edge matching***. It may seem that the light-edge matching does not perform any useful transformation during coarsening. However, the average degree of G_{i+1} produced by LEM is significant higher than that of G_i. Graphs with high average degree are easier to partition using certain heuristics such as Kernighan-Lin [3].

Heavy Clique Matching (HCM)   A ***clique*** of an unweighted graph $G = (V, E)$ is a fully connected subgraph of G. Consider a set of vertices U of V ($U \subset V$). The subgraph of G induced by U is defined as $G_U = (U, E_U)$, such that E_U consists of all edges $(v_1, v_2) \in E$ such that both v_1 and v_2 belong in U. Looking at the cardinality of U and E_U we can determined how close U is to a clique. In particular, the ratio $2|E_U|/(|U|(|U|-1))$ goes to one if U is a clique, and is small if U is far from being a clique. We refer to this ratio as *edge density*.

The *heavy clique matching* scheme computes a matching by collapsing vertices that have high edge density. Thus, this scheme computes a matching whose edge density is maximal. The motivation behind this scheme is that subgraphs of G_0 that are cliques or almost cliques will most likely not be cut by the bisection. So, by creating multinodes that contain these subgraphs, we make it easier for the partitioning algorithm to find a good bisection. Note that this scheme tries to approximate the graph coarsening schemes that are based on finding highly connected components [5, 15, 7].

As in the previous schemes for computing the matching, we compute the heavy clique matching using a randomized algorithm. Note that HCM is very similar to the HEM scheme. The only difference is that HEM matches vertices that are only connected with a heavy edge irrespective of the contracted edge-weight of the vertices, whereas HCM matches a pair of vertices if they are both connected using a heavy edge and if each of these two vertices have high contracted edge-weight.

3.2 Partitioning Phase

The second phase of a multilevel algorithm is to compute a minimum edge-cut bisection P_m of the coarse graph $G_m = (V_m, E_m)$ such that each part contains roughly half of the vertex weight of the original graph.

A partition of G_m can be obtained using various algorithms such as (a) spectral bisection [33, 2, 18], (b) geometric bisection [28] (if coordinates are available), and (c) combinatorial methods [25, 8, 9]. Since the size of the coarser graph G_m is small (*i.e.*, $|V_m| < 100$), this step takes a small amount.

We implemented three different algorithms for partitioning the coarse graph. The first algorithm uses the spectral bisection [33], and the other two use graph growing heuristics. The first graph-growing heuristic (GGP) randomly selects a vertex v and grows a region around it in a breadth-first fashion until half of the vertex-weight has been included. The second graph-growing heuristic (GGGP) also starts from a randomly selected vertex v but it includes vertices that lead to the smaller increase in the edge-cut. Since the quality of the partitions obtained by GGP and GGGP depends on the choice of v, a number of different partitions are computed starting from different randomly selected vertices and the best is used as the initial partition. In the experiments in Section 4.1 we selected 10 vertices for GGP and 5 for GGGP. We found all of these partitioning schemes to produce similar partitions with GGGP consistently performing better.

3.3 Uncoarsening Phase

During the uncoarsening phase, the partition P_m of the coarser graph G_m is projected back to the original graph, by going through the graphs $G_{m-1}, G_{m-2}, \ldots, G_1$. Since

each vertex of G_{i+1} contains a distinct subset of vertices of G_i, obtaining P_i from P_{i+1} is done by simply assigning the vertices collapsed to $v \in G_i$ to the partition $P_{i+1}[v]$.

Even though P_{i+1} is a local minima partition of G_{i+1}, the projected partition P_i may not be at a local minima with respect to G_i. Since G_i is finer, it has more degrees of freedom that can be used to improve P_i, and decrease the edge-cut. Hence, it may still be possible to improve the projected partition of G_{i-1} by local refinement heuristics. For this reason, after projecting a partition, a partition refinement algorithm is used. The basic purpose of a partition refinement algorithm is to select two subsets of vertices, one from each part such that when swapped the resulting partition has smaller edge-cut. Specifically, if A and B are the two parts of the bisection, a refinement algorithm selects $A' \subset A$ and $B' \subset B$ such that $A \backslash A' \cup B'$ and $B \backslash B' \cup A'$ is a bisection with a smaller edge-cut.

A class of algorithms that tend to produce very good results are those that are based on the Kernighan-Lin (KL) partition algorithm [25, 6, 20]. The KL algorithm is iterative in nature. It starts with an initial partition and in each iteration it finds subsets A' and B' with the above properties. If such subsets exist, then it moves them to the other part and this becomes the partition for the next iteration. The algorithm continues by repeating the entire process. If it cannot find two such subsets, then the algorithm terminates.

The KL algorithm we implemented is similar to that described in [6] with certain modifications that significantly reduce the run time. The KL algorithm, computes for each vertex v a quantity called *gain* which is the decrease (or increase) in the edge-cut if v is moved to the other part. The algorithm then proceeds by repeatedly selecting a vertex v with the largest gain from the larger part and moves it to the other part. After moving v, v is marked so it will not be considered again in the same iteration, and the gains of the vertices adjacent to v are updated to reflect the change in the partition. The algorithm terminates when the edge-cut does not decrease after x number of vertex moves. Since, the last x vertex moves did not decrease the edge-cut they are undone. The choice of $x = 50$ works quite well for all our graphs.

The efficient implementation of the above algorithm relies on the method used to compute the gains of successive finer graphs and the use of appropriate data structure to store these gains. Our algorithm computes the gains of the vertices during the projection of the partition. In doing so, it utilizes the computed gains for the vertices of the coarser graph and it only needs to compute the gains of the vertices that are along the boundary of the partition. The data structure used to store the gains is a hash table that allow insertions, updates, and extraction of the vertex with maximum gain in constant time. Details about the implementation of the KL algorithm can be found in [22].

In the next section we describe three different refinement algorithms that are based on the KL algorithm but differ in the time they require to do the refinement.

Kernighan-Lin Refinement The idea of Kernighan-Lin refinement (KLR) is to use the projected partition of G_{i+1} onto G_i as the initial partition for the Kernighan-Lin algorithm. The KL algorithm has been found to be effective in finding locally optimal partitions when it starts with a fairly good initial partition [3]. Since the projected partition is already a good partition, KL substantially decreases the edge-cut within a small number of iterations. Furthermore, since a single iteration of the KL algorithm stops as soon as x swaps are performed that do not decrease the edge-cut, the number of vertices swapped in each iteration is very small. Our experimental results show that a single iteration of KL terminates after only a small percentage of the vertices have been swapped (less than 5%), which results in significant savings in the total execution time of this refinement algorithm.

Greedy Refinement   Since we terminate each pass of the KL algorithm as soon as no further improvement can be made in the edge-cut, the complexity of the KLR scheme described in the previous section is dominated by the time required to insert the vertices into the appropriate data structures. Thus, even though we significantly reduced the number of vertices that are swapped, the overall complexity does not change in asymptotic terms. Furthermore, our experience shows that the largest decrease in the edge-cut is obtained during the first pass. In the greedy refinement algorithm (GR), we take advantage of that by running only a single iteration of the KL algorithm [3]. This usually reduces the total time taken by refinement by a factor of two to four (Section 4.1).

Boundary Refinement   In both the KLR and GR algorithms, we have to insert the gains of all the vertices in the data structures. However, since we terminate both algorithms as soon as we cannot further reduce the edge-cut, most of this computation is wasted. Furthermore, due to the nature of the refinement algorithms, most of the nodes swapped by either the KLR and the GR algorithm are along the boundary of the cut, which is defined to be the vertices that have edges that are cut by the partition.

In the boundary refinement algorithm, we initially insert into the data structures the gains for only the boundary vertices. As in the KLR algorithm, after we swap a vertex v, we update the gains of the adjacent vertices of v not yet being swapped. If any of these adjacent vertices become a boundary vertex due to the swap of v, we insert it into the data structures if the have positive gain. Notice that the boundary refinement algorithm is quite similar to the KLR algorithm, with the added advantage that only vertices are inserted into the data structures as needed and no work is wasted.

As with KLR, we have a choice of performing a single pass (boundary greedy refinement (BGR)) or multiple passes (boundary Kernighan-Lin refinement (BKLR)) until the refinement algorithm converges. As opposed to the non-boundary refinement algorithms, the cost of performing multiple passes of the boundary algorithms is small, since only the boundary vertices are examined.

To further reduce the execution time of the boundary refinement while maintaining the refinement capabilities of BKLR and the speed of BGR one can combine these schemes into a hybrid scheme that we refer to it as BKLGR. The idea behind the BKLGR policy is to use BKLR as long as the graph is small, and switch to BGR when the graph is large. The motivation for this scheme is that single vertex swaps in the coarser graphs lead to larger decrease in the

edge-cut than in the finer graphs. So by using BKLR at these coarser graphs better refinement is achieved, and because these graphs are very small (compared to the size of the original graph), the BKLR algorithm does not require a lot of time. For all the experiments presented in this paper, if the number of vertices in the boundary of the coarse graph is less than 2% of the number of vertices in the original graph, refinement is performed using BKLR, otherwise BGR is used.

4   Experimental Results

We evaluated the performance of the multilevel graph partitioning algorithm on a wide range of matrices arising in different application domains. The characteristics of these matrices are described in Table 1. All the experiments were performed on an SGI Challenge, with 1.2GBytes of memory and 200MHz Mips R4400. All times reported are in seconds. Since the nature of the multilevel algorithm discussed is randomized, we performed all experiments with fixed seed.

Matrix Name	Order	Nonzeros	Description
BCSSTK28 (BC28)	4410	107307	Solid element model
BCSSTK29 (BC29)	13992	302748	3D Stiffness matrix
BCSSTK30 (BC30)	28294	1007284	3D Stiffness matrix
BCSSTK31 (BC31)	35588	572914	3D Stiffness matrix
BCSSTK32 (BC32)	44609	985046	3D Stiffness matrix
BCSSTK33 (BC33)	8738	291583	3D Stiffness matrix
BCSPWR10 (BSP10)	5300	8271	Eastern US power network
BRACK2 (BRCK)	62631	366559	3D Finite element mesh
CANT (CANT)	54195	1960797	3D Stiffness matrix
COPTER2 (COPT)	55476	352238	3D Finite element mesh
CYLINDER93 (CY93)	45594	1786726	3D Stiffness matrix
FINAN512 (FINC)	74752	335872	Linear programming
4ELT (4ELT)	15606	45878	2D Finite element mesh
INPRO1 (INPR)	46949	1117809	3D Stiffness matrix
LHR71 (LHR)	70304	1528092	3D Coefficient matrix
LSHP3466 (LS34)	3466	10215	Graded L-shape pattern
MAP (MAP)	267241	937103	Highway network
MEMPLUS (MEM)	17758	126150	Memory circuit
ROTOR (ROTR)	99617	662431	3D Finite element mesh
S38584.1 (S33)	22143	93359	Sequential circuit
SHELL93 (SHEL)	181200	2313765	3D Stiffness matrix
SHYY161 (SHYY)	76480	329762	CFD/Navier-Stokes
TROLL (TROL)	213453	5885829	3D Stiffness matrix
WAVE (WAVE)	156317	1059331	3D Finite element mesh

Table 1: Various matrices used in evaluating the multilevel graph partitioning and sparse matrix ordering algorithm.

4.1   Graph Partitioning

As discussed in Sections 3.1, 3.2, and 3.3, there are many alternatives for each of the three different phases of a multilevel algorithm. It is not possible to provide an exhaustive comparison of all these possible combinations without making this paper unduly large. Instead, we provide comparison of different alternatives for each phase after making a reasonable choice for the other two phases.

Matching Schemes   We implemented the four matching schemes described in Section 3.1 and the results for a 32-way partition for some matrices is shown in Table 2. These schemes are (a) random matching (RM), (b) heavy edge matching (HEM), (c) light edge matching (LEM), and (d) heavy clique matching (HCM). For all the experiments, we used the GGGP algorithm for the initial partition phase and the BKLGR as the refinement policy during the uncoarsening phase. For each matching scheme, Table 2 shows the edge-cut, the time required by the coarsening phase (CTime), and the time required by the uncoarsening phase (UTime). UTime is the sum of the time spent in partitioning the coarse graph (ITime), the time spent in refinement (RTime), and the time spent in projecting the partition of a coarse graph to the next level finer graph (PTime).

	RM	HEM	LEM	HCM
BCSSTK31	14489	84024	412361	115471
BCSSTK32	184236	148637	680637	153945
BRACK2	75832	53115	187688	69370
CANT	817500	487543	1633878	521417
COPTER2	69184	59135	208318	59631
CYLINDER93	522619	286901	1473731	354154
4ELT	3874	3036	4410	4025
INPRO1	205525	187482	821233	141398
ROTOR	147971	110988	424359	98530
SHELL93	373028	237212	1443868	258689
TROLL	1095607	806810	4941507	883002
WAVE	239090	212742	745495	192729

Table 3: The edge-cut for a 32-way partition when no refinement was performed, for the various matching schemes.

In terms of the size of the edge-cut, there is no clear cut winner among the various matching schemes. The value of 32EC for all schemes are within 10% of each other. Out of these schemes, RM does better for 2 matrices, HEM does better for six matrices, LEM for three, and HCM for one.

The time spent in coarsening does not vary significantly across different schemes. But RM requires the least amount of time for coarsening, while LEM and HCM require the most (upto 38% more time than RM). This is not surprising since RM looks for the first unmatched neighbor of a vertex (the adjacency lists are randomly permuted). On the other hand, HCM needs to find the edge with the maximum edge density, and LEM produces coarser graphs that have vertices with higher degree than the other three schemes; hence, LEM requires more time to both find a matching and also to create the next level coarser graph. The coarsening time required by HEM is only slightly higher (upto 10% more) than the time required by RM.

Comparing the time spent during uncoarsening, we see that both HEM and HCM require the least amount of time, while LEM requires the most. In some cases, LEM requires as much as 7 times more time than either HEM or HCM. This can be explained by results shown in Table 3. This table shows the edge-cut of 32-way partition when no refinement is performed (*i.e.*, the final edge-cut is exactly the same as that found in the initial partition of the coarsest graph). Table 3 shows that the edge-cut of LEM on the coarser graphs is significantly higher than that for either HEM or HCM. Because of this, all three components of UTime increase for LEM relative to those of the other schemes. The ITime is higher because the coarser graph has more edges, RTime increases because a large number of vertices need to be swapped to reduce the edge-cut, and PTime increases because more vertices are along the boundary; which requires more computation [22]. The time spent during uncoarsening for RM is also higher than the time re-

quired by the HEM scheme by upto 50% for some matrices for somewhat similar reasons.

From the discussion in the previous paragraphs we see that UTime is much smaller than CTime for HEM and HCM, while UTime is comparable to CTime for RM and LEM. Furthermore, for HEM and HCM, as the problem size increases UTime becomes an even smaller fraction of CTime. As discussed in introduction, this is of particular importance when the parallel formulation of the multilevel algorithm is considered.

As the experiments show, HEM is a good matching scheme that results in good initial partitions, and requires little refinement. Even though it requires slightly more time than RM, it produces consistently smaller edge-cut. We selected the HEM as our matching scheme of choice because of its consistent good behavior.

Initial Partition Algorithms   As described in Section 3.2, a number of algorithms can be used to partition the coarse graph. We have implemented the following algorithms: (a) spectral bisection (SBP), (b) graph growing (GGP), and (c) greedy graph growing (GGGP). Due to space limitations we do not report the results here but they can be found in [22]. In summary, the results in [22] show that GGGP consistently finds smaller edge-cuts than the other schemes at slightly better run time. Furthermore, there is no advantage in choosing spectral bisection for partitioning the coarse graph.

Refinement Policies   As described in Section 3.3, there are different ways that a partition can be refined during the uncoarsening phase. We evaluated the performance of five refinement policies, both in terms of how good partitions they produce and also how much time they require. The refinement policies that we evaluate are (a) Greedy refinement (GR), (b) Kernighan-Lin refinement (KLR), (c) boundary Greedy refinement (BGR), (d) boundary Kernighan-Lin refinement (BKLR), and (e) the combination of BKLR and BGR (BKLGR).

The result of these refinement policies for partitioning graphs corresponding to some of the matrices in Table 1 in 32 parts is shown in Table 4. These partitions were produced by using the heavy-edge matching (HEM) during coarsening and the GGGP algorithm for initially partitioning the coarser graph.

A number of interesting conclusions can be drawn out of Table 4. First, for each of the matrices and refinement policies, the size of the edge-cut does not vary significantly for different refinement policies. For each matrix the edge cut of every refinement policy is within 15% of the best refinement policy for that particular matrix. On the other hand, the time required by some refinement policies does vary significantly. Some policies require up to 20 times more time than others. KLR requires the most time while BGR requires the least.

Comparing GR with KLR, we see that KLR performs better than GR for 8 out of the 12 matrices. For these 8 matrices, the improvement is less than 5% on the average; however, the time required by KLR is significantly higher than that of GR. Usually, KLR requires two to three times more time than GR.

Comparing the GR and KLR refinement schemes against their boundary variants, we see that the time required by the

	RM			HEM			LEM			HCM		
	32EC	CTime	UTime	32EC	CTime	UTime	32EC	CTime	UTime	32EC	CTime	UTime
BCSSTK31	44810	5.93	2.46	45991	6.55	1.95	42261	7.65	4.90	44491	7.48	1.92
BCSSTK32	71416	9.21	2.91	69361	10.26	2.34	69616	12.13	6.84	71939	12.06	2.36
BRACK2	20693	6.86	3.41	21152	7.54	3.33	20477	7.90	4.40	19785	8.07	3.42
CANT	323.0K	20.34	8.99	323.0K	22.39	5.74	325.0K	27.14	23.64	323.0K	26.19	5.85
COPTER2	32330	5.18	2.95	30938	6.39	2.68	32309	6.94	5.05	31439	7.25	2.73
CYLINDER93	198.0K	16.49	5.25	198.0K	18.65	3.22	199.0K	21.72	14.83	204.0K	21.61	3.24
4ELT	1826	0.82	0.76	1894	0.91	0.78	1992	0.92	0.95	1879	1.08	0.74
INPRO1	78375	10.40	2.90	75203	11.56	2.30	76583	13.46	6.25	78272	13.34	2.30
ROTOR	38723	12.94	5.60	36512	14.31	4.90	37287	15.51	8.30	37816	16.59	5.10
SHELL93	84523	36.18	10.24	81756	40.59	8.94	82063	46.02	16.22	83363	48.29	8.54
TROLL	317.4K	67.75	14.16	307.0K	74.21	10.38	305.0K	93.44	70.20	312.8K	89.14	10.81
WAVE	73364	20.87	8.24	72034	22.96	7.24	70821	25.60	15.90	71100	26.98	7.20

Table 2: Performance of various matching algorithms during the coarsening phase. *32EC* is the edge-cut of a 32-way partition, *CTime* is the time spent in coarsening, and *RTime* is the time spent in refinement.

	GR		KLR		BGR		BKLR		BKLGR	
	32EC	RTime	32EC	RTime	32EC	RTime	32EC	RTime	32EC	RTime
BCSSTK31	45267	1.05	46852	2.33	46281	0.76	45047	1.91	45991	1.27
BCSSTK32	66336	1.39	71091	2.89	72048	0.96	68342	2.27	69361	1.47
BRACK2	22451	2.04	20720	4.92	20786	1.16	19785	3.21	21152	2.36
CANT	323.4K	3.30	320.5K	6.82	325.0K	2.43	319.5K	5.49	323.0K	3.16
COPTER2	31338	2.24	31215	5.42	32064	1.12	30517	3.11	30938	1.83
CYLINDER93	201.0K	1.95	200.0K	4.32	199.0K	1.40	199.0K	2.98	198.0K	1.88
4ELT	1834	0.44	1833	0.96	2028	0.29	1894	0.66	1894	0.66
INPRO1	75676	1.28	75911	3.41	76315	0.96	74314	2.17	75203	1.48
ROTOR	38214	4.98	38312	13.09	36834	1.93	36498	5.71	36512	3.20
SHELL93	91723	9.27	79523	52.40	84123	2.72	80842	10.05	81756	6.01
TROLL	317.5K	9.55	309.7K	27.4	314.2K	4.14	300.8K	13.12	307.0K	5.84
WAVE	74486	8.72	72343	19.36	71941	3.08	71648	10.90	72034	4.50

Table 4: Performance of five different refinement policies. All matrices have been partitioned in 32 parts. *32EC* is the number of edges crossing partitions, and *RTime* is the time required to perform the refinement.

boundary policies is significantly less than that required by their non-boundary counterparts. The time of BGR ranges from 29% to 75% of the time of GR, while the time of BKLR ranges from 19% to 80% of the time of KLR. This seems quite reasonable, given that BGR and BKLR are simpler versions of GR and KLR, respectively. But surprisingly, BGR and BKLR lead to better edge-cut (than GR and KLR, respectively) in many cases. BGR does better than GR in 6 out of the 12 matrices, and BKLR does better than KLR in 10 out the 12 matrices. Thus, the quality of the boundary refinement policies is similar if not better than their non-boundary counterparts.

Even though BKLR appears to be just a simplified version of KLR, in fact they are two distinct schemes. In each scheme, a set of vertices from the two parts of the partition is swapped in each iteration. In BKLR, the set of vertices to be swapped from either part is restricted to be only along the boundary, whereas in the KLR it can potentially be any subset. BKLR performs better in conjunction with the HEM coarsening scheme, because for HEM the first partition of the coarsest graph is quite good (consistently better than the partition that can be obtained for other coarsening schemes such as RM and LEM), and it does not change significantly with each uncoarsening phase. Note that by restricting each iteration of KL on the boundary vertices, more iterations are needed for the algorithm to converge to a local minima. However, these iterations take very little time. Thus, BKLR provides the very precise refinement that is needed by HEM.

For the other matching schemes, and for LEM in particular, the partition of the coarse graph is far from being close to a local minima when it is projected in the next level finer graph, and there is room for significant improvement not just along the boundary. This is the reason why LEM requires the largest refinement time among all the matching schemes, irrespective of the refinement policy. Since boundary refinement schemes consider only boundary vertices, they may miss sequences of vertex swaps that involve non boundary vertices and lead to a better partition. To compare the performance of the boundary refinement policies against their non-boundary counterparts, for both RM and LEM, we performed another set of experiments similar to those shown in Table 4. For the RM coarsening scheme, BGR outperformed GR in 5 matrices, and BKLR outperformed KLR only in 5 matrices. For the LEM coarsening scheme, BGR outperformed GR only in 4 matrices and BKLR outperformed KLR only in 3 matrices.

Comparing BGR with BKLR we see that the edge-cut is better for BKLR for 11 matrices, and they perform similarly for the remaining matrix. Note that the improvement performed by BKLR over BGR is relatively small (less than 4% on the average). However, the time required by BKLR is always higher than that of BGR (in some cases upto four times higher). Again we see here that marginal improvements in the partition quality come at a significant increase in the refinement time. Comparing BKLGR against BKLR we see that its edge-cut is on the average within 2% of that of BKLR, while its runtime is significantly smaller than that of BKLR and somewhat higher than that of BGR.

In summary, when it comes to refinement policies, a relatively small decrease in the edge-cut usually comes at a significant increase in the time required to perform the refinement. Both the BGR and the BKLGR refinement policies require little amount of time and produce edge-cuts that are fairly good when coupled with the heavy-edge matching scheme. We believe that the BKLGR refinement policy strikes a good balance between small edge-cut and fast execution.

4.2 Comparison with Other Partitioning Schemes

The multilevel spectral bisection (MSB) [2] has been shown to be an effective method for partitioning unstructured problems in a variety of applications. The MSB algorithm coarsens the graph down to a few hundred vertices using random matching. It partitions the coarse graph using spectral bisection and obtains the Fiedler vector of the coarser graph. During uncoarsening, it obtains an approximate Fiedler vector of the next level fine graph by interpolating the Fiedler vector of the coarser graph, and computes a more accurate Fiedler vector using the SYMMLQ. The MSB algorithm computes the Fiedler vector of the graph using this multilevel approach. This method is much faster than computing the Fiedler vector of the original graph directly. Note that MSB is a significantly different scheme than the multilevel scheme that uses spectral bisection to partition the graph at the coarsest level. We used the MSB algorithm in the Chaco [19] graph partitioning package to produce partitions for some of the matrices in Table 1 and compared them against the partitions produced by our multilevel algorithm that uses HEM during coarsening phase, GGGP during partitioning phase, and BKLGR during the uncoarsening phase.

Figure 1 shows the relative performance of our multilevel algorithm compared to MSB. For each matrix we plot the ratio of the edge-cut of our multilevel algorithm to the edge-cut of the MSB algorithm. Ratios that are less than one indicate that our multilevel algorithm produces better partitions than MSB. From this figure we can see that for almost all the problems, our algorithm produces partitions that have smaller edge-cuts than those produced by MSB. In some cases, the improvement is as high as 60%. For the cases where MSB does better, the difference is very small (less than 1%). However the time required by our multilevel algorithm is significantly smaller than that required by MSB. Figure 4 shows the time required by the MSB algorithm relative to that required by our multilevel algorithm. Our algorithm is usually 10 times faster for small problems, and 15 to 35 times faster for larger problems. The relative difference in edge-cut between MSB and our multilevel algorithm decreases as the number of partitions increases. This is a general trend, since as the number of partitions increase both schemes cut more edges, to the limiting case in which $|V|$ partitions are used in which case all $|E|$ edges are cut.

One way of improving the quality of MSB algorithm is to use the Kernighan-Lin algorithm to refine the partitions (MSB-KL). Figure 2 shows the relative performance of our multilevel algorithm compared against the MSB-KL algorithm. Comparing Figures 1 and 2 we see that the Kernighan-Lin algorithm does improve the quality of the MSB algorithm. Nevertheless, our multilevel algorithm still produces better partitions than MSB-KL for many problems. However, KL refinement further increases the run time of the overall scheme as shown in Figure 4; thus, increases the gap in the run time of MSB-KL and our mul-

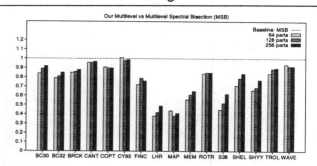

Figure 1: Quality of our multilevel algorithm compared to the multilevel spectral bisection algorithm. For each matrix, the ratio of the cut-size of our multilevel algorithm to that of the MSB algorithm is plotted for 64-, 128- and 256-way partitions. Bars under the baseline indicate that the multilevel algorithm performs better.

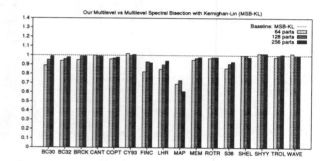

Figure 2: Quality of our multilevel algorithm compared to the multilevel spectral bisection algorithm with Kernighan-Lin refinement. For each matrix, the ratio of the cut-size of our multilevel algorithm to that of the MSB-KL algorithm is plotted for 64-, 128- and 256-way partitions. Bars under the baseline indicate that our multilevel algorithm performs better.

tilevel algorithm.

The graph partitioning package Chaco implements its own multilevel graph partitioning algorithm that is modeled after the algorithm by Hendrickson and Leland [20, 19]. This algorithm, which we refer to as Chaco-ML, uses random matching during coarsening, spectral bisection for partitioning the coarse graph, and Kernighan-Lin refinement every other coarsening level during the uncoarsening phase. Figure 3 shows the relative performance of our multilevel algorithms compared to Chaco-ML. From this figure we can see that our multilevel algorithm usually produces partitions with smaller edge-cut than that of Chaco-ML. For some problems, the improvement of our algorithm is between 10% to 50%. Again for the cases where Chaco-ML does better, it is only marginally better (less than 2%). Our algorithm is usually two to six times faster than Chaco-ML (Figure 4). Most of the savings come from the choice of refinement policy (we use BKLGR) which is usually four to six times faster than the Kernighan-Lin refinement implemented by Chaco-ML. Note that we are able to use BKLGR without much quality penalty only because we use the HEM coarsening scheme. In addition, the GGGP used in our method for partitioning the coarser graph requires much less time than the spectral bisection which is used in Chaco-ML.

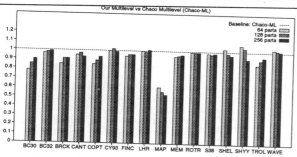

Figure 3: Quality of our multilevel algorithm compared to the multilevel Chaco-ML algorithm. For each matrix, the ratio of the cut-size of our multilevel algorithm to that of the Chaco-ML algorithm is plotted for 64-, 128- and 256-way partitions. Bars under the baseline indicate that our multilevel algorithm performs better.

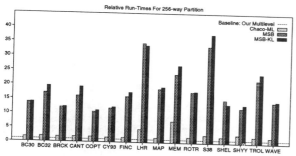

Figure 4: The time required to find a 256-way partition for Chaco-ML, MSB, and MSB-KL relative to the time required by our multilevel algorithm.

4.3 Sparse Matrix Ordering

The multilevel graph partitioning algorithm can be used to find a fill reducing ordering for a symmetric sparse matrix via recursive nested dissection. Let S be the vertex separator and let A and B be the two parts of the vertex set of G that are separated by S dering, A is ordered first, B second, while the vertices in S are numbered last. Both A and B are ordered by recursively applying nested dissection ordering. In our multilevel nested dissection algorithm (MLND) a vertex separator is computed from an edge separator by finding the minimum vertex cover [31]. The minimum vertex cover has been found to produce very small vertex separators.

Overall quality of a fill reducing ordering depends on whether or not the matrix is factored on a serial or parallel computer. On a serial computer, a good ordering is the one that requires the smaller number of operations during factorization. The number of operations required is usually related to the number of nonzeros in the Cholesky factors. The fewer nonzeros usually lead to fewer operations. However, since the number of operations is the square of the number of nonzeros, similar fills may have different operation counts. For this reason, all comparisons in this section are only in terms of the number of operations. On a parallel computer, a fill reducing ordering, besides minimizing the operation count, should also increase the degree of concurrency that can be exploited during factorization. In general, nested dissection based orderings exhibit more concurrency during factorization than minimum degree or-

derings [10, 27] that have been found to be very effective for serial factorization.

The minimum degree [10] ordering heuristic is the most widely used fill reducing algorithm that is used to order sparse matrices for factorization on serial computers. The minimum degree algorithm has been found to produce very good orderings. The multiple minimum degree algorithm [27] is the most widely used variant of minimum degree due to its very fast runtime.

The quality of the orderings produced by our multilevel nested dissection algorithm compared to that of MMD is shown in Figure 5. For our multilevel algorithm, we used the HEM scheme during coarsening, the GGGP scheme for partitioning the coarse graph and the BKLGR refinement policy during the uncoarsening phase. Looking at this figure we see that our algorithm produces better orderings for 11 out of the 18 test problems. For the other seven problems MMD does better. However, for many of these 7 matrices, MMD does only slightly better than MLND. The only exception is BCSPRW10 for which all nested dissection schemes perform poorly.

However, for the matrices arising in finite element domains, MLND does consistently better than MMD, and is some cases by a large factor (two to three times better for CANT, ROTR, SHEL, and WAVE). Also, from Figure 5 we see that MLND does consistently better as the size of the matrices increases and as the matrices become more unstructured. When all 18 test matrices are considered, MMD produces orderings that require a total of 702 billion operations, whereas the orderings produced by MLND require only 293 billion operations. Thus, the ensemble of 18 matrices can be factored roughly 2.4 times faster if ordered with MLND.

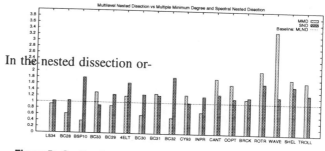

Figure 5: Quality of our multilevel nested dissection relative to the multiple minimum degree, and the spectral nested dissection algorithm. The matrices are displayed in increasing number of equations. **Bars above the baseline indicate that the MLND algorithm performs better.**

However, another, even more important, advantage of MLND over MMD, is that it produces orderings that exhibit significantly more concurrency than MMD. The elimination trees produced by MMD (a) exhibit little concurrency (long and slender), and (b) are unbalanced so that subtree-to-subcube mappings lead to significant load imbalances [26, 9, 14]. One the other hand, orderings based on nested dissection produce orderings that have both more concurrency and better balance [24, 16]. Therefore, when the factorization is performed in parallel, the better utilization of the processors can cause the ratio of the run time of parallel factorization algorithms running ordered using

MMD and that using MLND to be substantially higher than the ratio of their respective operation counts.

The MMD algorithm usually takes two to three times less time to order the matrices in Table 1 than the time required by MLND. However, efforts to parallelize the MMD algorithm have had no success [11]. In fact, the MMD algorithm appears to be inherently serial in nature. On the other hand, the MLND algorithm is amenable to parallelization. In [23] we present a parallel formulation of our MLND algorithm that achieves a speedup of 57 on 128-processor Cray T3D.

Spectral nested dissection (SND) [32] is a widely used ordering algorithm for ordering matrices for parallel factorization. As in the case of MLND, the minimum vertex cover algorithm was used to compute a vertex separator from the edge separator. The quality of the orderings produced by our multilevel nested dissection algorithm compared to that of the spectral nested dissection algorithm is also shown in Figure 5. From this figure we can see that MLND produces orderings that are better than SND for 17 out of the 18 test matrices. The total number of operations required to factor the matrices ordered using SND is 378 billion which is 30% more than the of MLND. Furthermore, as discussed in Section 4.2, the runtime of SND is substantially higher than that of MLND. Also, SND cannot be parallelized any better than MLND; therefore, it will always be slower than MLND.

References

[1] Stephen T. Barnard and Horst Simon. A parallel implementation of multilevel recursive spectral bisection for application to adaptive unstructured meshes. In *Proceedings of the seventh SIAM conference on Parallel Processing for Scientific Computing*, pages 627–632, 1995.

[2] Stephen T. Barnard and Horst D. Simon. A fast multilevel implementation of recursive spectral bisection for partitioning unstructured problems. In *Proceedings of the sixth SIAM conference on Parallel Processing for Scientific Computing*, pages 711–718, 1993.

[3] T. Bui and C. Jones. A heuristic for reducing fill in sparse matrix factorization. In *6th SIAM Conf. Parallel Processing for Scientific Computing*, pages 445–452, 1993.

[4] Tony F. Chan, John R. Gilbert, and Shang-Hua Teng. Geometric spectral partitioning (draft). Technical Report In Preparation, 1994.

[5] Chung-Kuan Cheng and Yen-Chuen A. Wei. An improved two-way partitioning algorithm with stable performance. *IEEE Transactions on Computer Aided Design*, 10(12):1502–1511, December 1991.

[6] C. M. Fiduccia and R. M. Mattheyses. A linear time heuristic for improving network partitions. In *In Proc. 19th IEEE Design Automation Conference*, pages 175–181, 1982.

[7] J. Garbers, H. J. Promel, and A. Steger. Finding clusters in VLSI circuits. In *Proceedings of IEEE International Conference on Computer Aided Design*, pages 520–523, 1990.

[8] A. George. Nested dissection of a regular finite-element mesh. *SIAM Journal on Numerical Analysis*, 10:345–363, 1973.

[9] A. George and J. W.-H. Liu. *Computer Solution of Large Sparse Positive Definite Systems*. Prentice-Hall, Englewood Cliffs, NJ, 1981.

[10] A. George and J. W.-H. Liu. The evolution of the minimum degree ordering algorithm. *SIAM Review*, 31(1):1–19, March 1989.

[11] Madhurima Ghose and Edward Rothberg. A parallel implementtaion of the multiple minimum degree ordering heuristic. Technical report, Old Dominion University, Norfolk, VA, 1994.

[12] J. R. Gilbert and E. Zmijewski. A parallel graph partitioning algorithm for a message-passing multiprocessor. *Internation Journal of Parallel Programming*, (16):498–513, 1987.

[13] John R. Gilbert, Gary L. Miller, and Shang-Hua Teng. Geometric mesh partitioning: Implementation and experiments. In *Proceedings of International Parallel Processing Symposium*, 1995.

[14] Anshul Gupta, George Karypis, and Vipin Kumar. Highly scalable parallel algorithms for sparse matrix factorization. Technical Report 94-63, Department of Computer Science, University of Minnesota, Minneapolis, MN, 1994. Submitted for publication in *IEEE Transactions on Parallel and Distributed Computing*. Available on WWW at URL ftp://ftp.cs.umn.edu/users/kumar/sparse-cholesky.ps.

[15] Lars Hagen and Andrew Kahng. A new approach to effective circuit clustering. In *Proceedings of IEEE International Conference on Computer Aided Design*, pages 422–427, 1992.

[16] M. T. Heath, E. G.-Y. Ng, and Barry W. Peyton. Parallel algorithms for sparse linear systems. *SIAM Review*, 33:420–460, 1991. Also appears in K. A. Gallivan et al. *Parallel Algorithms for Matrix Computations*. SIAM, Philadelphia, PA, 1990.

[17] M. T. Heath and P. Raghavan. A Cartesian nested dissection algorithm. Technical Report UIUCDCS-R-92-1772, Department of Computer Science, University of Illinois, Urbana, IL 61801, 1992. To appear in *SIAM Journal on Matrix Analysis and Applications*, 1994.

[18] Bruce Hendrickson and Rober Leland. An improved spectral graph partitioning algorithm for mapping parallel computations. Technical Report SAND92-1460, Sandia National Laboratories, 1992.

[19] Bruce Hendrickson and Rober Leland. The chaco user's guide, version 1.0. Technical Report SAND93-2339, Sandia National Laboratories, 1993.

[20] Bruce Hendrickson and Rober Leland. A multilevel algorithm for partitioning graphs. Technical Report SAND93-1301, Sandia National Laboratories, 1993.

[21] Zdenek Johan, Kapil K. Mathur, S. Lennart Johnsson, and Thomas J. R. Hughes. Finite element methods on the connection machine cm-5 system. Technical report, Thinking Machines Corporation, 1993.

[22] G. Karypis and V. Kumar. Multilevel graph partitioning schemes. Technical report, Department of Computer Science, University of Minnesota, 1995. Available on WWW at URL ftp://ftp.cs.umn.edu/users/kumar/mlevel_serial.ps.

[23] G. Karypis and V. Kumar. Parallel multilevel graph partitioning. Technical report, Department of Computer Science, University of Minnesota, 1995. Available on WWW at URL ftp://ftp.cs.umn.edu/users/kumar/mlevel_parallel.ps.

[24] George Karypis, Anshul Gupta, and Vipin Kumar. A parallel formulation of interior point algorithms. In *Supercomputing 94*, 1994. Available on WWW at URL ftp://ftp.cs.umn.edu/users/kumar/interior-point.ps.

[25] B. W. Kernighan and S. Lin. An efficient heuristic procedure for partitioning graphs. *The Bell System Technical Journal*, 1970.

[26] Vipin Kumar, Ananth Grama, Anshul Gupta, and George Karypis. *Introduction to Parallel Computing: Design and Analysis of Algorithms*. Benjamin/Cummings Publishing Company, Redwood City, CA, 1994.

[27] J. W.-H. Liu. Modification of the minimum degree algorithm by multiple elimination. *ACM Transactions on Mathematical Software*, 11:141–153, 1985.

[28] Gary L. Miller, Shang-Hua Teng, and Stephen A. Vavasis. A unified geometric approach to graph separators. In *Proceedings of 31st Annual Symposium on Foundations of Computer Science*, pages 538–547, 1991.

[29] B. Nour-Omid, A. Raefsky, and G. Lyzenga. Solving finite element equations on concurrent computers. In A. K. Noor, editor, *American Soc. Mech. Eng*, pages 291–307, 1986.

[30] R. Ponnusamy, N. Mansour, A. Choudhary, and G. C. Fox. Graph contraction and physical optimization methods: a quality-cost trade-off for mapping data on parallel computers. In *International Conference of Supercomputing*, 1993.

[31] A. Pothen and C-J. Fan. Computing the block triangular form of a sparse matrix. *ACM Transactions on Mathematical Software*, 1990.

[32] Alex Pothen, H. D. Simon, and Lie Wang. Spectral nested dissection. Technical Report 92-01, Computer Science Department, Pennsylvania State University, University Park, PA, 1992.

[33] Alex Pothen, Horst D. Simon, and Kang-Pu Liou. Partitioning sparse matrices with eigenvectors of graphs. *SIAM Journal of Matrix Analysis and Applications*, 11(3):430–452, 1990.

Optimal Circuit-Switched Routing in Hypercubes

Ausif Mahmood[1], Donald J. Lynch[2] and Roger B. Shaffer[2]

[1]Computer Engineering, University of Bridgeport, CT 06601 (email:mahmood@cse.bridgeport.edu)
[2]Washington State University at Tricities, Richland, WA 99352 (email:lynchd/rshaffer@beta.tricity.wsu.edu)

Abstract -- *In circuit-switched routing, the path between a source and its destination has to be established by incrementally reserving links before the data transmission can begin. If the routing algorithm is not carefully designed, deadlocks can occur in reserving these links. Deadlock-free algorithms based on dimension ordered routing, such as the E-cube, exist. Recently, adaptive, minimum-distance routing algorithms, such as the Turn Model and the UP Preference algorithms, have been reported. In this paper, we present a new class of adaptive, provably deadlock-free, minimum-distance routing algorithms. We prove that the algorithms developed here are optimally adaptive in the sense that any further flexibility in communication will result in deadlock. We show that the Turn Model is actually a member of our new class of algorithms that does not perform as well as other algorithms within the new class. We present both an analytical comparison of the flexibility and balance in routing provided by various algorithms, as well as a comparison based on uniform and non-uniform traffic simulations. The Extended UP Preference algorithm developed in this paper is shown to have improved performance with respect to existing algorithms.*

1. Introduction

Hypercubes, i.e., binary n-cube interconnection system, have been extensively studied in recent years [1-6]. Hypercubes have a regular structure with high connectivity and low diameter, allowing flexible and efficient communication between any two processors in the system, and the ability to tolerate faults in the system. These advantages have led to numerous hypercube multiprocessor implementations.

There are two methods of message communication in interconnection networks such as a hypercube: packet switched and circuit-switched schemes. In packet switched communication, a message is broken into packets and each packet is sent individually. In contrast, the circuit-switched scheme completely reserves the path from a source to its destination before message transmission begins. The links in the path are released only after the message transmission is completed. The circuit-switched scheme is preferred for transmission of long messages, and in applications where the

communication delay should be minimum after the source-destination (*s-d*) path has been established. There are two strategies for reserving links in the *s-d* path, *backtrack-and-retry*, and *reserve-and-hold*.

In the backtrack-and-retry strategy, when a partial path from a source node to an intermediate node, cannot be advanced towards the destination node because the required outgoing links from this intermediate node are busy, the entire reserved path is released. This strategy may lead to indefinite starvation of some message requests. In contrast, the reserve-and-hold strategy waits at an intermediate node until the partial reserved path can be advanced towards the destination node by reserving an appropriate outgoing link. Many commercially available hypercube systems (e.g., Intel's IPSC-860) use the reserve-and-hold strategy in implementing the circuit-switched mode of communication.

The reserve-and-hold strategy can lead to a deadlock state in reserving links. In the deadlock state, there is a set of messages for which no partial paths can be advanced because the required links are each reserved by other partial paths in the set. Various routing algorithms have been designed in the past to guarantee deadlock-free circuit-switched communication in a hypercube. In this paper, we focus on minimum-distance, adaptive, circuit-switched routing algorithms using the reserve-and-hold strategy.

2. Existing Hypercube Routing Algorithms

One popular, deadlock-free, circuit-switched routing method used in hypercubes is known as the *E-cube* algorithm [7]. It reserves the links in a strictly increasing (or decreasing) order of the dimensions in which the binary representation of the source and destination nodes differ. This algorithm is minimal in that the allowed paths take the smallest possible number of links between a source and its destination i.e., shortest Hamming distance. *E-cube* is considered to be oblivious or non-adaptive because there is only one allowed source-destination (*s-d*) path. Under *E-cube* routing in a hypercube, if a message is to be passed from node 2 to node 5, the only allowed path is $2 \rightarrow 3 \rightarrow 1 \rightarrow 5$. The *E-cube* routing algorithm has the advantage of simplicity. However, since there is only one prescribed path for a source-destination pair, much of the hypercube's flexibility is not utilized.

Dally and Seitz [8] introduced the concept of a channel dependency graph and showed that an algorithm for wormhole[a] routing can be deadlock-free only if its associated channel dependency graph is acyclic. They also introduced a methodology to develop deadlock-free routing algorithms. In their scheme, physical channels corresponding to cycles are split into a group of virtual channels, the virtual channels are ordered, and routing is restricted to visit channels in decreasing order to eliminate cycles in the channel dependency graph.

An adaptive, minimal routing algorithm has been presented in [9]. This algorithm has the drawback that it causes the node whose address is all ones to become a hotspot, creating a critical problem in system performance [6]. Another algorithm presented in [10] allows $k+1$ simultaneous disjoint paths from a source to a destination node that is a Hamming distance k from the source. A new algorithm, developed in this paper, achieves greater flexibility than any of the existing minimal routing algorithms, especially as the hypercube size increases. A number of non-minimal wormhole routing algorithms have been proposed in [12-15]. A non-minimal algorithm has the potential to cause infinite length paths known as livelocks and thus some mechanism is needed to avoid these. Although it has been shown that the chaos router reduces network congestion [13-14], the livelock prevention implementation has been difficult for circuit-switched applications [6].

In this paper, we limit our study to the class of minimal routing algorithms. In a minimal algorithm, when a message is routed from a source s to a destination d, the path chosen must involve only the dimensions in which s and d differ. Each link in the path changes a specific bit x_i in dimension i. If bit x_i changes from 0 to 1, then it is considered an up transition U_i. Similarly, a down transition D_i changes x_i from 1 to 0. Thus there are two sets of required transitions for every s-d pair; the set of up transitions $UT(s,d)$ and the set of down transitions $DT(s,d)$. These sets are formally described below.

$$UT(s,d) = \{i \mid 0 \leq i \leq (n-1), s_i=0, d_i=1\}$$
$$DT(s,d) = \{i \mid 0 \leq i \leq (n-1), s_i=1, d_i=0\}$$

Of the known existing minimal circuit-switched routing algorithms, the *Turn Model* [11] and the *UP* (or *DOWN*) *Preference* [6] algorithms have previously shown the greatest flexibility in providing deadlock-free paths from a source to its destination. To show how our new algorithms provide improved performance over the above algorithms, we describe the important features of each in the following paragraphs.

The *TURN Model* algorithm [17] is developed by analyzing the directions in which packets can turn in a network and the cycles they form. By removing the turns that cause cycles in the associated channel dependency graph, a deadlock-free routing is obtained. The *Turn Model* based routing algorithm operates in two phases:

PHASE 1 (down phase): *All down transitions required in the s-d path can be traversed in any order.*

PHASE 2 (up phase): *The up transitions are completed in any order.*

The *UP Preference* algorithm [6] is specified in terms of the order of the up and down transitions between any *s-d* pair. The algorithm specifies that any up transition U_i can be made at any point in the order of required transitions, but a down transition D_i can be made only after all lower dimension transitions have been completed. Table 1 shows an example of all possible paths from node 3 to node 4 and whether or not each path is allowed by the *UP Preference* algorithm. For comparison, Table 1 also shows whether or not a path is allowed by the *Turn Model* routing algorithm.

Table 1. An Example Showing Allowed Paths in the *Turn Model* and *UP Preference* Algorithms

PATH by node order	PATH by transition order	Allowed by *Turn Model*	Allowed by *Up Preference*
$3 \rightarrow 2 \rightarrow 0 \rightarrow 4$	$D_0 \rightarrow D_1 \rightarrow U_2$	Yes	Yes
$3 \rightarrow 2 \rightarrow 6 \rightarrow 4$	$D_0 \rightarrow U_2 \rightarrow D_1$	No	Yes
$3 \rightarrow 7 \rightarrow 6 \rightarrow 4$	$U_2 \rightarrow D_0 \rightarrow D_1$	No	Yes
$3 \rightarrow 1 \rightarrow 0 \rightarrow 4$	$D_1 \rightarrow D_0 \rightarrow U_2$	Yes	No
$3 \rightarrow 1 \rightarrow 5 \rightarrow 4$	$D_1 \rightarrow U_2 \rightarrow D_0$	No	No
$3 \rightarrow 7 \rightarrow 5 \rightarrow 4$	$U_2 \rightarrow D_1 \rightarrow D_0$	No	No

It has been proven in [6] that the *UP Preference* algorithm is deadlock-free because it yields acyclic channel dependency graph. Note that the *UP Preference* scheme is not optimal in terms of flexibility in allowed paths. For example, in a 3-D hypercube, the path $3 \rightarrow 1 \rightarrow 5 \rightarrow 4$ is not allowed by the *UP Preference* rule. However, if this is added to the set of allowed paths, the resulting channel dependency graph is still acyclic. As the next section shows, a new class of routing algorithms developed in this paper improves upon the algorithms of [6]. In fact, the set of paths allowed by one of the new algorithms is a superset of those allowed by the *UP Preference* rules. Thus we designate this new algorithm as the *Extended UP Preference* algorithm. Not all members of our new class of algorithms perform equally well. In particular, the *Turn Model* algorithm is a member of this new class which performs poorly because it creates congestion in routing traffic between *s-d* pairs, and it allows fewer paths in a hypercube than, for example, the *Extended Up Preference* algorithm.

---

[a] Wormhole routing is an extension of circuit-switching which allows incremental reserving and relinquishing of links and buffers in a network.

3. *Extended UP Preference* Algorithm

The *Extended UP Preference* algorithm is based on extending the deadlock-free routing in a 2-D hypercube, as defined by the *UP Preference* rule, to general *n*-D hypercubes. That is, the *Extended UP Preference* algorithm uniformly applies the *UP Preference* rule to each of the 2-D subcubes traversed in an *s-d* path within a larger hypercube. This results in a deadlock-free routing with much more flexibility than the original *UP Preference* algorithm. The reason for treating the overall routing in terms of 2-D subcube traversals is that a 2-D cube is the smallest network that contains physical loops, and the cycles in the channel dependency graph are related to these physical loops.

The channel dependency graph *D* defined in [8], for a given interconnection network *I* and routing function *R*, is a directed graph, $D = G(C,E)$. The vertices of *D* are the channels of *I*. The edges of *D* are the pairs of channels connected by *R* as,

$$E = \{(c_i, c_j) \mid R(c_i, n) = c_j \text{ for some } n \in N\}$$

where c_i and c_j are the channels in *I*, and *N* is the set of processing nodes in the interconnection network. Figure 1 shows the channel dependency graph for a 2-D hypercube without any restrictions in routing messages from a source to a non-adjacent destination. Each vertex in Figure 1 corresponds to a physical link or channel in the 2-D hypercube and is identified by the nodes that it connects. For example, the vertex labeled 1,3 corresponds to the channel from node 1 to node 3. An edge in the graph in Figure 1 exists if that path is allowed by the routing criterion, e.g., the edge labeled $1 \rightarrow 3 \rightarrow 2$ indicates that the path $1 \rightarrow 3 \rightarrow 2$ is allowed by the routing algorithm.

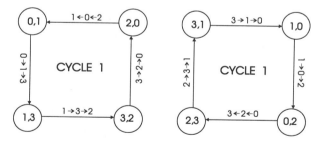

Figure 1. Channel Dependency Graph for the 2-D Hypercube with Unrestricted Routing

Since there are two cycles created in the channel dependency graph shown in Figure 1, *unrestricted* routing can deadlock. In order to remove the potential for deadlocks, at least one edge must be removed from each of the two cycles. If the top edges, labeled $2 \rightarrow 0 \rightarrow 1$ and $3 \rightarrow 1 \rightarrow 0$, are removed, then each of the two cycles in Figure 1 is broken. This results in one possible deadlock-free routing scheme in the 2-D

hypercube. Incidentally, this particular deadlock-free routing is the same as that allowed by the *UP Preference* rule in a 2-D hypercube.

Notice that there are sixteen possible deadlock-free routing solutions, depending upon which combination of edges is removed from the two cycles in the channel dependency graph for a 2-D hypercube. By uniformly applying one of these deadlock-free routing solutions to all the 2-D subcubes in a larger hypercube, a deadlock-free routing algorithm can be obtained. Section 3.1 elaborates on the various members of this new class of algorithms. One member of our new class of algorithms is obtained when the *UP Preference* rule on a 2-D hypercube is *extended* by uniformly applying it to all the 2-D subcubes in an *n*-D hypercube. The resulting routing is deadlock-free and has greater flexibility than the original *UP Preference* algorithm of [6].

Extended-UP Preference Routing Criterion

The *Extended-UP (Ex-UP) Preference* rule is equivalent to applying the standard *UP Preference* rules uniformly to all 2-D subcubes in a larger hypercube. Thus in the sequence of up and down links (transitions) traversed in the *s-d* path, the *UP Preference* rule must be satisfied for any two consecutive transitions in the path. For example, the path $3 \rightarrow 1 \rightarrow 5 \rightarrow 4$ $(D_1 \rightarrow U_2 \rightarrow D_0)$ which was not allowed by the *UP Preference* rule, is allowed under the *Ex-UP Preference* scheme because the consecutive 2-D transitions $D_1 \rightarrow U_2$ and $U_2 \rightarrow D_0$ are both individually allowed under *UP Preference,* and thus collectively $D_1 \rightarrow U_2 \rightarrow D_0$ (path $3 \rightarrow 1 \rightarrow 5 \rightarrow 4$) is legal under the *Ex-UP Preference* rule. The *Ex-UP Preference* rule is formally stated below.

For an *s-d* pair of nodes (X,Y), suppose a path has been setup to an intermediate node *I* as $(X, \ldots H, I, \ldots Y)$. In establishing the path from *H* to *I*, if a transition in dimension *prev_dim* is taken, then the following are set.

> If $I = X$, then *previous_transition* = UP, *prev_dim* = -1
> If an up transition occurred in the link from *H* to *I*
>   *previous_transition* = UP
>   else *previous_transition* = DOWN
> If (*previous_transition* = UP)
>   *lock_dimension* = -1 else *lock_dimension* = *prev_dim*

The following rules determine if the next link of dimension > *lock_dimension* can be reserved from *I*.

- A *j* dimension link from *I* can be reserved if $j \in UT(I,Y)$
- A *j* dimension link from *I* can be reserved if $j \in DT(I,Y)$ AND [$\{\exists m \mid m > j$ AND $m \in UT(I,Y)\}$ OR $\{\forall m, lock\_dimension \leq m < j$ AND $m \notin (UT(I,Y) \cup DT(I,Y))\}$]
- A *j* dimensional link from *I* can not be reserved if $j \notin (UT(I,Y) \cup DT(I,Y))$

Note that the *Extended* concept does not necessarily restrict a down transition in dimension i if the transitions in dimensions less than i remain, in order to proceed to the destination. Such a down transition can be taken in the *Ex-UP Preference scheme* (but not in *UP Preference*) if an up transition of dimension greater than i is remaining. In order to prevent deadlocks, all transitions whose dimensions are less than i are considered to be locked once a down transition in dimension i is taken. When an up transition of dimension higher than the *lock_dimension* occurs, it unlocks, i.e., allows all remaining transitions in any dimension. Theorem 1 formally demonstrates that the *Ex-UP Preference* scheme results in a deadlock-free routing.

Theorem 1: If an algorithm based on the *Ex-UP Preference* criteria is used to route messages in an n-D hypercube ($n \geq 2$), no deadlocks in routing can occur.

Proof: We use induction to prove this theorem.

<u>Base case</u>: $n = 2$. In a 2-D hypercube, the *Ex-UP Preference* rule disallows the paths $2 \rightarrow 0 \rightarrow 1$ and $3 \rightarrow 1 \rightarrow 0$. Since, the resulting channel dependency graph (Figure 2) is acyclic, the *Ex-UP Preference* rule based routing in a 2-D hypercube is deadlock-free.

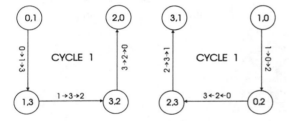

Figure 2. Channel Dependency Graph for the 2-D Hypercube under *Ex-UP Preference* Rule

<u>Inductive step</u>: We assume that the theorem holds for hypercubes of dimension 2,3, ... n, and show that it holds for a hypercube of dimension $n+1$.

Consider that the hypercube of dimension $n+1$ is partitioned along dimension n into two subcubes, N_0 and N_1, each of dimension n. All processor nodes in N_0 have addresses with a 0 in the nth dimension bit (i.e., $(n+1)^{th}$ bit). Similarly, the processor nodes in N_1 have addresses with a 1 in the most significant bit. Let C_0^n and C_1^n be the channel dependency graphs for N_0 and N_1 respectively corresponding to the *Ex-UP Preference* rule. By the induction hypothesis, C_0^n and C_1^n are acyclic.

When N_0 and N_1 are connected to form the $(n+1)$-D hypercube, there are 2^{n+1} additional links (channels) formed between the two n-D subcubes. Half of these, i.e., the up links from N_0 to N_1 are denoted as the set

$UL_{n+1} = \{(0,2^n), (1,2^n + 1), ... (2^n - 1, 2^{n+1} -1)\}$.

The other half corresponds to the down links from N_1 to N_0 forming the set of links,

$DL_{n+1} = \{(2^n,0), (2^n + 1,1), ... (2^{n+1} - 1, 2^n - 1)\}$.

The channel dependency graph for the $(n+1)$-D hypercube based on the *Ex-UP Preference* routing scheme is shown in Figure 3, where channel dependencies internal to C_0^n and C_1^n are omitted. The links (channels) in UL_{n+1} can appear as intermediate nodes in the channel dependency graph for a path connecting a processor node in N_0 to a processor node in N_1. However, since a path beginning in N_1 cannot continue as a path in N_0 under the *Ex-UP Preference* rule, the channels in DL_{n+1} cannot appear as intermediate nodes in the channel dependency graph for a path connecting a processor node in N_1 to a processor node in N_0. This condition results from the fact that a down transition in dimension n (bit $(n+1)$) can only be the last transition in the *Ex-UP Preference rule* as otherwise, it will lock all lower dimensions, and there is no higher up transition left to unlock these. Hence the channel dependency graph for the $(n+1)$-D hypercube is acyclic and deadlock-free. ∎

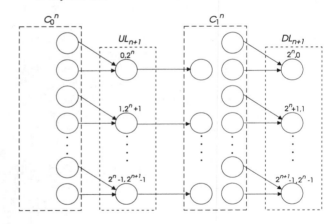

Figure 3. Channel Dependency Graph for the $(n+1)$-D Hypercube under *Ex-UP Preference* Rule

3.1 The Class of *Extended* Algorithms

Since the *Extended* concept applies a deadlock-free solution obtained for a 2-D hypercube to all the 2-D subcubes traversed in an *s-d* path in a larger hypercube, many deadlock-free routing algorithms can be devised. The unrestricted routing in a 2-D hypercube results in two independent cycles in the channel dependency graph, as shown in Figure 1. A deadlock-free routing algorithm in a 2-D hypercube must disallow these cycles by removal of one edge from each. Thus there are 16 possible deadlock-free routing schemes in a 2-D hypercube.

As shown earlier, the removal of the top edges in the two cycles in Figure 1 by disallowing paths $2\rightarrow0\rightarrow1$ and $3\rightarrow1\rightarrow0$ (shown graphically in Figure 4a), is equivalent to the *UP Preference* algorithm of [6] in a 2-D hypercube. If the bottom two edges ($1\rightarrow3\rightarrow2$ and $0\rightarrow2\rightarrow3$) in Figure 1 are removed, the *DOWN Preference* solution results. Similarly, the removal of the right and left edges in Figure 1 results in *Reverse UP Preference* (disallowing paths $3\rightarrow2\rightarrow0$ and $1\rightarrow0\rightarrow2$) and *Reverse DOWN Preference* (disallowing paths $0\rightarrow1\rightarrow3$ and $2\rightarrow3\rightarrow1$) algorithms, respectively. A *Reverse UP Preference* algorithm allows up transitions in any order, but a down transition can be taken only if all the higher transitions have been completed. Similarly, the *Reverse DOWN Preference* algorithm allows an up transition only if all higher dimension transitions have been completed. The disallowed paths in these four types of algorithms are graphically shown in Figure 4. All other *s-d* paths in Figure 4 are allowed.

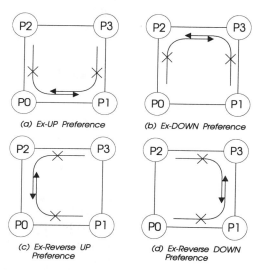

Figure 4. Disallowed Paths in the *Extended UP/DOWN Preference* Group

Figure 4 demonstrates that all the four different algorithms are closely related since they have the same pattern of disallowed paths. Further, by processor relabeling, one case can be obtained from the other. For example, if the processor address bits are all inverted in the *UP Preference* case, the *DOWN Preference* scheme results. Because of the close relationship of these four 2-D hypercube routing solutions, their performance when extended in a uniform manner to a larger hypercube will also be similar to each other. Therefore, only the *Extended UP Preference* case will be studied in detail.

Following similar reasoning, the remaining 12 deadlock-free solutions in the 2-D hypercube are also divided in three groups with four solutions in each group. One interesting case occurs when the bottom edge in cycle 1 and the left edge in cycle 2 of Figure 1 are removed by disallowing paths $1\rightarrow3\rightarrow2$ and $2\rightarrow3\rightarrow1$ respectively. This solution is equivalent to the *TURN Model* for a hypercube (i.e., *P-cube* of [11]). In this case, the communication between processor 1 and processor 2 is forced to go through processor 0. Under heavy traffic, this will cause a larger message flow through node 0 resulting in an artificial hot spot. For larger hypercubes, the *Turn Model* algorithm forces heavy traffic through nodes with low address labels and minimum traffic through nodes with high address labels, as will be seen in Sections 4 and 5. Section 4 also shows that the *TURN Model* algorithm is less flexible than the *Extended Up Preference* algorithm in terms of the total number of allowed paths.

3.2 Optimality of *Extended* Algorithms

We show that the *Extended* class of routing algorithms results in an optimal flexibility in the number of allowed paths in the sense that the addition of any new path to the allowed set of paths by an extended algorithm will result in deadlocks.

Theorem 2: For an *n*-D hypercube with $n\geq2$, the *Extended* algorithms are optimal in terms of the number of paths allowed while remaining deadlock-free.

Proof: Case 1, 2-D hypercube: All *s-d* paths of length 1 (link to neighbors) are allowed in a 2-D hypercube since they cannot cause a deadlock. For *s-d* paths of length 2, the unrestricted routing causes 2 independent cycles in the channel dependency graph as was shown in Figure 1. The *Extended* rule removes only one edge from each cycle to make the routing deadlock-free (Figure 2). Thus the algorithm provides optimal flexibility in routing in a 2-D hypercube.

Case 2, *n*-D hypercube ($n>2$): All paths of length l links ($l > 2$), are composed of a sequence of 2 link paths. Allowing a length l path ($l > 2$) implies that all of the 2-link paths it contains are also allowed. A length l path is disallowed in an *Extended* algorithm if any 2 link paths it contains are disallowed. Thus if a disallowed length l path ($l > 2$) is added to the set of allowed paths, then this path contains at least one disallowed 2-link path. This will allow the routing to deadlock as all the 2-D subcubes are already optimal in terms of allowed 2-link paths. ∎

As an example, in a 3-D hypercube, the path $3\rightarrow7\rightarrow5\rightarrow4$ is disallowed in the *Ex- UP Preference* rule. Allowing the path $3\rightarrow7\rightarrow5\rightarrow4$ will imply that the 2-link path $7\rightarrow5\rightarrow4$ is also allowed but that will cause a deadlock in the 2-D subcube where the most significant bit is always 1 (the path $7\rightarrow5\rightarrow4$ is disallowed by the *UP Preference* rule in this 2-D subcube). Further, it has also been proved recently in [16] that the *Turn Model* is optimal in the sense that it restricts the minimum number of turns to avoid deadlock. We have earlier shown that the *Turn Model* is one of the members of our new *Extended* class of algorithms. It is not useful in practice since it creates bottlenecks in routing traffic.

4. Analytical Comparison of Algorithms

One measure of the adaptivity provided by a minimum-distance routing algorithm is the count of all the allowed x-hop s-d paths in an n-D hypercube where $(2 \leq x \leq n)$. The 1-hop path count is not interesting in the comparison since every routing algorithm allows all possible 1-hop paths i.e., a path to the neighboring nodes. As a single quantitative measure of the adaptivity provided by a minimum-distance routing algorithm, we define the flexibility of an algorithm to be

$$Flexibility_{alg} = \frac{\left(\frac{N_{2h-alg}}{N_{2h-Ecube}} + \frac{N_{3h-alg}}{N_{3h-Ecube}} + \dots + \frac{N_{nh-alg}}{N_{nh-Ecube}} \right)}{n-1}$$

$$(1)$$

where N_{ih-alg} and $N_{ih-Ecube}$ are the number of allowed paths of length i-hops in a given algorithm and the *E-cube* algorithm, respectively. The flexibility as defined in (1) is a count of all allowed x-hop paths $(2 \leq x \leq n)$ in an algorithm, normalized by the path counts of the *E-cube* algorithm. The flexibility provided by the *E-cube* algorithm according to equation (1) is thus 1, indicating that the *E-cube* algorithm provides a single path from a source node to a destination node. A higher flexibility count is desirable for an algorithm since a greater number of alternate s-d paths would lessen congestion under heavy or uneven traffic.

In order to calculate the flexibility provided by an algorithm according to (1), counts of distinct allowed x-hop s-d paths must be determined. These counts can be computed recursively. The general formula for the x-hop path count in an n-D hypercube is given by (2).

$$N_{xh-alg}^{n\text{-}D} = N_{xh-alg}^{(n-1)\text{-}D} \cdot \left(\frac{2n}{n-x} \right)$$

$$(2)$$

where $(2 \leq x \leq n-1)$ and $n > 2$.

The superscripts in equation (2) indicate the dimension of the hypercube. Equation (2) states that for any minimum-distance routing algorithm, the x-hop count in an n-D hypercube can be obtained by multiplying the x-hop count in $(n-1)$-D hypercube by $\left(\frac{2n}{n-x} \right)$. Note that in unrestricted routing, the number of n-hop s-d paths (where destination is the complement of the source) in

an n-D hypercube is $2^n \cdot n!$. It can be verified that the number of $(n-1)$-hop s-d paths is $(2^n \cdot n!)/1!$, $(n-2)$-hop s-d paths are $(2^n \cdot n!)/2!$, and so on. In general, the x-hop s-d path count in unrestricted routing is given by (3).

$$N_{xh-unrest}^{n\text{-}D} = \frac{2^n \cdot n!}{(n-x)!}$$

$$(3)$$

Thus the ratio of x-hop path counts in an n-D and $(n-1)$-D hypercube is given by,

$$\frac{N_{xh-unrest}^{n\text{-}D}}{N_{xh-unrest}^{(n-1)\text{-}D}} = \frac{2n}{(n-x)}$$

$$(4)$$

Similar reasoning shows that for any restricted minimum-distance routing algorithm which scales uniformly according to the dimension of the hypercube, the increase in the number of allowed x-hop s-d paths from $(n-1)$-D to n-D hypercube will also be given by the factor of $2n/(n-x)$.

Returning to equation (2), the starting point for building the recursion is the count of all the allowed 2-hop s-d paths in a 2-D hypercube. Since equation (2) is valid only up to $(n-1)$-hop counts, the expressions for n-hop s-d path counts allowed by different algorithms are needed to compute their flexibility. We discuss the n-hop count expressions for each of the algorithms of interest in the following paragraphs.

An n-hop s-d path in a minimum-distance routing algorithm traverses all dimensions in linking a source to its destination. Thus for each of the 2^n nodes in an n-D hypercube, each source node has only one destination node that is n hops away. The n-hop path count is then the aggregate of all the paths allowed by an algorithm in establishing a link from each node in the hypercube to its complement node. The *E-cube* algorithm, being non-adaptive, allows only one such s-d path per node. Thus there is a total of 2^n n-hop s-d paths in an n-D hypercube for the *E-cube* algorithm denoted by $N_{nh-Ecube}^{n\text{-}D}$.

$$N_{nh-Ecube}^{n\text{-}D} = 2^n$$

$$(5)$$

The total number of allowed n-hop s-d paths in the *UP Preference* and the *Turn Model* is similar and given by equation (6).

$$N_{nh-UP-Preference}^{n\text{-}D} = N_{nh-Turn}^{n\text{-}D} = (n+1)!$$

$$(6)$$

The expression for the n-hop s-d path counts allowed in the *Ex-UP Preference* algorithm is given recursively by (7).

$$N_{nh-Ex-UP-Pref}^{n\text{-}D} = 2 \cdot N_{nh-Ex-UP-Pref}^{(n-1)\text{-}D} + \sum_{i=1}^{n-1} \left[\frac{(n-1)!}{(i-1)!(n-i)!} \right] N_{ih-Ex-UP-Pref}^{i\text{-}D}$$

$$(7)$$

where $N_{1h-UP-Preference}^{0\text{-}D} = 1$.

The n-hop s-d path count in unrestricted routing (which can deadlock) is given by equation (8).

$$N_{nh-unrest}^{n-D} = (n!)(2^n) \qquad (8)$$

Flexibility, as defined by equation (1), can now be computed for different algorithms in this study. The $\log_2$ plot of the flexibility of different algorithms versus hypercube sizes is shown in Figure 5. The *Ex-UP Preference* algorithm is approximately 100 times more adaptive than the *UP Preference* and *Turn Model* algorithms for a hypercube size of 20. The *E-cube* algorithm has a flexibility of 1 for all hypercube sizes, and so is not shown in Figure 5. The flexibility provided by unrestricted routing (which does deadlock) is also plotted in Figure 5 for comparison with the deadlock-free routing algorithms.

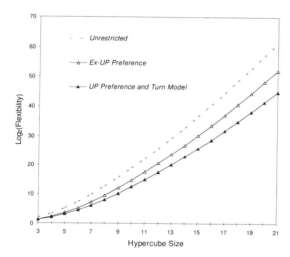

Figure 5. Comparison of Flexibility of Different Minimum-Distance Routing Algorithms

Although the *UP Preference* and the *Turn Model* based algorithms provide the same count of x-hop s-d paths and thus the same flexibility as defined by equation (1), their routing paths are different. As discussed in Section 3, the *Turn Model* forces more messages through certain nodes and creates artificial hotspots in routing. Thus the flexibility count alone is not a complete indicator of an algorithm's performance. A good algorithm should provide high flexibility and also balance the traffic through different nodes in the hypercube.

In order to quantitatively measure traffic balance under a routing algorithm, we analyze the Intermediate Node Traffic Count (*INTC*) through each node in the hypercube. Suppose the allowed paths between an s-d pair P-Q are, P→X→Y→Q and P→X→W→Q. Because the message has to pass through node X to get from P to Q, the *INTC* for X is 1. However, to proceed further towards Q, the message could either go through

Y or W. Hence the *INTC* for Y and W is 0.5. Similarly, if the routing algorithm allowed an additional s-d path from P to Q as P→L→M→Q, the *INTC* would become 2/3 for node X, and 1/3 for nodes L, M, W and Y.

By considering all s-d pairs for a given size hypercube and enumerating the paths allowed by a routing algorithm, the *INTC* through each node can be computed. The standard deviation of *INTC* for all nodes is then a good measure of the *balance* in routing provided by an algorithm. Table 2 lists the flexibility (as given by equation (1)) and the standard deviation of *INTC* for hypercube sizes of up to 7. The desirable attributes in an adaptive routing algorithm are high flexibility and low standard deviation of *INTC*. Table 2 shows that the *Ex-UP Preference* has the highest flexibility and a relatively low standard deviation of *INTC*. Although *E-cube* has 0 standard deviation, its performance will be limited as its flexibility is 1. The mean of *INTC*, depends only on the hypercube size, and is independent of the algorithm.

Table 2. Flexibility and Standard Deviation of *INTC* for Different Algorithms

Hyper-cube Size	UP Preference (Flex., S.D.)	Turn Model (Flex., S.D.)	Extended-UP Preference (Flex., S.D.)	Unrestricted (deadlocks) (Flex., S.D.)
3	(2.25, 2.29)	(2.25, 3.51)	(2.38, 2.19)	(4, 0)
4	(4, 7.84)	(4, 12.45)	(4.71, 7.18)	(10.67, 0)
5	(8.62, 23.47)	(8.62, 38.1)	(11.98, 20.6)	(38, 0)
6	(22.65, 64.9)	(22.6, 107)	(38.86, 54.5)	(174.40, 0)
7	(71.4, 170.4)	(71.4, 286)	(156, 137.4)	(985.33, 0)

Despite the fact that the *UP Preference* and the *Turn Model* routing algorithms have identical flexibility counts, the standard deviation of *INTC* for *Turn Model* is much higher indicating its poor performance. The traffic simulations in the next section confirm this result.

5. Simulation Study of Routing Algorithms

A discrete event simulator was developed to study the performance of various routing algorithms on a hypercube multiprocessor. The simulator models the hypercube by two unidirectional links between all neighboring nodes, i.e., one outgoing link and one incoming link, to and from each neighbor, respectively. Each node contains a message queue and an associated router controller implementing the routing algorithm being studied. It is assumed that it takes one time unit to reserve a link if it is available. Once all links between a source and its destination have been established, the message transmission takes place, consuming an amount of time proportional to the size of the message. All of the links involved in a transmission are released when the message transmission is completed.

In our study, the message transmission times can vary between 100 and 900 time units. Both a uniform distribution of message lengths (with a mean of 500), and a Poisson distribution of message lengths (with a mean of 250 time units) are included in the simulation model. The Poisson distribution allows fewer long messages, relatively, than the uniform distribution. It has been used in related simulation studies of routing schemes e.g., [6] and represents practical applications in a circuit-switched hypercube. In gathering statistics, 16,000 messages are generated in an 8-D hypercube. Of these, the first 3,000 and the last 3,000 messages are ignored in order to insure that routing performance results are independent of initial loading and final unloading effects. The choice of an 8-D hypercube and 16,000 total messages allows the simulation to execute in a reasonable amount of CPU time with stable results.

The simulator keeps track of all nodes that are busy in originating a message in every simulation time unit. The percent of busy nodes averaged over the simulation time interval is defined in this work as the "traffic load". This load can be adjusted indirectly by changing the periodic message injection rate into the system. For circuit switched communication, the setup time i.e., the time involved to reserve the links between an s-d pair is directly proportional to the overall communication throughput achievable. The average message size in the system affects this setup time since the links in the reserved path are occupied for that duration of time. For a given traffic type, the simulator records the setup time for each message. The average setup time for the 10,000 messages is then normalized by dividing it with the average message size.

Figures 6 and 7 show the traffic simulation results for the bit-reverse and bit-transpose traffic, respectively. In the bit-reverse traffic, a source node with binary address $(x_{n-1}, x_{n-2}, ... x_0)$ sends a message to the destination node with address $(x_0, x_1, ... x_{n-1})$. In the bit-transpose traffic, the destination is selected to be the node with address $(x_{n/2-1}, x_{n/2-2}, ... x_0, x_{n-1}, x_{n-2}, ... x_{n/2})$. Both bit-reverse and bit-transpose traffics occur in many practical computations and can cause worst-case behavior in oblivious routers for hypercubes [17]. Figures 6 and 7 show the superior performance of the *Ex-UP Performance* algorithm as compared to other algorithms studied. The *Ex-UP Preference* algorithm performs even better relatively when the message lengths are uniformly distributed or when a larger mean for message lengths is used.

Uniform traffic, bit complement traffic, geometric traffic were also studied for the different algorithms. The *Ex-UP Preference* algorithm performs better than other algorithms for all these traffic types. The relative results are similar to those of Figures 6 and 7.

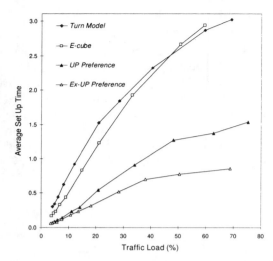

Figure 6. Comparison of Algorithms Under Bit-Reverse Traffic with Poisson Distribution of Message Lengths.

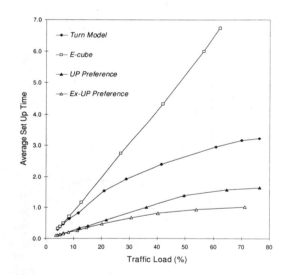

Figure 7. Comparison of Algorithms Under Bit-Transpose Traffic with Poisson Distrib. of Message Lengths.

6. Conclusions

A new class of minimum-distance deadlock-free routing algorithms has been introduced in this paper. A member of this new class, termed as the *Ex-UP Preference* algorithm, is shown to be optimal in terms of the number of allowed s-d paths with superior performance in comparison to existing algorithms. An existing algorithm, known as the *Turn Model,* also belongs to our new class of algorithms. However, it does not perform as well as other members of the new class, because it creates congestion in routing the traffic.

An analytical measure for judging the performance of adaptive minimum-distance routing algorithms has also been developed in this paper. It takes into account the flexibility provided in routing as well as the balance of traffic load through different nodes in a hypercube network. This measure, when applied to the new algorithms developed in this paper and other existing algorithms, indicates the relatively high flexibility of the new *Ex-UP Preference* algorithm. Further, the *Ex-UP Preference* algorithm has low standard deviation of *INTC*, indicating a good balance in routing the traffic in a hypercube. Various types of traffic were simulated on a hypercube network employing different circuit switched routing algorithms. In all traffic simulations, the *Ex-UP Preference* algorithm achieves the lowest *s-d* path setup time of all existing algorithms, supporting the analytical comparisons.

Finally, although we have focused on the hypercube network in this paper, the approach followed here to develop optimal, minimum-distance, deadlock-free routing algorithms can be applied to any recursively defined network. By removing the fewest paths needed to eliminate cycles in the channel dependency graph in the smallest dimension network with physical loop(s), a deadlock-free routing can be obtained. By uniformly extending this routing to higher dimension networks, a deadlock-free routing algorithm can be obtained. Of the many possible, deadlock-free solutions, one that shows the best balance in routing the traffic should be chosen. Our future work lies in a similar direction, and also at extending the ideas presented here to wormhole routing.

Acknowledgments: This research was funded in part by a grant from the National Science Foundation Center for Design of Analog and Digital Integrated Circuits under grant #CDADIC 92-4.

We are thankful to Prof. C. S. Raghavendra for his suggestions in improving this paper.

7. REFERENCES

[1] Seitz, C. L., "The Cosmic Cube," *Communications of the ACM*, 28-1, July 1985, pp. 22-33.

[2] Saad, Y. and Shultz, M. H., "Topological Properties of Hypercubes," *Journal of Parallel and Distributed Computing*, 6, 1989, pp. 115-135.

[3] Choi, S. B., and Somani, A. K., "Rearrangeable Circuit-Switched Hypercube Architectures for Routing Permutations," *Journal of Parallel and Distributed Computing*, vol. 19, no. 2, October 1993, pp. 125-130.

[4] Chiu, G-M, and Wu, S-P, "Fault Tolerant Routing Strategy in Hypercube Systems," *Proc. of the 24th International Symposium on Fault Tolerant Computing*, June, 1994, pp. 382-391.

[5] Kim, J., and Das, C. R., "Hypercube Communication Delay with Wormhole Routing," *IEEE Transactions on Computers*, vol. 43, no. 7, July, 1994, pp. 806-814.

[6] Chiu, G-M, Chalasani, S., and Raghavendra, C. S., "Flexible Routing Criteria for Circuit-Switched Hypercubes," *Journal of Parallel and Distributed Computing*," 22, 1994, pp. 279-294.

[7] Sullivan, H. and Brashkow, T. R., "A Large Scale Homogeneous Machine," *Proc. of the Fourth Annual Symposium on Computer Architecture*," 1977, pp. 105-124.

[8] Dally, W. J., and Seitz, C. L., "Deadlock-Free Message Routing in Multiprocessor Interconnection Networks," *IEEE Transactions on Computers*, vol. c-36, no. 5, May, 1987, pp. 547-553.

[9] Konstantinidou, S., "Adaptive, Minimal Routing in Hypercubes," *Proc. of the 6th MIT Conference on Advanced Research in VLSI*, 1990, pp. 139-153.

[10] Li, Q., "A Dual-Channel Binary Hypercube Network," *Proc. of the 13th International Conference on Computers and Communications*, 1994, pp. 268-274.

[11] Glass, C. J. and Ni, L. M., "The *Turn Model* for Adaptive Routing," *Proc. of the 19th Symposium on Computer Architecture*, May 1992, pp. 278-287.

[12] Kim, J. and Shin, K. G., "Deadlock-Free Fault Tolerant Routing in Injured Hypercubes," *IEEE Transactions on Computers*, vol. 42, no. 9, Sept. 1993, pp. 1078-1088.

[13] Konstantinidou, S. and Snyder, L., "The Chaos Router: A Practical Application of Randomization in Network Routing," *Proc. of the 2nd Annual Symposium on Parallel Algorithms and Architectures*, 1990, pp. 21-30.

[14] Bolding, K., Fulgham, M. L. and Snyder, L., "The Case for Chaotic Adaptive Routing," *Technical Report* CSE-94-02-04, University of Washington, Seattle, 1994.

[15] Ngai, J. N. and Seitz, C. L. "A Framework for Adaptive Routing in Multicomputer Networks," *Proc. of the 1989 ACM Symposium on Parallel Algorithms and Architectures*, 1989, pp. 1-9.

[16] Lin, C-C and F-C Lin, "Minimal Turn Restrictions for designing Deadlock-Free Adaptive Routing," *Proc. of the 6th IEEE Symposium on Parallel and Distributed Processing*, Oct. 1994, pp. 680-687.

[17] Fulgham, M. L., "Performance of Chaos and Oblivious Routers Under Non-uniform Traffic," *Technical Report* CSE-93-06-01, University of Washington, Seattle, 1993.

Unicasting in Faulty Hypercubes Using Safety Levels

Jie Wu

Department of Computer Science and Engineering
Florida Atlantic University
Boca Raton, FL 33431
jie@cse.fau.edu

Abstract – *We consider a distributed unicasting algorithm for hypercubes with faulty nodes (including disconnected hypercubes) using the safety level concept. The safety level of each node in an n-dimensional hypercube is an approximated measure of the number and distribution of faulty nodes in the neighborhood and it can be easily calculated through n − 1 rounds of information exchange among neighboring nodes. Optimal unicasting between two nodes is guaranteed if the safety level of the source node is no less than the Hamming distance between the source and the destination. The feasibility of an optimal or suboptimal unicasting can be easily determined at the source node by comparing its safety level, together with its neighbors' safety levels, with the Hamming distance between the source and the destination. The proposed scheme is also the first attempt to address the unicasting problem in disconnected hypercubes.*

1   Introduction

With its numerous attractive features, the binary hypercube has been one of the dominating topological structures for distributed-memory systems. Efficient interprocessor communication is a key to the performance of a hypercube system. *Unicasting* is a one-to-one communication between a source and a destination. Unicasting in fault-free hypercubes has been extensively studied and surveys of unicasting scheme are presented in [3], [6].

As the number of processors in a hypercube system increases, the probability of processor failure also increases. There has been a number of fault-tolerant unicasting schemes proposed in previous work [1], [4], [7]. Most of these schemes assume that each node knows either only the neighbors' status or the status of all the nodes. A model that uses the former assumption is called *local-information-based*, while a model that uses the later assumption is called *global-information-based*. Local-information-models use a weaker but a more reasonable assumption; however, local information can only be used to achieve local optimization and most of approaches based on this

model are heuristic in nature. Therefore, the length of a routing path is unpredictable in general and global optimization, such as time and traffic in routing, is impossible. A global-information-based model can obtain an optimal or suboptimal result; however, it requires a complex process that collects global information. Normally global information is presented in a tabular format and it is not easy to use.

We propose here a novel concept called *safety level* which is an integer associated with each node in the system. Safety level is a concise representation of the distribution of faulty nodes in the system. It is also considered as a special type of *limited global information*, a compromise between local-information and global-information based approaches. Basically, each node in an *n*-cube is assigned a safety level k, where $0 \le k \le n$, and this node is called k-safe. A k-safe node indicates that there exists at least one Hamming distance path (i.e. the optimal path) from this node to any node within k distance. The safety level of each node can be calculated using a simple $(n-1)$-round iterative algorithm which is independent of the number and distribution of faults in the hypercube. An optimal unicasting between two nodes is guaranteed if the safety level of the source node is no less than the Hamming distance between these two nodes. The unicasting proposed in this paper can also be used in disconnected hypercubes, where nodes in the hypercube are separated into two or more parts. The feasibility of an optimal or suboptimal unicasting can by easily determined at the source node by comparing its safety level, together with its neighbors' safety levels, with the Hamming distance between the source and the destination.

In this paper, we also show that the safety level concept covers a larger set of safe nodes than both Lee-Hayes' [5] and Wu-Fernandez' [10] definitions and it can be used in unicasting in various faulty hypercubes (including disconnected hypercubes) more effectively. We prove that under both Lee-Hayes' and Wu-Fernandez' safe node definitions, the safe node set is empty for any disconnected hypercube; that is, the unicasting algorithms proposed by Lee-Hayes [5] and

Chiu-Wu [2] are not applicable to disconnected hypercubes. In separate papers, we have shown that the safety level concept can also be used to achieve optimization or suboptimization in multicasting [11] and broadcasting [9] in a faulty hypercube.

2 Notation and Preliminaries

The n-dimensional hypercube (or n-cube) Q_n is a graph having 2^n nodes labeled from 0 to $2^n - 1$. Two nodes are joined by an edge if their addresses, as binary integers, differ in exactly one bit position. More specifically, every node a has address $a_{n-1}a_{n-2}\cdots a_0$ with $a_i \in \{0,1\}$, $0 \le i \le n-1$, and a_i is called the ith bit (also called the ith dimension) of the address. We denote node a^i the neighbor of a along dimension i. Symbol $\oplus$ denotes the bitwise exclusive OR operation on binary addresses of two nodes. Let $e^k = e_{n-1}e_{n-2}\cdots e_0$ where $e_k = 1$ and $e_j = 0, \forall j \neq k$. For example, $1101 \oplus e^2 = 1001$. Clearly $a \oplus e^i$ represents setting or resetting the ith bit of a. The distance between two nodes s and d is equal to the Hamming distance between their binary addresses, denoted by $H(s,d)$.

A path connecting two nodes s and d is termed *optimal path* (also called *Hamming distance path*) if its length is equal to the Hamming distance between these two nodes. Clearly, $s \oplus d$ has value 1 at $H(s,d)$ bit positions corresponding to $H(s,d)$ distinct dimensions. These $H(s,d)$ dimensions are called *preferred dimensions* and the corresponding nodes are termed *preferred neighbors*. The remaining $n - H(s,d)$ dimensions are called *spare dimensions* and the corresponding nodes are *spare neighbors*. Clearly, an optimal path is obtained by using links at each of these $H(s,d)$ preferred dimensions in some order. For example, suppose $s = 0101$ and $d = 1011$ then $s \oplus d = 0101 \oplus 1011 = 1110$. Therefore, dimensions 3, 2, 1 are preferred dimensions and dimension 0 is a spare dimension; that is, nodes 1101, 0001, and 0111 are preferred neighbors and node 0100 is a spare neighbor of node $s = 0101$.

3 Safety Levels

In our approach, limited global information is captured in the safety level associated with each node. In a given n-cube, the safety level of each node ranges from 0 to n. The safety level associated with a node is an approximation of the number and distribution of faulty nodes in the neighborhood, rather than just the number of faulty nodes. Let $S(a) = k$ be the safety status of node a, where k is referred to as the level of safety, and a is called k-safe. A faulty node is 0-safe which corresponds to the lowest level of safety, while an n-safe node (also called a *safe node*) corresponds to

the highest level of safety. A node with k-safe status is called *unsafe* if $k \neq n$.

Definition 1: *The safety level of a faulty node is* 0. *For a nonfaulty node a, let $(S_0, S_1, S_2, ..., S_{n-1})$, $0 \le S_i \le n$, be the nondecreasing safety level sequence of node a's n neighboring nodes in an n-cube, such that $S_i \le S_{i+1}$, $0 \le i \le n-2$. The safety level of node a is defined as: if $(S_0, S_1, S_2, ..., S_{n-1}) \ge (0, 1, 2, ..., n-1)$[1], then $S(a) = n$ else if $(S_0, S_1, S_2, ..., S_{k-1}) \ge (0, 1, 2, ..., k-1) \land (S_k = k-1)$ then $S(a) = k$.*

The following iterative algorithm (GS) calculates the safety level of each node in an n-cube. For simplicity, we show here only the synchronous version of GS, although it can be implemented asynchronously. We assume that all nonfaulty nodes in Q_n have n as their initial safety levels. N is the set of all the nonfaulty nodes in Q_n. The selection of Δ, the number of iterations used in GS, will be discussed in the next section.

Algorithm GLOBAL_STATUS (GS)
{ Initially all nonfaulty nodes are n-safe,
    faulty nodes are 0-safe, and *round* = 1.}
  while round $\le \Delta$
    parbegin
      NODE_STATUS(a), $\forall a \in N$
    parend;
    round := round +1.

Procedure NODE_STATUS(a)
  at node a determine the nondecreasing status
  sequence of neighboring nodes $(S_0, S_1, S_2, ..., S_{n-1})$;
  if $(S_0, S_1, S_2, ..., S_{n-1}) \ge (0, 1, 2, ..., n-1)$
    then mark a as n-safe (or safe);
  if $(S_0, S_1, S_2, ..., S_{k-1}) \ge (0, 1, 2, ..., k-1) \land (S_k = k-1)$
    then mark a as k-safe.

Figure 1 shows the safety level of each node in a faulty 4-cube with four faulty nodes (represented as black nodes). Based on the safety level definition, the safety levels of all the nodes that have two (or more) faulty neighbors will be changed to 1 after the first round, as in the case for nodes $0001, 0010, 0111, 1011, 1101$ in Figure 1. That is, the effect of 0-safe status of faulty nodes will first propagate to their neighbors, then neighbors' neighbors and so on. For example, after the second round the safety levels of node 0101 change to 2, because this node has three 1-safe neighbors. The safety level of every node remains stable after two rounds and each value represents the safety level of the corresponding node.

Theorem 2: *The GS algorithm identifies a k-safe ($k \neq n$) node of an n-cube in k rounds, i.e., at the kth round this node reaches a stable status.*

---

[1] $seq_1 \ge seq_2$ if and only if each element in seq_1 is greater or equal to the corresponding element in seq_2.

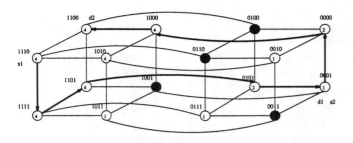

Figure 1: A 4-cube with four faulty nodes

Proof: We prove this theorem using mathematical induction on k, the safety level of a node. When $k = 1$, clearly there are at least two faulty (0-safe) neighboring nodes for any 1-safe node. This node can identify its safety level in one step. Suppose for all k-safe nodes, where $k \leq i$, exactly k rounds are required for these nodes to stabilize their status. Based on Definition 1 an $(i + 1)$-safe node can identify its status once all its neighbors, which have safety levels lower than $i + 1$, have the stable status. By the induction assumption, exactly after the ith round all those neighbors are stabilized, and it takes one extra step for this $(i + 1)$-safe node to be stabilized. □

Corollary : *To identify the status of all the nonfaulty nodes in any faulty hypercube (which might be a disconnected hypercube), the number of rounds (Δ in GS) is $n - 1$, where n is the dimension of the faulty hypercube.*

Definition 2 (Lee and Hayes [5]) *A nonfaulty node is unsafe if and only if there are at least two unsafe or faulty neighbors.*

Definition 3 (Wu and Fernandez [10]) *A nonfaulty node is unsafe if and only if one of the following conditions is true: (a) There are two faulty neighbors, or (b) There are at least three unsafe or faulty neighbors.*

In general, it is difficult, if not impossible, to compare the size of safe node sets because each safe node set depends on the distribution of faulty nodes. However, it is clear that, for each distribution of faulty nodes, the safe node set obtained using the definition in this paper contains the set using the definition in [10], which in turn contains the set using the definition in [5]. The safety level defined here provides more accurate information than the previous ones. Surprising, it takes fewer rounds ($n - 1$ by the above Corollary) to determine the safety level of each node than using both of the other definitions, which require $O(n^2)$ rounds of information exchanges in the worst case.

Theorem 3 [9]: *In a faulty n-cube with fewer than n faulty nodes, each nonfaulty but unsafe node has a safe neighbor.*

Theorem 4: *If the safety level of a node is k ($0 < k \leq n$), then there is at least one Hamming distance path from this node to any node within k Hamming distance.*

Proof: We prove this theorem using the mathematical induction on j, the distance between the source and the destination.

When $j = 1$, the source node can clearly reach all the neighbors, faulty and nonfaulty. Assume that this theorem holds true for all $j < i$. When $j = i$, based on the hypercube property that there are j node-disjoint optimal paths between two nodes separated by j Hamming distance, there are j preferred dimensions of a source node to any destination node which is j distance away. Based on the safety level definition and assuming that the safety level is $k (\geq j)$, the nondecreasing safety level sequence of source node's neighboring nodes satisfies the following condition: $(S_0, S_1, S_2, ..., S_{k-1}) \geq (0, 1, 2, ..., k-1) \wedge (S_k = k - 1)$. Therefore, among $j (\leq k)$ preferred neighbors, there is at least one neighbor whose safety level is at least $j - 1$. Since the distance between this neighbor and the destination node is $j - 1$ and based on the induction assumption, there exists an optimal path from the source node to the destination node. □

4 Unicasting in Faulty Hypercubes

The result of Theorem 4 provides a simple way to identify an optimal path between two nodes: An optimal path is generated by selecting a preferred neighbor with the highest safety level at each routing step. We consider two unicasting algorithms: one is an optimal algorithm in which a message is guaranteed to be forwarded to the destination node along an optimal path and the other is a suboptimal algorithm in which a message is forwarded to the destination node along a path with a length of the Hamming distance between the source and the destination plus two. The selection between the optimal and the suboptimal algorithms can be decided locally at the source node, using the following information: (1) The Hamming distance of the source s and the destination d. This can be done by computing $H(s, d) = |s \oplus d|$. The preferred neighbor sets and the spare neighbor sets are obtained based on $s \oplus d$. (2) The safety level of the source node $S(s)$. (3) The safety levels of the neighboring nodes of the source node $S(s^0)$, $S(s^1)$, $S(s^2)$, ..., $S(s^{n-1})$ for neighbors along dimensions, $0, 1, 2, ..., n - 1$. Note that safety levels of neighbors are available at each node and they are updated at each application of GS.

A *navigation vector*, $N = s \oplus d$, is introduced which is calculated at the source node and is passed to a selected neighbor after resetting or setting the corresponding bit, depending on whether this neighbor is a preferred one or a spare one. In the proposed unicasting algorithm, a preferred neighbor (of the source) that satisfies certain safety level requirement is selected and the corresponding algorithm is optimal. More specifically, the safety level requirement for optimal unicasting is as follows: the safety level of the source is at least equal to the Hamming distance between the source and the destination or a preferred neighbor's safety level is at least equal to the Hamming distance minus one. If there is no such a preferred neighbor, then a spare neighbor whose safety level is at least equal to the Hamming distance plus one (the condition for suboptimal unicasting) is selected and the corresponding algorithm is suboptimal. At each intermediate node, a preferred neighbor with the highest safety level is selected (for both optimal and suboptimal algorithms). Each intermediate node knows its preferred and spare neighbors upon receiving the unicast message with a navigation vector. A unicasting completes when the navigation vector becomes zero; that is, each bit is zero.

At the source node s with unicast message m and destination d, and at any intermediate node a with unicast message m and navigation vector N:

Algorithm Unicasting_at_source_node
$N = s \oplus d$; $H = H(s, d)$;
if $(S(s) \geq H) \vee (\exists i(S(s^i) \geq H - 1 \wedge N(i) = 1))$
then Optimal_unicasting:
    send $(m, N \oplus e^i)$ to s^i,
    where $S(s^i) = \max\{S(s^j)|N(j) = 1\}$
else if $\exists i(S(s^i) \geq H + 1 \wedge N_i = 0)$
    then Suboptimal_unicasting:
        send $(m, N \oplus e^i)$ to s^i,
        where $S(s^i) = \max\{S(s^j)|N(j) = 0\}$
    else failure

Algorithm Unicasting_at_intermediate_node
if $N = 0$
then stop
else send $(m, N \oplus e^i)$ to a^i,
    where $S(a^i) = \max\{S(a^j)|N(j) = 1\}$

Consider a unicasting example in the faulty 4-cube of Figure 1, where $s_2 = 0001$ and $d_2 = 1100$ are the source and the destination. In this case, the safety level of the source (which is 1) is less than the Hamming distance between the source and the destination (which is 3). However, there are two preferred neighbors (0000 and 0101) whose safety levels are 2 (which is the Hamming distance minus one). Therefore, optimal unicasting is still possible by selecting one of these two preferred neighbors, say 0000. The corresponding routing path $0001 \rightarrow 0000 \rightarrow 1000 \rightarrow 1100$ is shown in Figure 1. An optimal unicasting from node $s_1 = 1110$ to $d_1 = 0001$ is also shown in Figure 1.
Theorem 5: *Suppose the Hamming distance between the source and the destination is j for a given unicas-*

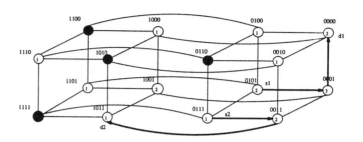

Figure 2: Unicasting in a disconnected 4-cube

ting. When the source is $k(\geq j)$-safe or there is an $l(\geq j-1)$-safe preferred neighbor of the source node, optimality is guaranteed using the proposed unicasting algorithm for any unicasting. When there is a $m(\geq j+1)$-safe spare neighbor of the source node, at least suboptimality is guaranteed.

This theorem can be directly derived based on Theorem 4 and the proposed unicasting algorithm.

The proposed unicasting algorithm can be applied to disconnected hypercubes. To our best knowledge, our approach is the first one that addresses routing in disconnected hypercubes. We first show an example of a disconnected hypercube and see how the proposed algorithm works in this case. Figure 2 shows a disconnected 4-cube with four faulty nodes: 0110, 1010, 1100, and 1111. Clearly, any unicasting initiated at node 1110 will fail. The source node detects this by checking its neighbors' safety levels and its own safety level. However, unicasting is possible if it is initiated from the other part of the partition. For example, in a unicasting with source $s_1 = 0101$ and destination $d_1 = 0000$, the Hamming distance between the source and the destination is 2 and the safety level of the source is 2. Therefore, optimal unicasting is possible and the corresponding path is shown in Figure 2. An example of suboptimal unicasting from $s_2 = 0111$ (with a safety level of 1) to $d_2 = 1011$ is also shown in Figure 2.

Theorem 6: *For Wu and Fernandez's safe node definition, the safe node set is empty for any disconnected hypercube.*

Proof: Assume that the nonfaulty node set of a given faulty n-cube is partitioned into two disjoint sets V and W. Moreover, any two nodes, one from each of V and W, are disconnected in the faulty n-cube.

Select a node v from V and assume that its address is 0000 (otherwise we can always remap nodes to make the address of v 0000). We then classify nodes in the n-cube into n levels. A node with i one bits in its address belongs to level i. Note that any node at level i has

i neighbors at level $i - 1$. Assume that level k is the lowest (closest to node v: 0000) level among the nodes in W, then k should be larger then 1; otherwise W and V are connected. Randomly select a node w from W at level k, there are at least $k \geq 2$ neighbors at level $k-1$. Based on the fact that W and V are disconnected and k is the lowest level among the nodes in W, all k neighbors at level $k - 1$ are faulty. Therefore, node w is unsafe. Since node w is randomly selected, all nodes at level k that belong to W are unsafe.

At level $k + 1$, again we randomly select a node w' in W. Node w' has $k + 1 \geq 3$ neighbors at level k, among them some are unsafe nodes in W and the rest are faulty. Based on the Wu-Fernandez's definition (Definition 3) of safe node, a node is unsafe if there are three unsafe or faulty neighbors, all the nodes at level $k + 1$ that are in W are unsafe. The same argument can be applied level by level until reaching the highest level among the nodes in W. In this way, we prove that all the nodes in W are unsafe. By interchanging the role of V and W, we can prove that all the nodes in V are unsafe.

A disconnected hypercube with multiple disjoint parts can be considered as a result of a series of partitions of a disconnected hypercube with two disjoint parts. Clearly, the above conclusion applies to any disconnected hypercube with multiple disjoint parts. □

Because for any given faulty hypercube the safe node set based on Wu-Fernandez' definition covers the one based on Lee-Hayes' definition and based on the result of Theorem 6, we conclude that the unicasting algorithms proposed by Lee-Hayes [5] and Chiu-Wu [2] are not applicable to any disconnected hypercube.

5 Conclusions

We have proposed a distributed unicasting algorithm for faulty hypercubes. The algorithm uses limited global information captured by a safety level associated with each node. The safety level can be calculated through simple $(n - 1)$ rounds of information exchanges among neighboring nodes in an n-cube. A source node can easily decide to perform either an optimal or a suboptimal unicasting, based on its safety level, its neighbors' safety levels, and the Hamming distance between the source and the destination, A source node can also identify cases when optimal and suboptimal paths are blocked by faulty nodes and when the corresponding unicasting tries to forward a message to another part in a disconnected faulty hypercube. The proposed approach is the first attempt to address reliable unicasting in disconnected hypercubes.

The proposed routing algorithm can also be used together with other heuristic and/or greedy routing algorithms, such as *randomized routing* [8] and *depth-first routing* [1]. Such a combination is especially efficient and useful in a system with many faults, which result in a relatively small percentage of safe nodes in the system. When the source and neighbors' safety levels are too low to use the proposed routing algorithm, a heuristic or greedy routing algorithm is first applied until an intermediate node with a sufficiently high safety level is reached; then the proposed routing algorithm is used to guide the message to the destination through an optimal path.

References

[1] M. S. Chen and K. G. Shin. Adaptive fault-tolerant routing in hypercube multicomputers. *IEEE Trans. on Computers*. 39, (12), Dec. 1990, 1406-1416.

[2] G. M. Chiu and S. P. Wu. Fault-tolerant routing strategy in hypercube multicomputers. *Proc. of the 24th International Symposium on Fault-Tolerant Computing*. June 1994.

[3] P. T. Gaughan and S. Yalamanchili. Adaptive routing protocols for hypercube interconnection networks. *Computer*. 26, (5), May 1993, 12-24.

[4] J. M. Gordon and Q. F. Stout. Hypercube message routing in the presence of faults. *Proc. of the 3rd Conference on Hypercube Concurrent Computers and Applications*. Jan. 1988, 251-263.

[5] T.C. Lee and J.P. Hayes. A fault-tolerant communication scheme for hypercube computers. *IEEE Transactions on Computers*. 41, (10), Oct. 1992, 1242-1256.

[6] L. M. Ni and P. K. McKinley. A survey of routing techniques in wormhole networks. *Computer*. 26, (2), Feb. 1993, 62-76.

[7] C. S. Raghavendra, P. J. Yang, and S. B. Tien. Free dimensions - an effective approach to achieving fault tolerance in hypercubes. *Proc. of the 22nd International Symposium on Fault-Tolerant Computing*. 1992, 170-177.

[8] L. Valiant. A scheme for fast parallel communication. *SIAM Journal on Computing*. 34, (1), May 1992, 350-361.

[9] J. Wu. Broadcasting in injured hypercubes using limited global information. TR-CSE-92-39, Dept. of Computer Science and Engineering, Florida Atlantic University, Nov. 1992.

[10] J. Wu and E. B. Fernandez. Reliable broadcasting in faulty hypercube computers. *Microprocessing and Microgramming*. 39, 1993, 43-53.

[11] J. Wu and K. Yao. Fault-tolerant multicasting in hypercubes using limited global information. To appear in *IEEE Transactions on Computers*.

Efficient Multi-Packet Multicast Algorithms on Meshes with Wormhole and Dimension-Ordered Routing

Luc De Coster[*†]   Natalie Dewulf[†]   Ching-Tien Ho[‡]

[†]K.U. Leuven – ESAT
Kard. Mercierlaan 94
B-3001 Heverlee, Belgium
luc.decoster@esat.kuleuven.ac.be

[‡]IBM Almaden Research Center
650 Harry Road
San Jose, CA 95120
ho@almaden.ibm.com

Abstract. We give efficient multicast algorithms on d-dimensional meshes with wormhole and dimension-ordered routing and with 1-port and asynchronous communication model. Previous algorithms either perform poorly for long messages or work only for the broadcast case, not multicast. In this paper, we develop some fundamental theories regarding the congestion-free property for a class of multicast algorithms that we define on meshes in the asynchronous environment. The class of algorithms contains the binomial tree algorithm and the Hamiltonian path algorithm as special cases. Moreover, it contains a new algorithm, which is optimal within the class and generally performs better than the combination of the two previous algorithms. We also implemented the new algorithm on the Intel Delta parallel system. The experimental result shows an improvement of 27% is obtained for multicasting a 128 kbyte message within 64 nodes using our algorithm than the best previous algorithm.

1   Introduction

Several parallel systems based on 2D/3D mesh or torus topologies are available, such as the Intel Delta and Paragon, MIT J-Machine, and Cray T3D. Many recent parallel systems of mesh, torus or hypercube topologies use wormhole and dimension-ordered routing. Efficient communication between processors is critical for good performance. Most multiprocessors support the point-to-point communication, or *unicast*, in hardware between any two, not necessarily adjacent, processors. Some multiprocessors also support

hardware *broadcast*. A general primitive which covers both unicast and broadcast is *multicast*, in which broadcast is performed within an arbitrary subset of processors. Besides covering the two extremes, multicast in its general form is also a very important primitive used in various applications [10].

There have been much previous work on broadcast and multicast on meshes and tori with various communication models. Most of the recent results assumed wormhole routing (e.g., [1, 2, 4, 6, 7, 11, 12, 14, 15]). Many of them also assumed dimension-ordered routing. Efficient *broadcast* algorithms on meshes, based on hybrids of algorithms, were given in [1, 2, 7]. However, these broadcast algorithms cannot be generalized to multicast algorithms without losing efficiency.

Efficient *multicast* algorithms for short messages on meshes were given in [11, 12]. The multicast algorithm by Lee [11] is not optimal. The algorithm by McKinley et al. [12] is based on a binomial tree constructed from the dimension-ordered list of the multicast set. The algorithm is optimal when the message size is of one unit. However, for a long message, this algorithm generally performs worse than a pipelined-based algorithm. In fact, to our knowledge, we do not know any prior multicast algorithm on meshes that works efficiently for any message size. In this paper, we give efficient *multicast* algorithms for an arbitrary message size on a d-dimensional mesh with wormhole and dimension-ordered routing.

First, we define the best[1] shape of the tree based on some simple recursion. Second, we order the multicast set according to a dimension ordering, starting from the source node, in a wraparound manner. Then, we map this list of nodes to the tree according to the *pre-order* traversal. We will define an appropriate schedul-

---

*L. De Coster is supported by Belgian National Fund for Scientific Research.

[1]Best within a class of algorithms defined later.

ing of receiving and sending packets at each node in the tree. Thus, each tree corresponds to an algorithm. We will show that the class of algorithms defined through the preorder mapping and packets scheduling is congestion-free (even in the asynchronous environment). Note that the class of algorithms that we define contains many known algorithms as special cases, such as the binomial-tree algorithm, the Hamiltonian-path algorithm and many multiple-level hybrid algorithms. The algorithm, which corresponds to the tree with the best shape, called the Fibonacci algorithm, generally outperforms all these previously known algorithms. Finally, due to the space limitation, we omit all the proofs and many illustrative examples and figures. See [5] for details.

2 Preliminaries

We use K to denote the message size (in bytes or any appropriate unit) to be multicast. All sets are assumed to be *ordered*. The *multicast set*, denoted S, is the (ordered) set of nodes involved in the multicast including the source node. Throughout, we let $S = \{D_0, D_1, \cdots, D_{n-1}\}$ in some order specified later and let the source node in S be D_s. We assume a d-dimensional mesh with wormhole and dimension-ordered routing, with one-port and asynchronous communication model, and without intermediate-node intercept capability. Let τ be the initial start-up time (overhead) and let t_c be the data transmission time per byte. We model the communication complexity for sending a K-byte message to another node as $T = \tau + K t_c$ when there is no congestion, regardless of the distance [8, 13].

In a multicast algorithm, a non-destination cannot be an end node of one routing step (and, for instance, use this node as a source node for later steps). It should be noted that a node cannot get a copy of its "by-passing" message, as this is the hardware constraint of most current machines

For the purpose of defining two notions of congestion-free, we define two different communication models: *synchronous* and *asynchronous*. The notions were first defined in [12]. In a *synchronous* communication model, the communications are synchronized step by step for the whole network, i.e., not just between a sender and its corresponding receiver. While in an *asynchronous* communication model, for the sake of defining message congestion, a very general communication behavior of a real system is covered as follows. (1) A communication, which occupies a path from a

source node to a destination node, can be arbitrarily long in time. (2) A sender of a communication can have an arbitrary delay after the communication completes, i.e., the path is released, before proceeding to the next send. (3) A receiver of a communication can have an arbitrary delay after the communication completes before proceeding to the next receive.

A communication algorithm is *stepwise congestion-free* if all edges are congestion-free in the *synchronous* communication model. A communication algorithm is *depth congestion-free* if all edges are congestion-free in the *asynchronous* communication model. A *snapshot* of a communication algorithm is the communication graph taken at any time instance.

3 Theory for a Class of Multicast Algorithms on Meshes

In this section, we develop the fundamental theory regrading congestion property for a class of multicast algorithms on meshes with ascending routing.

Definition 1 Let M be a d-dimensional mesh of form $\ell_0 \times \ell_1 \times \cdots \times \ell_{d-1}$. Each node can be assigned an address $(a_0, a_1, \cdots, a_{d-1})$ where $0 \le a_i < \ell_i$ for all $0 \le i < d$. $V(M)$ and $E(M)$ denote the set of nodes and edges, respectively, of the mesh M.

Definition 2 Given any two nodes $A = (a_0, a_1, \cdots, a_{d-1})$ and $B = (b_0, b_1, \cdots, b_{d-1})$ in a mesh M, the *ascending path* from A to B, denoted $P_+(A, B)$, is defined as follows:

$$A = (a_0, a_1, \cdots, a_{d-1}) \xrightarrow{0} (b_0, a_1, \cdots, a_{d-1}) \xrightarrow{1}$$

$$(b_0, b_1, a_2, \cdots, a_{d-1}) \xrightarrow{2} \cdots \xrightarrow{d-1} (b_0, b_1, \cdots, b_{d-1}) = B.$$

Here, we use the notation "$s \xrightarrow{i} d$" to denote the shortest path along the i-th dimension from s to d, where it is assumed that the addresses of s and d only differ in dimension i, if $s \ne d$.

We say that a routing is *ascending*, if the routing path between any two nodes, say from A to B, follows the ascending path $P_+(A, B)$.

Definition 3 ([12]) Given any two nodes $A = (a_0, a_1, \cdots, a_{d-1})$ and $B = (b_0, b_1, \cdots, b_{d-1})$, we say that A precedes B with respect to the *ascending order*, denoted $A <_+ B$, if and only if there exists a j, where $0 \le j < d$, such that $a_j < b_j$ and $a_i = b_i$ for all $j < i < d$.

Definition 4 We say $S = \{D_0, D_1, \cdots, D_{n-1}\}$ is an *ascending-ordered* set (or a set of *ascending-order*) if

$$D_0 <_+ D_1 <_+ \cdots <_+ D_{n-1}.$$

Definition 5 Given an ordered set $S = \{D_0, D_1, \cdots, D_{n-1}\}$, the *ascending-routed Hamiltonian path* of a set S, denoted $HP_+(S)$, is the set of paths defined as

$$\{P_+(D_i, D_{i+1}) | 0 \leq i \leq n-2\}.$$

Note that S is not necessarily an ascending-ordered set. The term "ascending-routed" is taken from the fact that each path $P_+(D_i, D_{i+1})$ in the set, which traverses two adjacent nodes in S, follows the ascending routing. Similarly, an *ascending-routed Hamiltonian cycle* of a set can be defined as follows.

Definition 6 Given an ordered set $S = \{D_0, D_1, \cdots, D_{n-1}\}$, The *ascending-routed Hamiltonian cycle* of a set S, denoted $HC_+(S)$, is the set of paths defined as

$$HP_+(S) \cup \{P_+(D_{n-1}, D_0)\}.$$

We say that two paths are edge-disjoint if and only if all of their composing edges are mutually edge-disjoint.

Theorem 1 *Let S be an ascending-ordered set and S' be some rotation of S. Then, all paths in $HC_+(S')$ are mutually edge-disjoint.*

We now first define a class of multicast algorithms $\mathcal{A}$ with respect to a rotation of an ascending-ordered set S on a d-dimensional mesh M with ascending routing. We then show, based on Theorem 1, that all algorithms belong to this class $\mathcal{A}$ are *depth* congestion-free.

For convenience, define an ordered set $S' = \{D'_0, D'_1, \cdots, D'_{n-1}\}$, where $D'_i = D_{(s+i) \bmod n}$ for all $0 \leq i \leq n-1$. We will packetize K into m packets, for a carefully chosen parameter m, each of size $\leq \lceil K/m \rceil$.

1. If $|S'| = 1$ then stop. Otherwise, partition the set S' into two subsets $S_1 = \{D'_0, D'_1, \cdots, D'_{h-1}\}$ and $S_2 = \{D'_h, D'_{h+1}, \cdots, D'_{n-1}\}$. (Here, different values of h will result in different algorithms.)

2. The source node D'_0 sends all the m packets to D'_h one by one. When the source node finishes sending m packets, it repeats Step 1 with $S' = S_1$ and $n = h$.

3. When node D'_h, the leader of S_2, receives the first packet, it initiates a multicast by repeating Step 1 with $S' = S_2$ and $n = n - h$. Note that the receiving of successive packets from node D'_0 continues, concurrently with the repetition of Step 1.

Note that the parameter m and the choice of h at each level (i.e., each time Step 1 is applied) determine a unique algorithm in this class. Clearly, both the binomial tree algorithm (where $h = \lceil n/2 \rceil$ and $m = 1$) and the pipelined Hamiltonian path algorithm (where $h = 1$ and optimal m derived in [8]) are special cases in this class.

Another alternative way to describe the class of algorithms $\mathcal{A}$ is to use a tree. Given a tree of n nodes, the ordered set of n nodes in S' is one-to-one mapped to the tree nodes by *preorder* traversal. The packets scheduling of each tree node is as follows. Each tree node, upon receiving a packet from its parent, starts propagating them to its rightmost child packet by packet in a pipelined manner. When it finishes sending all m packets to the rightmost child, it proceeds to the second rightmost child, and so on. Note that the partitioning of S' into S_1 and S_2 above corresponds to the partitioning of a tree into its rightmost subtree (S_2) and the remaining part (S_1). That is, the choice of h at all levels collectively determines a unique tree shape.

Consider S' as a logical linear array laid out from left to right. It is clear that all the communications (i.e., data flows) are from left to right. Note that a node can start propagating part of the multicast message it receives packet by packet in a pipelined manner. Following the divide-and-conquer definition of the algorithm, it can be easily shown that for any given snapshot of any algorithm in the class $\mathcal{A}$, there is no "overlapping" among the active paths on S'. The next lemma gives a more formal definition of this property.

Lemma 1 *For any given snapshot of any algorithm in the class $\mathcal{A}$, if there are two active paths (i, j) and (i', j') then either $i < j \leq i' < j'$ or $i' < j' \leq i < j$.*

Based on this lemma and Theorem 1, it can be shown that any algorithm in this class $\mathcal{A}$ is depth congestion-free.

Theorem 2 *Any algorithm in this class $\mathcal{A}$ is depth congestion-free.*

4 Multicast on Meshes

We now present the *Fibonacci tree* algorithm, which is based on the recursion of the generalized Fibonacci numbers. The Fibonacci tree algorithm is optimal within the class of algorithms $\mathcal{A}$. First, we need a definition.

Definition 7 Let $N(t, m)$ be the maximum number of nodes (including the source node) that can receive m multicast packets from the source node in t rounds based on an algorithm in the class $\mathcal{A}$.

With this definition, we can derive a simple recursion as follows.

Theorem 3

$$N(t, m) = \begin{cases} N(t - m, m) + N(t - 1, m), & if\ t \geq m \\ 1, & otherwise. \end{cases}$$

For any given m, the sequence $N(*, m)$ is in fact a generalized Fibonacci sequence. For instance, when $m = 1$, $N(t, 1) = 2^t$. When $m = 2$, $N(t, 2) = N(t - 1, 2) + N(t - 2, 2)$; $N(*, 2) = 1, 1, 2, 3, 5, 8, 13, ...$, the well-known Fibonacci sequence.

For the purpose of comparison, we will compare our Fibonacci tree algorithm against the better one of two naive multicast algorithms: *Hamiltonian path* algorithm and *binomial tree* algorithm. Note that both algorithms are also implemented with the depth congestion-free property by observing the class of algorithms $\mathcal{A}$. Also, pipelining is applied to the Hamiltonian path algorithm. (See [5] for details.)

Figure 1 gives a comparison among the two naive algorithms and the Fibonacci tree algorithm for multicasting a 64 kbyte message within a 64 node set. The measured times on the Intel Delta mesh are plotted on the top as a function of the logarithm of the number of packets. The estimated times are plotted similarly on the bottom. As expected, the Fibonacci tree algorithm gives the best performance for all m. The optimal m of the Fibonacci tree algorithm for the given n and K occurs around $2^4 = 16$ packets, each of 4 kbytes. Compared to the better one of the two naive algorithms (which occurs at $m = 1$ using the binomial tree algorithm), the Fibonacci tree algorithm improves by about 17% with our measurement for this case.

Figure 2 shows the percentage of improvement of the Fibonacci tree algorithm with respect to the better one of the two naive algorithms. We also implement the three algorithms on Delta for the data point $n = 64$ and $K = 128$ kbytes and measure an improvement of 27% over the better of the two naive ones.

5 Concluding Remark

We have shown that a class of multicast algorithms on meshes can be performed with depth congestion-free property even in the asynchronous environment. The idea is first to sort the multicast set according to the ascending or descending ordering, then to map the multicast set, starting from the source node in a wraparound manner, to any multicast tree with pre-order traversal. We have also given a new algorithm—the Fibonacci tree algorithm based on a recursion like generalized Fibonacci numbers. We also derived detailed theoretical comparisons among these algorithms, and confirmed them with experimental results on the Intel Delta. In particular, the Fibonacci tree algorithm improved the better of the two previous algorithms by 27% for multicasting a 128 kbyte message within 64 nodes. Finally, all multicast algorithms in the class $\mathcal{A}$ are deadlock-free even when there are multiple multicast instances possibly mixed with multiple unicast instances [3]. The main reason is that our multicast algorithms are implemented as user-level unicasts, i.e., each "packet" is in fact a user-level unicast message rather than a wormhole-level flit.

References

[1] M. Barnett, S. Gupta, D. G. Payne, L. Shuler, R. van de Geijn and J. Watts. Interprocessor Collective Communication Library (InterCom). In *Proceedings of the Scalable High-Performance Computing Conference*, pp. 357–364, May 1994.

[2] M. Barnett, D. G. Payne, R. van de Geijn and J. Watts. *Broadcasting on Meshes with Wormhole Routing*. Manuscript, 1994.

[3] R. V. Boppana, S. Chalasani and C. S. Raghavendra. On multicast wormhole routing in multicomputer networks, In *Proceedings of the Sixth IEEE Symposium on Parallel and Distributed Processing*, pp. 722–729, October 1994.

[4] J. Bruck, L. De Coster, N. Dewulf, C.-T. Ho and R. Lauwereins. On the design and implementation of broadcast and global combine using the postal model. In *Proceedings of the Sixth IEEE Symposium on Parallel and Distributed Processing*, pp. 594–602, October 1994.

[5] L. De Coster, N. Dewulf and C.-T. Ho. *Efficient Multi-Packet Multicast Algorithms on Meshes with Wormhole Routing and Dimension-Ordered*

Routing. IBM Research Report RJ 9937, February 1995.

[6] E. Fleury and P. Fraigniaud, Multicasting in Meshes, in *Proceedings of the 1994 International Conference on Parallel Processing*, Vol. III, pp. 151–158, August, 1994.

[7] S. E. Hambrusch, F. Hameed and A. Khokhar. *A Study of Coarse-Grained Communication Operations on Mesh Architectures.* Technical report, Purdue University, May 1994.

[8] C.-T. Ho and M.T. Raghunath. Efficient communication primitives on hypercubes. *Journal of Concurrency: Practice and Experience*, 4(6):427–457, September 1992.

[9] S. L. Johnsson and C.-T. Ho. Spanning graphs for optimum broadcasting and personalized communication in hypercubes. *IEEE Trans. Computers*, 38(9):1249–1268, September 1989.

[10] Y. Lan, A.-H. Esfahanian, and L. M. Ni. Multicast in hypercube multiprocessors. Technical report, Dept. of Computer Science, Michigan State Univ., July 1987.

[11] T. C. Lee. Conflict-free multicast in circuit-switched multiprocessor systems. Technical report, IBM RC 17567, September 1991.

[12] P. K. McKinley, H. Xu, A. Esfahanian and L. M. Ni, Unicast-Based Multicast Communication in Wormhole-Routed Networks, in *Proceedings of the 1992 International Conference on Parallel Processing*, Vol. II, pp. 10–19, August 1992.

[13] L. M. Ni and P. K. McKinley. A survey of wormhole routing techniques in direct networks. *IEEE Computers*, 26(2):62–76, February 1993.

[14] D. K. Panda, S. Singhal and P. Prabhakaran. Multidestination message passing mechanism confirming to base wormhole routing scheme. *Proceedings of Parallel Routing and Communication Workshop*, May 1994.

[15] D. F. Robinson, P. K. McKinley and B. H. C. Cheng, Optimal Multicast Communication in Wormhole-Routed Torus Networks, *Proceedings of the 1994 International Conference on Parallel Processing*, Vol. I, pp. 134–141, August 1994.

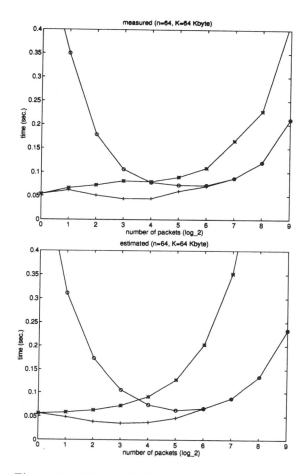

Figure 1: Measured (top) and estimated (bottom) times of the Hamiltonian path ('o'), the binomial tree ('*'), and the Fibonacci tree ('+') algorithms.

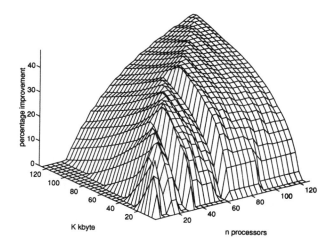

Figure 2: Improvement of the Fibonacci tree algorithm compared with the better one of the two naive algorithms.

PERFORMANCE ASSESSMENT OF LARGE ATM SWITCHING NETWORKS WITH PARALLEL SIMULATION TOOL

W. Liu, G. Petit[+] and E. Dirkx
Vrije Universiteit Brussel, Dept INFO
Pleinlaan 2
B-1050 Brussels, Belgium
email: erik@info.vub.ac.be
[+]Traffic Technology Dept., Bell Telephone Co.
F. Wellesplein 1
B-2018 Antwerp, Belgium
email: gpet@alcbel.be

Abstract -- *ATM switches should deliver satisfactory performance for various types of services. This paper describes a parallel simulation tool which can be used to evaluate the performance of ATM switching networks down to the sub-cell level and its implementation on a MIMD parallel computer. A combination of general purpose and problem specific algorithms results in a high computational efficiency, portability and scalability.*

1. INTRODUCTION

Asynchronous Transfer Mode (ATM) has been accepted widely due to advantages of efficiency and flexibility in a context of multirate, multimedia services. Many switch architectures have been proposed since ATM has been selected as transfer mode of Broadband Integrated Services Digital Network (BISDN) by the International Telecommunication Union (ITU). The performance of a switching network must be analysed by evaluation techniques so as to make sure that the designed system is able to guarantee the Quality Of Service (QOS) for all connections. On the other hand, such a network should be implemented in a cost effective way to facilitate chip design.

The functional range of analytic methods is restricted to a high degree of abstraction because of unrealistic assumptions and many approximations that are necessary to keep a model tractable. Simulation techniques are more flexible and can be applied to wide areas of applications in an arbitrary degree of detail. Nonetheless, it is very difficult to obtain reliable results from a simulation of a large ATM switching system without the support of huge computing power and efficient algorithm. Our solution is to decompose the network under study into a collection of executable components and execute them in parallel on a multicomputer system. The implementation of this

strategy on a MIMD parallel computer resulted in a general purpose, yet highly efficient software tool.

2. PERFORMANCE MODELLING

The large ATM switching networks are generally constructed by interconnecting switching modules as multistage interconnection network. Their architectures may differ from each other in the aspects of basic switching blocks, buffer location, interconnection of switching components, single or multiple paths, self- or label- routing, contention resolution mechanism, blocking or nonblocking, growing capability, fault-tolerance, etc. The ATM switches of interest to our work are scalable architectures based on modular approach. The network consists of multiple switching modules which can be arranged in several stages on demand. The internal buffer is used at switching element (SE) to solve cell output contentions. Output queueing and the shared buffer are applied for the incoming cells in a SE due to the high performance and efficient utilization of buffer memory [1]. A cell self-routing mechanism is supported so that each SE can make a very fast routing decision simply by inspecting the routing tag. Multiple paths are allowable for the cells belonging to the same connection. Hence, cell traffic load can be distributed over all available internal links. Multiple-path networks improve the performance and allow for the construction of very large switches from switch modules [2].

With the self-routing technique, a self-routing tag must be attached to each ATM cell according to its Virtual Path Identifier (VPI) and Virtual Channel Identifier (VCI). This tag points to the corresponding outlet for the cell destination. Together with the tag and other control information, an ATM cell has to be repacketized using an internal cell structure. This introduces conversion overhead. When arriving at a SE, a cell may be buffered temporarily to wait for the corresponding output links in the routing group to become free. The First-In First-Out (FIFO) discipline is applied to individual logical output queues.

According to the operating principles and the architectures, the ATM switching networks are modelled as a set of functional modules: the source, the line interface, the switching element, the routing, the cell resequencing, the data analysis module.

Cells loss rate is an important characteristic of an ATM network. The cell loss due to buffer overflow is closely related to the switch design. Cell congestion may cause buffer memory overflow, and thus result in the loss of incoming cells. A satisfactory size of buffer memory can only be determined by means of very detailed evaluation of the internal operation of the switching network.

Switching delay is another important performance index for multistage interconnection networks. The ATM technique multiplexes a diverse mix of sources on a single transmitting medium. The services with real time constraints, e.g. voice, are sensitive to the delay and delay variation. Again a measurement can be made based on very detailed evaluation of the internal operation of the switching network.

3. THE MPSR SWITCHES

The Multipath Self Routing (MPSR) switching fabric [3] is constructed with the topology of a Clos network. The architecture is extensible in a fully modular way, from a few up to 16384 external links. Its internal packet format is the Multi-Slot Cell (MSC), which consists of a train of equal-sized slots. The first slot contains a self-routing tag (SRT) and control information, the following slots contain the payload. Hence a MPSR switch is more flexible than a dedicated ATM switching system.

The fundamental building block of MPSR is called Integrated Switch Element (ISE), which is a 16*16 or 32*32 switching matrix integrated in a single VLSI chip. The ISE uses shared buffer: the MSCs to the same output port will be put in a logical queue linked by means of a list technique. Since all logical queues share the buffer memory in an ISE, each memory address can be allocated temporarily to any output port in case of demand. A Switch Module (SM) is a higher-level ATM switching matrix based on eight identical

ISEs divided in two sub-stages. Large switching networks are constructed from these SM Boards.

A Traffic Switching Unit (TSU) is formed by sixteen or thirty-two Termination LinK modules (TLK), and four Access Switches (AS), and may have up to 128 external links of 155 Mb/s. A switch plane is formed by a number of SMs arranged in stages, and its size may be up to 2048 links of 155 Mb/s. A MPSR switch may have a number of identical planes in parallel to increase the maximum allowable throughput. The largest MPSR switch consists of 128 TSUs and 16 switch planes, which involves three stages: the Access Switch (AS) stage, the Plane Switch 1 (PS1) stage, and the Plane Switch 2 (PS2) stage respectively. The full configuration of three-stage MPSR switch is shown in Figure 1.

4. CONCURRENT SIMULATION

In order to obtain detailed information about the performance indices of an ATM switch, the simulation must go down to very low level to measure the event occurrences. To simulate a large scale switching network based on this strategy, huge computing power and large memory space are required. Clearly, the most cost effective way is to carry out this type of simulations on MIMD parallel computers.

To implement a concurrent simulation, it is necessary to decompose the system under study into a set of processes that can been executed on several processors in parallel. The decomposition of an application can be done according to parallel heuristics such as pipelining, farming and/or data level parallelism [4]. Additionally several distributed simulation techniques have been proposed such as the Chandy-Misra algorithm [5] and the Time-Warp algorithm [6].

As cells traverse a switching network stage by stage towards their respective outputs, it is natural to treat such a switch fabric as a pipeline. This implies that individual stages of the switch may operate independently and simultaneously for a certain duration. Based on this heuristic, the SEs in a column form a stage of the pipeline, where arrival packets are flowing. Hence, a switch fabric is partitioned into a group of sequential processes with respect to the stages. Advantages of the partition involve functional consistency, balanced load, and low communication overhead.

The functional consistency stems from the fact that the SEs at a stage play the same routing strategy for the traffic. The SEs of a stage will either distribute incoming packets, or route incoming packets under a specific routing mode. Thus, this partitioning conforms to the manipulation relations of the switching networks.

Balanced load is achieved by distributing the processes with relatively equivalent granularity to a group of identical processors. The activating time of each process depends on the amount of data to be processed. For the ATM switching networks, in general, the quantity of arrival packets at each stage is almost equal (from a statistical viewpoint). Load balancing in different stages of the switching fabric therefore means a balanced execution time of the different processes at various stages of the multistage switch.

Interprocessor communications are overhead; the less they occur in a parallel environment, the higher the efficiency of the parallelism. For the simulation of multistage ATM switches, low communication overhead is assured by the partitioning mentioned earlier. To save data transmission and latency time, all data to be passed between stages at a given simulation clock time are sent as a single message.

ATM switches handle incoming traffic with very high speed, and in general there is a large amount of packets for each basic SE in any cycle. This implies that the arrival and departure events occur very frequently, even in moderate-size SEs. It is unlikely to exploit benefit from the conservative Chandy-Misra algorithm [5] for the simulation of large scale ATM switching networks due to the high density of event occurrences and communication overhead. The same reason also holds for Time Warp for this kind of application. There is no chance for a node to have some spare periods to schedule speculative events. With this approach, the cost of roll-back could be excessive for a model of an ATM switch with very heavy traffic and a lot of state information in each logical process.

Even though it is preferable to decompose a switching fabric according to stages, this partitioning may not suit large scale switching networks well because of the increasing computational and memory requirements per process. If the computation time of all processes is much more than the communication time, more processors would be needed to share the computation in order to reduce the simulation time. Also, the data structures can be distributed to more

processors. Therefore, further partitioning should be performed for very large scale switching networks.

To keep the load even and symmetrical on different processors, a substage could be divided into a power of 2 processes that can be mapped to the corresponding number of processors. This further partitioning is based on the situation that no message exchange is necessary between processors in vertical direction because each individual SE manages its shared buffer memory and packet streams independently. However, this further partitioning also increases the overhead of process communication. In order to maintain computing efficiency, such a partitioning should conform to the type of model under study.

5. SIMULATION PERFORMANCE

A 2048*2048 MPSR switch, with external ATM links operating at 155.52 Mb/s, has been simulated by means of the simulation tool on a transputer based multicomputer. A single simulation run for the switch takes about twelve days in total with offered load of 0.8 Erlang and five million time-slots. Using an independent uniform traffic pattern, about 1314 million ATM cells were generated during the whole simulation process. Figure 2 gives the configuration of the processor network for carrying out this simulation. From the statistical results, the size of a shared buffer memory is obtained which ensures the cell loss ratio is below a quantile of 10^{-6}. Also, the cell switching delay was derived at the same quantile. This means that the probability that the delay in the switch is larger than a certain value is smaller than 10^{-6}.

Even though the simulation run took a long time, the speedup is almost linear with the number of the processors. It was found that during execution of the simulation, computation time dominates communication time. This implies that the efforts to minimize the communication cost by overlapping the communications among different processors and communication with computation on individual processor have been very successful.

6. CONCLUSION

A concurrent simulation tool to evaluate the performance of a class of ATM switching networks has been developed and implemented. To speedup the simulation process, the problem was decomposed to a set of concurrent processes which are distributed over a network of processors. With a very high level of detail in the simulation, the benefits from parallelism are significant (and necessary). From the experiments, we can conclude that MIMD multicomputers are quite suitable to the problems of evaluating performance for large high speed ATM switching systems.

A current limitation of the simulation tool is that the number of processors required depends on the configuration of a switching fabric. For different switching networks, the user may have to adjust the map of processors to reflect the specific architecture of the switch to be simulated.

REFERENCES

[1] K.A. Lutz, "Considerations on ATM Switching Techniques," *Int. J. Digit. Analog Cabled Syst.*, (Vol. 1, 1988), pp.237-243.

[2] P. Newman, "ATM Technology for Corporate Networks," *IEEE Commun. Mag.*, (April, 1992), pp.90-101.

[3] M.A. Henrion, et al., "Switching Network Architecture for ATM based Broadband Communications," *ISS'90*, Stockholm, (May 1990).

[4] R.W. Hockney, and C.R. Jesshope, "*Parallel Computer 2*," Adam Hilger Ltd, Bristol, (1988), 625pp.

[5] K.M. Chandy, V. Holmes, and J. Misra, "Distributed simulation of networks," *Computer Networks*, (Feb. 1979), pp.105-113.

[6] D.R. Jefferson, "Virtual Time," *ACM Trans. Programming Languages and Systems*, (July 1985), pp.404-425.

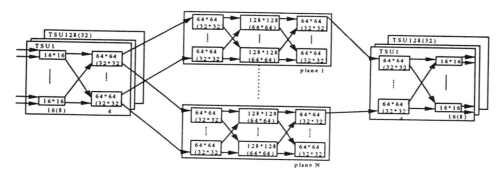

Fig. 1. The configuration of three-stage MPSR switch

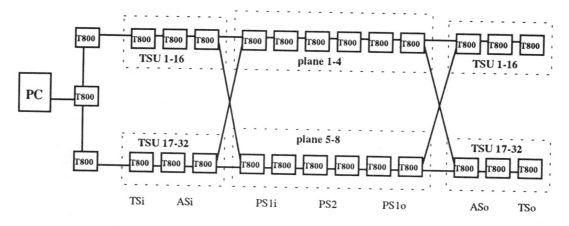

Fig. 2. Transputer configuration for the 2048*2048 MPSR switch

THE SYSTOLIC RECONFIGURABLE MESH

Mary M. Eshaghian
Dept. of Computer and Information Science
New Jersey Institute of Technology
Newark, NJ 07102
mary@cis.njit.edu

Russ Miller
Department of Computer Science
State University of New York at Buffalo
Buffalo, NY 14260
miller@cs.buffalo.edu

Abstract

In this paper, we introduce the Systolic Reconfigurable Mesh (SRM), which combines aspects of the reconfigurable mesh with that of systolic arrays. Every processor controls a local switch that can be reconfigured during every clock cycle in order to control the physical connections between its four bi-directional bus lines. Data is input from one side of the systolic reconfigurable mesh and output from another side, one column per unit time. Efficient algorithms are presented for intermediate-level vision tasks including histograming, connectivity, convexity and proximity.

1 Introduction

The reconfigurable mesh was originally proposed as a massively parallel computing model in the mid-1980s [6, 4, 8]. A review of the algorithmic literature [7] suggests that the major emphasis has been on *i*) fundamental problems, including sorting and arithmetic, *ii*) problems involving regularly structured data, such as matrices and images, in areas such as graph theory and image processing, and *iii*) geometric problems, where the amount of input data is sparse compared to the size of the reconfigurable mesh. Even though aspects of this machine have been incorporated into the communication of the MasPar machines and the Gated Connection Network of the Image Understanding Architecture [1], most of the algorithmic results in this area are of purely theoretical interest.

In this paper, we introduce a practical, scaled-down variant, of the reconfigurable mesh. Due to space limitations, the details of several of the algorithms will either be omitted or summarized.

2 The Systolic Reconfigurable Mesh

A *systolic reconfigurable mesh of size* n^2 consists of n^2 processors arranged in a two-dimensional grid overlaid with a grid-shaped reconfigurable bus. Every processor controls a local *switch* which can be reconfigured during every clock cycle to control the physical connections between its four bi-directional bus lines. The switches allow the broadcast bus to be divided into subbuses, providing smaller reconfigurable meshes. For a given set of switch settings, a *subbus*

refers to a broadcast bus over a maximally connected subset of processors. Without loss of generality, assume that data is input from the left side of the systolic reconfigurable mesh, one column of data per unit time, and eventually output from the right side, again, one column per unit time.

Given mn input items to be processed on a systolic reconfigurable mesh of size n^2, the computations on the systolic reconfigurable mesh fall naturally into three phases. The first phase can be classified as the *input and preprocessing phase*, where at time t, n data items are input and a (small) fixed number of operations are performed over some subset of the tn items that have been input thusfar. After completing the input and preprocessing phase following time m, the *static computing phase* is performed on the $O(n^2)$ items currently available in the systolic reconfigurable mesh. Finally, the *output and postprocessing phase* consists of pumping the data out, typically n items per unit time, and performing a (small) fixed number of operations on the remaining data, until all of the output has been generated. Given I input and preprocessing cycles, each consisting of $\bar{I}$ operations, a static phase consisting of $\bar{S}$ operations and O output and postprocessing cycles, each consisting of $\bar{O}$ operations, the total processing time is $T = \bar{I}IS\bar{O}O$. Naturally, one wants to minimize T. However, it is often desirable to eliminate S so as to improve the programmability of the systolic reconfigurable mesh and reduce the number of context switches that must be employed in changing between phases. Therefore, whenever possible, it is desirable to design an algorithm with the property that the running time is kept to a minimum and $S = 0$.

Variations of the systolic reconfigurable mesh exist, including the *bit model (BM)*, where individual processors can operate on a fixed number of bits of data in unit time, and the *word model (WM)*, in which individual processors can operate on a fixed number of $2 \log n$ bit words of data in unit time. For both variations, we assume that concurrent writes of a single bit (for BM) or word (for WM) of data to a subbus is permitted if the values are identical. We also assume that in unit time a bit (for BM) or word (for WM) of data can be broadcast to a subbus and read by all processors attached to that subbus. These assumptions are all reasonable and realizable in terms of designing and implementing a VLSI chip or a board that can be used in a workstation environment. Furthermore, every step of an algorithm designed for the word model can be simulated on a bit model variant in $O(\log N)$ time, where a word is $\log N$ bits wide.

3 Designing Algorithms

Denote a systolic reconfigurable mesh of size n^2 by S and a reconfigurable mesh of size n^2 by R. Suppose algorithm A requires t time on R. Then A can be simulated on S by first inputing the n^2 pieces of data in n time, then performing algorithm A on S in t time, then outputting the $O(n^2)$ final entries in n time. Therefore, the total time for simulating reconfigurable mesh algorithm A on a systolic reconfigurable mesh is $2n + t$. Suppose algorithm B requires t' time on a mesh of size n^2, where $t = \Omega(n)$. Then B can be simulated on S by first inputing the n^2 pieces of data in n time, then performing algorithm B on S in t' time, then outputting the $O(n^2)$ final entries in n time. That is, the total time for simulating mesh algorithm B on a systolic reconfigurable mesh is $2n + t'$. Note that in the above two cases, the time complexities of the static phase are t and t', respectively.

Therefore, if one were only concerned with an asymptotic analysis of algorithms on the systolic reconfigurable mesh, then one could simply simulate any reconfigurable mesh or mesh algorithm in $O(n)$ time. (Notice that the class of $O(n)$ time reconfigurable mesh algorithms properly includes the class of $O(n)$ time mesh algorithms.) However, our interest is in practical algorithms that can be efficiently performed on a realizable architecture designed as a companion board in a workstation targeted at image processing applications. That is, our goal is to design algorithms that not only eliminate the static phase, but which are complete in no more than $2n$ cycles. More generally, we are concerned with i) minimizing constants, ii) preserving the systolic nature of the systolic reconfigurable mesh and iii) eliminating the static stage of the algorithms.

3.1 Histogram

Consider the problem of creating a histogram of input values, where all such values are in the range of $1 \ldots n$. The results will be maintained in the last column of the systolic reconfigurable mesh. Specifically, the total number of occurrences for value i, $1 \leq i \leq n$, will be maintained in $SRM(i, n)$. The algorithm we present for the word model includes a sorting operation in every cycle. Each input and preprocessing cycle consists of the following.

1. Shift into the first column.

2. Sort these n data items in constant time [2, 3].

3. Every processor in the first column examines its (sorted) value and that of its neighbors. Each such processor then splits the column bus between itself and its neighbor if the data values are distinct.

4. In each subbus, the bottommost processor broadcasts its row position. The result of this segmented broadcast is that the topmost processor associated with every subbus knows the number of input data items with that value.

5. These partial results are then broadcast as follows. Perform the following for every processor in the first column that is a topmost processor on a subbus, as just defined. Without loss of generality, we describe the broadcast for such a processor $SRM(i, 1)$, which contains the number of occurrences over the current input set of size n for data value d, $1 \leq d \leq n$.

   (a) Connect the row bus in row i and broadcast this count to processor $SRM(i, d)$.

   (b) Connect the column bus in column d and broadcast the count from processor $SRM(i, d)$ to processor $SRM(d, d)$.

   (c) Connect the row bus in row d and broadcast the count from processor $SRM(d, d)$ to processor $SRM(d, n)$.

6. Processor $SRM(d, n)$ adds this count to its current running sum.

After n input and preprocessing cycles, as just described, the n final values can be output in 1 cycle, as the results are stored in the last column of the SRM. It is important to note that the static stage of processing has been eliminated and that the output and postprocessing stage consists of one simple output step. However, it is not clear whether or not the complexity of the input and preprocessing stage can be simplified by avoiding, for example, the complete sorting step associated with every column of input data.

Theorem 1 *The histogram problem for n^2 input items can be solved on the word-model systolic reconfigurable mesh of size n^2 using only a fixed number of broadcasts per input and preprocessing cycle, eliminating the static phase, and outputting the image in lockstep fashion with no additional work.* □

The histogram problem can be solved for the bit model while avoiding the sorting operation during every step.

Corollary 2 *The histogram problem for n^2 input items can be solved on the bit-model systolic reconfigurable mesh of size n^2 using $\Theta(\log N)$ broadcasts per input and preprocessing cycle, eliminating the static phase, and outputting the results upon completion.* □

3.2 Convex Hull

Consider the problem of marking the extreme points representing the convex hull of an $n \times n$ binary image. While the results for the histogram problem just presented were stored in a convenient predetermined location of the SRM, for this problem the results will be maintained along with the image. That is, the value of a pixel at position (i, j) will be maintained as part of a record that also includes a flag indicating whether or not pixel (i, j) is an extreme point of the image. It should be noted that this "marked" flag will be set to true after the completion of cycle $t < n$ (i.e., after t columns of the image have been input and preprocessed) if and only if pixel (i, j) is an extreme point of the restriction of the image to the rightmost $n \times t$ subimage.

Given the word model, each input and preprocessing cycle consists of the following.

1. Shift into the first column.

2. Perform bus splitting in column one of the SRM for the purpose of identifying the topmost and bottommost black pixels in that column (i.e., from the column of data just input).

3. Mark these pixels as extreme points and broadcast their coordinates to the entire SRM.

4. Every marked pixel uses the coordinates of these two new extreme points, its point and its previous extreme point in enumerated order to determine whether or not it is to remain marked as an extreme point.

5. The extreme points that precede these two new extreme points in the enumerated ordering are identified and their locations are broadcast to these new extreme points.

Notice that as the data flows from left to right in a truly systolic fashion, the marked field and preceding point location field for each pixel remains tied to the pixel data. Therefore, once all of the data has been input, the data may be output in a natural fashion. Again, it is very important to note that the static stage is nonexistent. Further, the output and postprocessing stage is reduced simply to outputting the highlighted/marked image.

Theorem 3 *Given an $n \times n$ binary image, and a word-model (bit-model) systolic reconfigurable mesh of size n^2, using only a small fixed ($\Theta(\log N)$) number of broadcasts per input and preprocessing cycle, eliminating the static phase, and outputting the image in lockstep fashion with no additional work, the convex hull of the image can be marked .* □

3.3 Nearest Neighbor

Consider the problem of determining for every black pixel in an $n \times n$ binary image, the coordinates of a nearest neighboring black pixel. By coupling the coordinates of a running nearest neighbor to every pixel, and exploiting the triangle inequality, the following can be obtained.

Theorem 4 *Given an $n \times n$ binary image, and a word-model (bit-model) systolic reconfigurable mesh of size n^2, using only a small fixed ($\Theta(\log N)$) number of broadcasts per input and preprocessing cycle, eliminating the static phase, and outputting the image in lockstep fashion with no additional work, a nearest neighboring black pixel can be identified for every black pixel in the image.* □

3.4 Minimum or Maximum

Consider the problem of finding the minimum (or maximum) value from a set of n^2 input values. The algorithm relies on creating a cross-product during every input and preprocessing phase to determine the running result. The static stage is nonexistent and the output and postprocessing stage simply consists of outputting one value.

Theorem 5 *Given an $n \times n$ binary image, and a word-model (bit-model) systolic reconfigurable mesh of size n^2, using only a small fixed ($\Theta(\log N)$) number of broadcasts per input and preprocessing cycle, eliminating the static phase, and outputting a single value, the minimum or maximum value can be determined.* □

3.5 Component Labeling

Consider the problem of labeling an $n \times n$ binary image. A simple approach would be to input the image in a natural fashion, apply a straightforward bit-polling algorithm to label the image [5] and then output the image in a natural fashion. Notice that the time for this algorithm is $2n + \log_2 n$ and that 2 context switches are required due to the inclusion of the static stage. It is possible to overlap the static stage with the input stage by doing the bit polling at every step as the image is inputed in a lock-step fashion. Then in $n \log_2 n$ time the image is labeled. The labeled image can be output by repeating the previous steps for another n steps (leading to $2n \log_2 n$ total running time) or by context switching to n simple shift steps (leading to $n \log_2 n + n$ total running time). Similarly, given an $n \times n$ adjacency matrix input, the $\log n$ bit polling algorithm of [1] can be applied to every input step so that in $n \log_2 n$ time the image is labeled. It becomes quite a challenge to produce an efficient algorithm that omits the static stage and also avoids repeating the $\log n$-time based labeling schemes in each step. In this section, we consider algorithms that are concerned with omitting the static stage and avoiding the $\Theta(\log n)$ time labeling during each step.

3.5.1 Algorithm 1:

For this algorithm and the next (section 3.5.2), the component labels assigned to the pixels will be of the form: (C_R, C_L, R_T), where C_R is the index of the rightmost column containing a pixel of the component, C_L is the index of the leftmost column containing a pixel of the component, and R_T is the index of the topmost row containing a pixel in the component. As with many of the previous algorithms, the component label record will remain coupled to the pixel value as the pixel travels through the systolic reconfigurable mesh. When a pixel is initially input to the SRM, the fields of the component label will be initialized to 0. These values will be updated throughout the course of the algorithm. The following is an outline of the first algorithm.

1. Take the input one slice at a time. For slice i do: Form a connected bus over every connected region; Every black pixel in column 1 of the SRM broadcasts the value i to be stored as C_L for its connected region; Shift right. (Notice that after cycle $t = n$, all n slices have been input and C_L has been identified for every connected component.)

2. For $t = 2n + i$, for row $i = 1$ to n do: Every upper boundary black pixel (white neighbor to its top) in row i broadcast the value i to be stored as R_T for its connected component. Notice that this value is stored only if $R_T = 0$. (Notice that after cycle $t = 2n$, the values of R_T have been properly determined.)

3. At cycle $2n + i$, the ith column of the image is output as follows: For each black pixel in column i, if $C_R = 0$ then broadcast the value of i as C_R to all its connected region including itself; Shift out. (Notice that after cycle $t = 3n$, the nth slice will be output, and all pixels that are output will have their correct final labels.)

Theorem 6 *Given an $n \times n$ digitized image, it can be labeled on the word-model systolic reconfigurable mesh of size n^2 using only a small fixed number of broadcasts per input and preprocessing cycle, and outputting the image in lockstep fashion with a small fixed number of broadcasts per output.* $\square$

3.5.2 Algorithm 2:

All the algorithms presented so far in this paper take the input from the left and output the results to the right. However, by inputting from the left and outputting to the top, the following theorem can be obtained.

Theorem 7 *Given an $n \times n$ digitized image, it can be labeled on the bit-model systolic reconfigurable mesh of size n^2 using only $\log n$ number of broadcasts per input and preprocessing cycle, eliminating the static phase, and outputting the image in lockstep fashion from the top.* $\square$

3.5.3 Algorithm 3:

This algorithm is quite different from some that have appeared earlier in this paper in that there is a nontrivial input and preprocessing phase, as well as a nontrivial output and postprocessing phase. The label will be represented as (C_L, C_R, T), where C_L is the index of the leftmost column containing a pixel of the component, C_R is the index of the rightmost column containing a pixel of the component, and the T field is used to break ties in the case of distinct figures with the same (C_L, C_R) values. C_L will be computed for every pixel during the input and preprocessing phase of the algorithm. C_R will be computed for every pixel during the output and postprocessing phase of the algorithm. Since a number of figures can occupy the same leftmost and rightmost column of the image, for such figures, a tie-breaking scheme will be used to determine T during the output and postprocessing step.

Each input and preprocessing cycle consists of the following.

1. In lockstep, shift the image to the right while inputting the next column of data into the first column of the SRM.

2. Initialize the component labels for all pixels now in column one of the SRM to $(0, 0, 0)$.

3. All processors that currently hold a black pixel of the image, connect their bus to all neighboring pixels that also maintain black pixels. The result is that there is a subbus over all figures with respect to the restriction of the image that has thusfar been input.

4. Exploiting the concurrent write capability, all processors in the first column of the SRM now broadcast their column label. All processors receiving such a value, store this in C_L, potentially replacing a previous value.

After n of these simple input and preprocessing cycles, all pixels know the C_L component of their label. The C_R and T components will be determined during the output and postprocessing phase. Each output and postprocessing cycle is concerned with those processors maintaining pixels in the last column of the SRM that have not previously received their final component labels. For such processors, the output and postprocessing cycle consists of the following.

1. All such processors that are currently responsible for a black pixel of the image, connect their bus to all neighboring pixels also maintaining black pixels. The result is that there is a subbus over all figures with respect to the restriction of the image that remains in the SRM.

2. Exploiting the concurrent write capability, all processors in the last column of the SRM now broadcast the column label of their pixel. All processors receiving a value store this in C_R. (Recall that this broadcast operation is only performed by processors in the last column of the SRM that are currently responsible for pixels which have not previously had their final labels determined.)

3. Unfortunately, multiple unique figures may now have the same component labels in terms of their C_L and C_R values. The labels are disambiguated as follows.

   (a) Every processor in the last column that contains a black pixel connects its bus to its northern processor if and only if that processor maintains a black pixel. In a similar fashion, every processor in the last column that contains a black pixel connects its bus to its southern processor if and only if that processor maintains a black pixel. Next, perform two bus broadcast operations in the last column of the SRM over these subbusses so as to identify the topmost and bottommost pixels in every connected component over the restriction of the image to this last column.

(b) Each such processor in the last column of the SRM that represents a pixel in the image at position (i, j) prepares a record (C_L, C_R, i, j, T). (Notice that all such records have identical values for j, which is included in the record for consistency.) Sort these, fewer than n, records using the sparse sorting result of [2] or [3]. After sorting, all records with identical (C_L, C_R) values will appear in contiguous positions in the last column ordered by their original position.

(c) For each top/bottom pair (which are in adjacent processors in the last column of the SRM) representing identical labels, the bottom processor will broadcast a bit and observe whether or not this bit is received by its mate. This operation is somewhat intricate, involving several row and column broadcasts. This is done by first moving the labels back to their original unsorted location, and then using the standard technique of odd-even bus splitting, the bits are broadcast to find the topmost-bottommost pair of each group having identical (C_L, C_R) fields.

(d) Finally, move the labels back, and perform a bus-splitting-and-broadcast operation over the ordered set of data in the last column so as to broadcast a processor ID to be used as the T value in disambiguating labels with identical (C_L, C_R) fields.

4. In lockstep, shift the image to the right so as to output the next column of data (pixel and label information).

It should be noted that the algorithm takes advantage of the concurrent write capability and is able to eliminate the static stage. However, the cost of the output and post-processing stage, while asymptotically constant, requires a non-trivial amount of work and is primarily of theoretical interest.

Theorem 8 *Given an $n \times n$ digitized image, it can be labeled on the word-model systolic reconfigurable mesh of size n^2 using only a fixed number of broadcasts per input and preprocessing cycle, eliminating the static phase, and outputting the image in lockstep fashion with using $O(1)$ sorting.* $\square$

3.5.4 Algorithm 4:

If every figure is restricted so as to be vertically convex, then the image can be quite simply labeled during the input and preprocessing phase.

Theorem 9 *Given an $n \times n$ digitized image, it can be labeled on the word-model systolic reconfigurable mesh of size n^2 using only a small fixed number of broadcasts per input and preprocessing cycle, eliminating the static phase, and outputting the image in lockstep fashion with no additional work.* $\square$

4 Conclusion

In this paper, we introduced a novel architecture, namely the *systolic reconfigurable mesh* (SRM). The motivation for this model is that it can be used, in practice, as an attached array processor board or chip, where the data will be piped in and out in a systolic fashion. Similarly, we concentrated on designing efficient algorithms for two variations of this model, namely, the bit-model and the word-model. We showed that a suitable algorithm for this model is one which (*i*) can be performed while the data moves in and out without any static stage, (*ii*) consists of simple operations for each step, and (*iii*) has low constants and minimal total execution time. Aiming to satisfy these properties, we presented several algorithms for intermediate-level vision tasks including histograming, connectivity, convexity, and proximity. For labeling an $n \times n$ image, we presented and discussed four different algorithms. Open problems include a generic labeling algorithm for the word-model and a convex hull algorithm that operates efficiently on an image containing multiple figures.

References

[1] Mary M. Eshaghian, K. Kim, G. Nash, and D. B. Shu. Implementation and application of a gated connection network in image understanding architecture. In H. Li and Q. Stout, editors, *Reconfigurable Massively Parallel Computers*, pages 64–87. Prentice Hall, 1991.

[2] J. -W. Jang and V. K. Prasanna. An optimal sorting algorithm on reconfigurable mesh. In *Proc. 6th International Parallel Processing Symposium*, pages 130–137. IEEE, 1992.

[3] J. -F. Jenq and S. Sahni. Histogramming on a reconfigurable mesh computer. In *Proc. 6th International Parallel Processing Symposium*, pages 425–432. IEEE, 1992.

[4] H. Li and M. Maresca. Polymorphic-torus network. *IEEE Trans. on Computers*, 38(9):1345–1351, September 1989.

[5] R. Miller, V. K. P. Kumar, D. Reisis, and Q. F. Stout. Meshes with reconfigurable buses. *Proc. 15th MIT Conference on Advanced Research in VLSI*, pages 163–178, March 1988.

[6] R. Miller, V. K. P. Kumar, D. I. Reisis, and Q. F. Stout. Parallel computations on reconfigurable meshes. *IEEE Trans. on Computers*, 42(6):678–692, June 1993.

[7] K. Nakano. A list of papers on algorithms for reconfigurable architectures. *Parallel Processing Letters*, 1995, to appear.

[8] C. C. Weems, S. P. Levitan, A. R. Hanson, and E. M. Riseman. The image understanding architecture. *International Journal of Computer Vision*, 2:251–182, 1989.

Sorting and Selection on Distributed Memory Bus Computers

S. Rajasekaran and S. Sahni
Department of CIS, University of Florida

Abstract. In this paper we study the problems of sorting and selection on the Distributed Memory Bus Computer (DMBC) recently introduced by Sahni. In particular we present: 1) An efficient algorithm for computing the sum of n bits; 2) An optimal $O(1)$ time sorting algorithm; 3) An optimal randomized logarithmic time integer sorting algorithm; and 4) An optimal randomized constant time selection algorithm. Our algorithms will run without change in performance on many related models as well. For example, these algorithms apply to the RMBM model of Vaidyanathan et al.

1 Introduction

Mesh connected computers are impressive models of computing owing to their numerous special features. In recent times, meshes with underlying buses have attracted the attention of many a researcher. The bus system could either be static or reconfigurable. Many attempts have been made to build computers based on these models and the results obtained have been promising. Mesh connected computers with bus systems are also interesting as theoretical models of computing. For example, n numbers can be sorted in $O(1)$ time using an $n \times n$ reconfigurable mesh. On the other hand, given only a polynomial number of processors, even the CRCW PRAM needs $\Omega(\frac{\log n}{\log\log n})$ time to sort n numbers.

Several variants of the reconfigurable mesh can be found in the literature. The DMBC model was proposed by Sahni [7] in an attempt to separate the cost of switching elements (and hence the I/O bandwidth) from that of full fledged processing elements. It is conceivable that the switching elements cost less than full powered processing elements. Often, many of the processors in a reconfigurable mesh get used only as switching elements. Thus it makes sense to separate processors from switching elements. A related model known as RMBM has been proposed by Thiruchelvan, Trahan, and Vaidyanathan [8]. This model also attempts to separate the I/O bandwidth from the processing power.

A DMBC of size (n, m) is a rectangular array where the first row consists of full fledged processing elements and the other rows comprise of switching elements. The rows are numbered 1 through m and the columns

1 through n. Think of the rows as broadcast buses. A row can be broken at any point using *segment* switches. There is a segment switch between any two columns in each row. These segment switches are locally controllable. Also, a processor can *fuse* the buses in any manner. For example, processor i can fuse the first 10 buses in column i. At any time, a processor can read from a bus or write into a bus and connect appropriately all its segment and fuse switches. More details about the model can be found in [7].

Several fundamental algorithms for the DMBC were given in [7]. In this paper we consider the problems of sorting (both general keys and integer keys) and selection. The following results are supplied: 1) An algorithm to add n bits in $O(1)$ time on an (n, n^ϵ) DMBC, for any fixed $\epsilon > 0$. An algorithm on an (n, n) DMBC has been given in [7]; 2) A randomized algorithm for sorting n integers in the range $[1, n(\log n)^c]$ (for any constant $c > 0$) on a DMBC $(\frac{n}{\log n}, n)$ that runs in time $O(\log n)$ with high probability. In contrast Thiruchelvan, Trahan, and Vaidyanathan [8] present a deterministic algorithm with a run time of $O(\log n \log^* n)$ on an RMBM of size $(\frac{n}{\log n \log^* n}, n)$. 3) A sorting algorithm that can sort n numbers in $O(1)$ time on a DMBC $(n^{1+\epsilon}, n)$ for any fixed $\epsilon > 0$. We show that this algorithm is optimal. The same algorithm also runs in time $O(1)$ on a PRAM with reconfigurable buses of size $(n^{1+\epsilon}, n^\delta)$, for any constants $\epsilon > 0$ and $\delta > 0$; and 4) A constant time randomized selection algorithm that employs an (n, n^ϵ) DMBC.

The rest of this paper is organized as follows: In section 2 we present some preliminaries and our bit sum algorithm. Our integer sorting algorithm is presented in section 3. Sections 4 and 5 contain our sorting and selection algorithms. Section 6 concludes the paper.

2 Preliminaries

In this section we provide some preliminary facts and results that will be employed in the paper.

We say a randomized algorithm uses $\widetilde{O}(f(n))$ amount of any resource (like time, space, etc.) if the amount of resource used is no more than $c\alpha f(n)$ with probability $\geq (1 - n^{-\alpha})$ for any α, c being a constant. We could also define $\widetilde{\Theta}(.)$, $\widetilde{o}(.)$, etc. in a similar manner.

By high probability we mean a probability of $\geq (1 - n^{-\alpha})$ for any constant $\alpha \geq 1$.

The following lemma has been proved by Sahni [7]:

Lemma 2.1 *We can add n bits in a unit of time using a DMBC $(3n, 2n)$. We can also add n bits in $O(1)$ time using a DMBC (n, n).*

We prove the following lemma (proof of which has been omitted):

Lemma 2.2 *The sum of n bits can be computed in $O(\frac{1}{\epsilon^2})$ time using a DMBC (n, n^ϵ).*

Our algorithm is based on a similar algorithm for the PARBUS model [3] and makes use of the technique of carry look ahead addition. However there are some crucial changes and new ideas in our algorithm.

3 Integer Sorting

In this section we consider the problem of sorting n numbers where each number is an integer in the range $[0, n(\log n)^c]$ for any fixed c. The algorithm is randomized and runs in time $\widetilde{O}(\log n)$ on a DMBC $(\frac{n}{\log n}, n)$.

We adopt the algorithm of Rajasekaran and Reif [6]. There are two basic ideas in the algorithm: 1) the radix sorting and 2) random sampling.

Radix Sorting. The idea is captured by the following lemma:

Lemma 3.1 *If n numbers in the range $[0, R]$ can be stable sorted using P processors in time T, then we can also stable sort n numbers in the range $[0, R^c]$ in $O(T)$ time using P processors, c being any constant.*

Summary. A summary of [6]'s algorithm is as follows: There are two phases in the algorithm (called *Coarse Sort* and *Fine Sort*). In Coarse Sort, the n given numbers are sorted with respect to their $\log n - 3 \log \log n$ LSBs. This algorithm is a non-stable sort. Followed by Coarse Sort, the numbers are stable sorted with respect to the remaining bits. This step is called Fine Sort. We follow these phases in our adoption also.

Fine Sort. If the n given numbers are in the range $[0, n(\log n)^c]$, in this phase we are interested in sorting these n numbers with respect to their $(c+3) \log \log n$ MSBs. That is, we are interested in sorting n $O(\log \log n)$-bit numbers. An optimal algorithm for this problem has been given in [6]. A similar algorithm has been employed for the RMBM model [8]. Since a DMBC can simulate an RMBM, the following lemma follows:

Lemma 3.2 *We can stable sort n $O(\log \log n)$-bit numbers on a $(\frac{n}{\log n}, n)$ DMBC in $O(\log n)$ time.*

Coarse Sort. This algorithm sorts n numbers in the range $[0, \frac{n}{\log^3 n}]$ in $\widetilde{O}(\log n)$ time using a DMBC $(\frac{n}{\log n}, n)$. The sort is non-stable. The idea is to count how many keys are there in the input of a given value and use hashing to rearrange the keys according to their values. Let $k_1, k_2, \ldots, k_n$ be the n given keys. Let $N(i)$ be the number of keys in the input whose value is i, for $1 \le i \le \frac{n}{\log^3 n}$. Important steps in the algorithm are: 1) Randomly sample $\frac{n}{\log n}$ keys from the input and sort this sample in $O(\log n)$ time using Cole's algorithm; 2) In the sorted sample, we can determine how many keys are there of value i, $1 \le i \le \frac{n}{\log^3 n}$. If $S(i)$ is the number of sample keys of value i, we obtain $A(i)$'s that are approximations to $N(i)$'s as follows. $A(i) = d \max\{S(i) \log n, \log n\}$ where d is a constant. $A(i)$'s thus obtained satisfy [6]: a) $A(i) \ge N(i)$ for each i and b) $\sum_i A(i) = O(n)$ with high probability; 3) Finally use hashing to rearrange the keys according to the sorted order.

Implementation on the DMBC. We assume the CRCW version of the DMBC. In particular we assume the following: At any given time, more than one processor can try to write to the same bus in which case a random processor succeeds. Step 1 of the above algorithm can be performed on a $(\frac{n}{\log n}, n)$ DMBC in $O(\log n)$ time using Cole's algorithm. Note that a DMBC can simulate an EREW PRAM of the same size. In step 2, approximations to $N(i)$'s can be computed in $O(1)$ time.

In step 3, hashing is done as follows. We use a slightly different scheme than that has been employed in [6]. If a key k_j is of value i, we say k_j belongs to bucket i. Bucket i will be assigned $\lceil \frac{A(i)}{\log n} \rceil$ successive processors. Processors are assigned in the order of bucket values, i.e., processors of bucket 2 will follow processors for bucket 1; processors for bucket 3 will follow processors of bucket 2; and so on. Any key that belongs to bucket i will be hashed onto a random processor assigned to bucket i.

To begin with, each processor π has $\log n$ keys that it has to hash onto other processors. At any given time processor π chooses a random remaining key k_j and tries to write it in bus i', where i' is the *id* of a random processor of the bucket that k_j belongs to. If π succeeds in this attempt, it will eliminate k_j from its queue and proceed with the remaining keys. If it does not, it will choose a random remaining key in the next step and proceed in a similar manner. It works until all its keys have been hashed onto appropriate processors. Also in every time step, each processor π will read from bus π and collect the key from the bus into its local queue.

We claim that all the processors will complete their tasks within time $\widetilde{O}(\log n)$. The proof is as follows. Consider any bucket i. Let m_i be the number of processors assigned to bucket i. This will mean that the number of keys in bucket i is $\leq m_i \log n$ with high probability. In step t of the algorithm, let n_t^i be the number of keys of bucket i (from among all the processors) that have not yet been hashed onto. Let $N_t = \sum_i n_t^i$.

Then, the expected value of n_t^i is $\frac{N_t}{n} N(i)$. The expected number of remaining keys of bucket i that will be chosen by their respective processors in time t is $\frac{n_t^i}{\log n} \frac{n}{N_t}$. That is, the expected number of keys of bucket i that will be chosen in time t is $\frac{N(i)}{\log n} \leq m_i$. This in turn means that the expected number of remaining keys of bucket i that will succeed in time t is $\Omega(\frac{N(i)}{\log n})$. Therefore, the expected number of keys (from among all buckets) that will succeed in time t is $\Omega(\frac{n}{\log n})$. Applying Chernoff bounds, the number of keys that will succeed in time t is $\Omega(\frac{n}{\log n})$ with high probability. Put together, all the processors will complete their hashings within time $\widetilde{O}(\log n)$.

After hashing has been completed, there will be $\widetilde{O}(\log n)$ keys in each processor. We perform a prefix computation followed by a routing so that there will be exactly $\log n$ keys in each processor. Prefix computation takes $O(\log n)$ time and the routing takes $\widetilde{O}(\log n)$ time.

Therefore we get the following theorem.

Theorem 3.1 *A DMBC $\left(\frac{n}{\log n}, n\right)$ can sort n numbers in $\widetilde{O}(\log n)$ time provided the numbers are in the range $[0, n(\log n)^c]$, c being any constant.*

4   General Sorting

We consider the problem of sorting n general keys in this section. The input is assumed to be from a linear order and nothing else is assumed about the keys. Several optimal sequential algorithms have been designed for sorting which run in time $O(n \log n)$, n being the number of keys to be sorted (see e.g., [2]). Optimal parallel sorting algorithms are also known for several models of computing. For instance Cole's algorithm runs in $O(\log n)$ time, given n EREW PRAM processors. Since a DMBC (n, n) can simulate an n-processor EREW PRAM, it follows that sorting can also be done in $O(\log n)$ time on an (n, n) DMBC. Thus we will focus our attention on the problem of constant time sorting on the DMBC.

Fact 4.1 *If sorting has to be performed in $O(1)$ time on a DMBC, then the number of processors has to be $\Omega(n^{1+\epsilon})$ for some constant $\epsilon > 0$. Also, the number of rows has to be $\Omega(n)$.*

Proof. Omitted.

The following lemma due to Sahni [7] will be employed:

Lemma 4.1 *We can sort n numbers in $O(1)$ time using an (n^2, n) DMBC.*

In this section we present an $O(1)$ time optimal sorting algorithm, i.e., an algorithm that uses a DMBC $(n^{1+\epsilon}, n)$ for any fixed $\epsilon > 0$. This algorithm is based on the column sort algorithm of Leighton [4].

Let $k_1, k_2, \ldots, k_n$ be the n given numbers. These numbers are thought of as forming a matrix M with $r = n^{2/3}$ rows and $s = n^{1/3}$ columns. There are 7 steps in the algorithm: 1) Sort the columns in increasing order; 2) Transpose the matrix preserving the dimesnsion as $r \times s$. I.e., pick the elements in column major order and fill the rows in row major order; 3) Sort the columns in increasing order; 4) Rearrange the numbers applying the reverse of the permutation employed in step 2; 5) Sort the columns in a way that adjacent columns are sorted in reverse order; 6) Apply two steps of odd-even transposition sort to the rows; and 7) Sort each column in increasing order. At the end of step 7, it can be shown that, the numbers will be sorted in column major order.

Implementation Details on DMBC. We will store the n given numbers in the first row of the DMBC with no more than one key per processor. At any given time each key will know which row and which column of the matrix M it belongs to. Whenever we need to sort the columns, we will make sure that the numbers belonging to the same column will be found in successive processors.

On a DMBC (n, n), note that any permutation can be performed in $O(1)$ time. This means that steps 2 and 4 can be performed in $O(1)$ time. Step 6 can be performed in $O(1)$ time as well as follows: Rearrange the numbers such that elements in the same row are in successive processors and apply two steps of the odd-even transposition sort. After this, move the keys to where they came from.

Next we describe how we implement steps 1,3,5, and 7. We first assume that we have a DMBC $(n^{5/3}, n)$. Later we will indicate how to reduce the size to $(n^{1+\epsilon}, n)$ for any $\epsilon > 0$.

Partition the DMBC into $n^{1/3}$ parts each of size $(n^{4/3}, n)$. Rearrange the n given numbers such that there are $n^{2/3}$ numbers in each part located in the first $n^{2/3}$ processors of each part. Think of each part as having a column of the matrix M. Now sort the numbers in each part (i.e., each column of M) using Sahni's algorithm (see lemma 4.1). This can be done in $O(1)$ time. This implies that steps 1,3,5, and 7 of column-sort can be performed in $O(1)$ time. Therefore it follows that n numbers can be sorted in $O(1)$ time on an $(n^{5/3}, n)$ DMBC.

We can reduce the size of DMBC to get (details in the full paper):

Theorem 4.1 *A DMBC $(n^{1+\epsilon}, n)$ can sort n numbers in $O(1)$ time, ϵ being any constant > 0.*

Sorting on a PRAM with Reconfigurable Buses. A PRAM with reconfigurable buses (call it PRAM-RB) of size (n, m) is nothing but a DMBC of size (n, m) where the n processing elements also have a shared common memory for communication. An integer sorting algorithm for this model is known. Using the sort ideas in [7] and the bit sum algorithm of section 3, n numbers may be sorted on a PRAM-RB of size (n^2, n^δ) in $O(1)$ time, for any fixed $\delta > 0$. (Proof can be found in the full paper).

5   Selection

Given a sequence of n numbers $k_1, k_2, \ldots, k_n$ and an $i \leq n$, the problem of selection is to identify the ith smallest of the n numbers. An elegant linear time sequential algorithm is known for selection (see e.g., [2]). Floyd and Rivest have given a simple linear time randomized algorithm for sequential selection [1].

Optimal parallel algorithms are also known for selection on various models of computing. Most of the parallel selection algorithms (both deterministic and randomized) make use of the technique of sampling.

In this section we show that selection can be done in $\widetilde{O}(1)$ time on a DMBC (n, n^ϵ) for any fixed $\epsilon > 0$. The basic idea is the following: 1) Pick a random sample S of size $q = o(n)$; 2) Choose two elements ℓ_1 and ℓ_2 from the sample whose ranks in S are $i\frac{q}{n} - \delta$ and $i\frac{q}{n} + \delta$ for some appropriate δ. One can show that these elements 'bracket' the element to be selected with high probability; 3) Eliminate all keys whose values are outside the range $[\ell_1, \ell_2]$; 4) Perform an appropriate selection from out of the remaining keys.

We implement the above scheme on the CRCW version of the DMBC. In particular we assume the following: At any given time, more than one processor can try to write to the same bus in which case one of them succeeds and we don't know which one. More details can be found in the full paper.

The theorem we prove is:

Theorem 5.1 *Selection can be performed in $\widetilde{O}(1)$ time on a DMBC of size (n, n^ϵ), for any fixed $\epsilon > 0$.*

6   Conclusions

In this paper we have presented efficient algorithms for sorting and selection on the DMBC. We have considered both integer sorting and general sorting problems. An interesting open problem is to obtain a matching deterministic selection algorithm. Also, it is not clear if we need n^ϵ rows in order to perform selection. However the processor bound of our selection algorithm is optimal. Also it is an open problem to reduce the processor bound of our $O(1)$ time sorting algorithm when applied to integers. For example, if we have to sort n bits, we could do so in $O(1)$ time on a DMBC (n, n^ϵ), for any fixed $\epsilon > 0$ (cf. lemma 2.2).

References

[1] R.W. Floyd, and R.L. Rivest, Expected Time Bounds for Selection, Communications of the ACM, Vol. 18, No.3, 1975, pp. 165-172.

[2] E. Horowitz and S. Sahni, *Fundamentals of Computer Algorithms*, Computer Science Press, 1978.

[3] J. Jang, H. Park, and V.K. Prasanna, A Fast Algorithm for Computing Histograms on a Reconfigurable Mesh, Proc. Frontiers of Massively Parallel Computing, 1992, pp. 244-251.

[4] T. Leighton, Tight Bounds on the Complexity of Parallel Sorting, IEEE Transactions on Computers, C-34(4), 1985, pp. 344-354.

[5] S. Rajasekaran, Mesh Connected Computers with Fixed and Reconfigurable Buses: Packet Routing, Sorting, and Selection, Proc. *First Annual European Symposium on Algorithms*, Oct. 1993.

[6] S. Rajasekaran and J.H. Reif, Optimal and Sublogarithmic Time Randomized Parallel Sorting Algorithms, SIAM Journal on Computing 18, 1989, pp. 594-607.

[7] S. Sahni, Data Manipulation on the Distributed Memory Bus Computer, to appear in Parallel Processing Letters, 1995.

[8] R.K. Thiruchelvan, J.L. Trahan, and R. Vaidyanathan, Sorting on Reconfigurable Multiple Bus Machines, Proc. International Prallel Processing Symposium, 1994.

Generalized Algorithm for Parallel Sorting on Product Networks

Antonio Fernández,*
MIT Laboratory for Computer Science
545 Technology Square
Cambridge, MA 02139

Nancy Eleser, and Kemal Efe[†]
Center for Advanced Computer Studies
University of Southwestern Louisiana
Lafayette, LA 70504

Abstract

If G is an arbitrary factor graph with N nodes, its r dimensional product contains N^r nodes. We present an algorithm which sorts N^r keys stored in r-dimensional product of any factor graph G in $O(r^2 S(N))$ time where $S(N)$ depends on G. We show that for any factor graph G, $S(N)$ is bounded above by $O(N)$, establishing an upper bound of $O(r^2 N)$ for the time complexity of sorting N^r keys on any product network. When r is fixed, this leads to the asymptotic complexity $O(N)$ to sort N^r keys, which is optimal for several instances of product networks. We also illustrate examples of product networks that can sort N^r keys in $O(r^2)$, or $O(N)$, or $O(Log^2 N)$ time.

Keywords: sorting, interconnection networks, product networks, algorithms, odd-even merge.

1 Introduction

In [1], Batcher presented two efficient sorting networks. Algorithms derived from these networks have been presented for a number of different parallel architectures, like the shuffle-exchange network [10], the grid [11, 5], the cube-connected cycles [8], and the mesh of trees [6].

In this paper we generalize Batcher's algorithm to merge N sorted sequences into a single sorted sequence. From this multiway merge operation we derive a sorting algorithm, and we show how to use this approach to obtain an efficient sorting algorithm for any homogeneous product network. Among the main results of this paper, we show that the time complexity of sorting N^r keys for any r dimensional N^r-node product graph is bounded above as $O(r^2 N)$. We also illustrate special cases of product networks with running times of $O(r^2)$, or $O(N)$, or $O(Log^2 N)$.

2 Definitions and Notations

Let G be a N-node connected graph. We define its r-dimensional homogeneous product as follows.

Definition 1 *Given a graph G with vertex set $V_G = \{0, 1, \cdots, (N-1)\}$ and arbitrary edge set E_G, the r-dimensional product of G, denoted PG_r, is the graph whose vertex set is $V_{PG} = \{0, 1, \cdots, (N-1)\}^r$ and edge set E_{PG} defined as follows: two vertices $x = x_r x_{r-1} \cdots x_1$, and $y = y_r y_{r-1} \cdots y_1$ are adjacent in PG_r if and only if both of the following conditions are true:*

1. *x and y differ in exactly one symbol position,*

2. *if i is the differing symbol index, then $(x_i, y_i) \in E_G$.*

Note that we can split the r-dimensional product network into N copies of $r-1$ dimensional product networks by erasing all the edges at an arbitrary dimension. When we do so, the uth copy obtained is denoted as $[u]PG_{r-1}^i$ where i is the dimension being erased. For example, if $i = r$ (i.e. the highest dimension) then we obtain N copies of $[u]PG_{r-1}^r$, each isomorphic to PG_{r-1}. Similarly, we can split the PG_r into N^2 copies of smaller product networks, each isomorphic to PG_{r-2} by erasing all the edges in two of the dimensions. In this case we extend the notation as $[u, v]PG_{r-2}^{i,j}$, where i and j denote the dimensions of edges being erased, and the pair $[u, v]$ uniquely identifies a copy obtained. This notation can be extended to erasing multiple dimensions in the obvious way, with the order of terms in square brackets corresponding to the order in superscripts.

When we focus on a k-dimensional subgraph of the r-dimensional product network, where $k < r$, it sometimes gets too long to write all the dimensions being erased, and it may be easier to just specify the dimensions of edges not erased. In particular, we will have occasion to refer to subgraphs obtained by erasing all dimensions but one, and thus the remaining subgraph will be isomorphic to the factor graph G. We will use $PG_1^{\{i\}}$ to denote such a subgraph, where i denotes the dimension of edges not erased. We also extend this notation similarly to above case, and use $PG_2^{\{i,j\}}$ to denote a two-dimensional subgraph with i and j indicating the dimension of edges not erased.

For an arbitrary factor graph G, vertex labels $0 \cdots N-1$ can define the ascending order of data when sorted. However we need to define an order for the nodes of PG_r, which will determine the final location

---

*This research has been partially funded by the CICYT/Spanish Ministry of Education and Science under grant pf94 04166960.

[†]This research has been partially funded by a grant from LEQSF.

of the sorted keys. The order defined is known as "snake" order.

Definition 2 *(Snake Order) for the r-dimensional product graph:*

1. *If $r = 1$, snake order corresponds to the order defined for G.*

2. *If $r > 1$, suppose that snake order has been already defined for PG_{r-1}. Then,*

   (a) *$[u]PG^r_{r-1}$ has the same order as PG_{r-1} if u is even, and reverse order if u is odd.*

   (b) *if $u < v$ then any value in $[u]PG^r_{r-1}$ precedes any value in $[v]PG^r_{r-1}$.*

The hamming weight of a vertex s is defined as $W(s) = \sum^r_{i=1} s_i$, where s_i is the symbol value at position i of the vertex label s. If the symbol at ith index position is the don't care symbol "*" then this symbol position is omitted when computing the Hamming weight. Depending on the parity of its Hamming weight, we say that a vertex is even or odd.

Similarly, we can define a Hamming weight for the $PG^{\{1\}}_1$ subgraphs of the product graph, by simply starting the summation from 2 when computing the Hamming weight. If the Hamming weight of a $PG^{\{1\}}_1$ is even, we say that it is an even subgraph. Otherwise it is an odd subgraph. We can also compute the Hamming weight for a $PG^{\{1,2\}}_2$ subgraph and say that it is even or odd.

Just like the definition of snake order for vertices, we can also define snake order for subgraphs $PG^{\{1\}}_1$ or $PG^{\{1,2\}}_2$ in the obvious way. This will be useful later.

Suppose that a sorted sequence is stored in some r-dimensional product network in snake order. The following lemma shows how to split the sequence into N subsequences such that the subsequence u contains every Nth term beginning with uth term.

Lemma 1 *Let S be a sorted sequence stored in some r-dimensional product network in snake order. By reversing the values at odd $PG^{\{1\}}_1$ subgraphs and then erasing the lowest-dimension edges from the product network, we obtain N copies of the product network with $r - 1$ dimensions. The uth copy (where $u \in \{0 \cdots N-1\}$) will contain every Nth term of the original sequence beginning with the uth value.*

Proof is omited due to space limitations. The interested reader can refer to [?].

3   Sorting Algorithm

The heart of the proposed sorting algorithm is the multi-way merge operation. The multiway merge algorithm combines N sorted sequences $A_i = (a_{i,0}, a_{i,1}, \ldots, a_{i,n-1})$, for $i = 0, \ldots, N-1$, into a single sorted sequence $J = (j_0, j_1, \ldots, j_{nN-1})$. For simplicity, we assume n to be a power of N, N^r, where $r > 1$.

To merge N sequences of N^r keys each, we initially assume the existence of an algorithm which can sort

N^2 keys. We make no assumption about the efficiency of this algorithm as yet. In Section 5 we discuss several possible ways to obtain efficient algorithms for this purpose. The purpose of this assumption is to maintain the generality of discussions, independent of the factor network used to build the product network.

3.1   Multiway Merge Algorithm

Here we consider how to merge N sorted sequences, $A_i = (a_{i,0}, a_{i,1}, \ldots, a_{i,n-1})$, for $i = 0, \ldots, N-1$, into a single large sorted sequence.

The merge operation consists of the following steps:

1. Distribute the keys of each sorted sequence A_i among N sorted subsequences $B_{i,j}$, for $i = 0, \ldots, N-1$ and $j = 0, \ldots, N-1$. The subsequence $B_{i,j}$ will have the form $(a_{i,j}, a_{i,j+N}, a_{i,j+2N}, \ldots, a_{i,j+(n-N)})$, for $i = 0, \ldots, N-1$ and $j = 0, \ldots, N-1$. This is equivalent to picking every Nth element of A_i starting with the jth element and putting them in $B_{i,j}$. Note that each subsequence $B_{i,j}$ is sorted since we put the elements in the same order as they appeared in A_i.

2. Merge the N subsequences $B_{i,j}$ into a single sorted sequence C_j, for $j = 0, \ldots, N-1$. This is done with a recursive call to the multiway merge process if the total number of keys in $B_{i,j}$ is at least N^2. If the number of keys in $B_{i,j}$ is N, a sorting algorithm for sequences of length N^2 is used to obtain a single sequence, because a recursive call to the merge process would not make much progress in this case (this point will be clear by the end of this subsection).

3. Interleave the sequences C_j into a single sequence $D = (d_0, d_1, \cdots, d_{N^{r+1}-1})$. The first N terms of the sequence D is obtained by reading the first element from each C_j, $j = 1 \cdots N$. The next set of N terms in D are obtained by reading the second value from each C_j, $j = 1 \cdots N$, etc.

   We prove below that D is now "almost" sorted; the potential dirty area (window of keys not sorted) has length no larger than N^2.

4. Clean the dirty area. To do so we start by dividing the sequence D into N^{r-1} subsequences of N^2 consecutive keys each. That is, the first N^2 terms of D are labeled as E_1, the next N^2 terms are labeled as E_2, and so on.

   We then independently sort the E_i subsequences in alternate orders by using the algorithm which we assumed available for sorting N^2 keys. E_i is transformed into a sequence F_i where F_i contains the keys of E_i sorted in non-decreasing order if i is even or in non-increasing order if i is odd, for $i = 0, \ldots, N^{r-1} - 1$.

   Now, we apply two steps of odd-even transposition between the sequences F_i, for $i = 0, \ldots, N^{r-1} - 1$. In the first step of odd-even transposition, each pair of sequences F_i and F_{i+1}, for i even, are compared element by element. Two sequences G_i and G_{i+1} are formed where $g_{i,k} = min(f_{i,k}, f_{i+1,k})$ and $g_{i+1,k} = max(f_{i,k}, f_{i+1,k})$. In the second step of the odd-even transposition, G_i and G_{i+1} for i odd are compared in a similar manner to form the sequences H_i and H_{i+1}.

Finally, we sort each sequence H_i in non-decreasing order, generating sequences I_i, for $i = 0, \ldots, N^{r-1} - 1$. The final sorted sequence J is the concatenation of the sequences I_i.

We need to show that the process described actually merges the sequences. To do so we use the well known zero-one principle.

Lemma 2 *When sorting an input sequence of zeroes and ones, the sequence D obtained after the completion of step 3 is sorted except for a dirty area which is never larger than N^2.*

Proof: Assume that we are merging sequences of zeroes and ones. Let z_i be the number of zeroes in sequence A_i, for $i = 0, \ldots, N - 1$. The rest of elements in A_i are ones. Step 1 breaks each sequence A_i into N subsequences $B_{i,j}$, $j = 0, \ldots, N - 1$. The number of zeroes in a subsequence $B_{i,j}$ is $\lfloor z_i/N \rfloor + q_{ij}$, where $q_{ij} = 1$ if $j \leq z_i \bmod N$. Observe that, for a given i, the sequences $B_{i,j}$ can differ from each other in their number of zeroes by at most one.

At the start of step 2, each column j is composed of the subsequences $B_{i,j}$ for $i = 0, \ldots, N - 1$. At the end of step 2, all the zeroes are at the beginning of each sequence C_j. The number of zeroes in each sequence C_j is the sum of the number of zeroes in $B_{i,j}$ for fixed j and $i = 0, \cdots, N - 1$. Thus, two sequences C_j can differ from each other by at most N zeroes. In step 4 we interleave the N sorted sequences into the sequence D by taking one key at a time from each sequence C_j. Since any two sequences C_j can differ in their number of zeroes by at most N, and since there are N sequences being interleaved, the length of the window of keys where there is a mixture of ones and zeroes is at most N^2. ∎

Lemma 3 *Step 4 cleans the dirty area.*

Proof: We know that the dirty area of the sequence D, obtained in step 4, has at most length N^2. If we divide the sequence D into consecutive subsequences, E_i, of N^2 keys each, the dirty area can either fit in exactly one of these subsequences or be distributed between two adjacent subsequences.

If the dirty area fits in one subsequence E_i, then after the initial sorting and the odd-even transpositions, the sequences H_i contain exactly the same keys as the sequences E_i, for $j = 0, \ldots, N^{r-1}$. Then, the last sorting in each sequence H_i and the final concatenation yield a sorted sequence J.

However, if the dirty area is distributed between two adjacent subsequences, E_i and E_{i+1}, we have two subsequences containing both zeroes and ones. After the first sorting, the zeroes are located at one side of F_i and at the other side of F_{i+1}.

One of the two odd-even transposition steps will not affect this distribution, while the other step is going to move zeroes from the second sequence to the first and ones from the first to the second. After these two steps, H_i is filled with zeroes or H_{i+1} is filled with ones. Therefore, only one sequence contains zeroes and ones combined. The last step of sorting will sort this sequence. Then the entire sequence J will be sorted. ∎

The reader can observe that at the end of Step 3, the dirty area will still have length N^2 even when we are merging N sequences of length N each. Thus, we do not make much progress when we apply the multiway merge process in this case. Here we assume the availability of a special sorting algorithm designed for the two dimensional version of the product network. In subsequent sections we discuss several methods to obtain such algorithms as we consider more specific product networks.

3.2 Application of Merging Algorithm to Sort

Using the above algorithm, and an algorithm to sort sequences of length N^2, it is easy to obtain a sorting algorithm to sort a sequence of length N^r, for $r \geq 2$.

First divide the sequence into subsequences of length N^2 and sort each subsequence independently. Then, apply the following process until only one sequence remains:

1. Group all the sorted sequences obtained into sets of N sequences each. (If we are sorting N^{r+1} keys, then initially there will be N^{r-2} groups, each containing N sequences of length N^2).

2. Merge the sequences in each group into a single sorted sequence using the algorithm shown in the previous section. If now there is only one sorted sequence then terminate. Otherwise go to Step 1.

4 Implementation in Homogeneous Product Networks

Here we mainly focus on the implementation of the multiway merge algorithm in PG_r in detail. The sorting algorithm trivially follows from the merge operation as described above. The initial scenario is N sorted sequences, of N^{r-1} keys each, stored in the N subgraphs $[u]PG_{r-1}^r$ in snake order. Before the sorting algorithm starts each processor holds one of the keys to be sorted. During the sorting algorithm, each processor needs enough memory to hold at most two values being compared.

Step 1: To explain how this step can be implemented, we refer to Lemma 1. By Lemma 1, if we reverse the order in "odd" $PG_1^{\{1\}}$ subgraphs, then we obtain the sequence $B_{u,v}$ in the subgraph $[u,v]PG_{r-2}^{r,1}$ sorted in snake order.

Reversing the order in a G subgraph can be performed by a permutation routing algorithm available for G.

Step 2 This step is implemented by merging together the sequences in subgraphs $[u,v]PG_{r-2}^{r,1}$ with the same u value into one sequence in $[v]PG_{r-1}^1$. If $r-1=2$, the merging is done by directly sorting with an algorithm for PG_2. If $r-1>2$, this step is done by a recursive call to the multiway merge algorithm.

Step 3 No movement of data is involved in this step, and we obtain a sequence sorted almost completely except for a small dirty area in the case of sorting zeroes and ones.

Step 4 This step cleans the dirty area. The $PG_2^{\{1,2\}}$ subgraphs are ordered by the snake order. In this step we independently sort the keys in $PG_2^{\{1,2\}}$ subgraphs, where the sorted order alternates in "consecutive" subgraphs. We now perform two steps of odd-even transposition between these subgraphs. In the first step, the nodes in the "odd" $PG_2^{\{1,2\}}$ subgraphs are compared with corresponding nodes in their "predecessor" subgraphs. The values are exchanged if the value in the predecessor subgraph is larger. In the second step of odd-even transposition, the values in the nodes of the "even" $PG_2^{\{1,2\}}$ subgraphs are compared (and possibly exchanged) with those of their predecessor subgraphs. A final sorting within each of the $PG_2^{\{1,2\}}$ subgraphs ends the merge process.

One point which needs to be examined in more detail here is that, depending on the factor graph G, the two elements that need to be compared and possibly exchanged with each other may or may not be adjacent in PG_r. If G has a hamiltonian path, then the nodes of G can be labeled in the order they appear on the hamiltonian path to define the sorted order for G. Then, the two steps of odd-even transposition sort is easy to implement since it involves communication between adjacent nodes in PG_r. If however G is not Hamiltonian (e.g. a complete binary tree), the two elements that need to be compared may not be adjacent, but they will always be in a common G subgraph. In this case permutation routing within G may be used to perform the compare-exchange step as follows: two nodes that need to compare their values send their values to each other. Then, depending on the result of comparison, each node can either keep its original value if the values were already in correct order, or they drop the original value and keep the new value if they were out of order.

4.1 Analysis of Time Complexity

To analyze the time taken by the sorting algorithm we will initially study the time taken by the merge process in a k-dimensional network. This time will be denoted as $M_k(N)$. Also let $S_2(N)$ denote the time required for sorting in PG_2 and $R(N)$ denote the time required for permutation routing in G.

Lemma 4 *Merging N sorted sequences of N^{k-1} keys in PG_k takes $M_k(N) = 2(k-2)S_2(N)+3(k-2)R(N)+ S_2(N)$ time steps.*

Proof: The time taken by step 1 of the merge process is just the time to reverse the order of the keys in $PG_1^{\{1\}}$-subgraphs. This process can be done with a permutation routing algorithm for G, that takes time $R(N)$. Step 2 is a recursive call to the merge procedure for $k-1$ dimensions, and hence will take $M_{k-1}(N)$ time. Step 3 does not take any computation time. Finally, step 4 takes the time of one sorting in PG_2, two permutation routings in G (for the steps of odd-even transposition), and one more sorting in PG_2.

Therefore, the value of $M_k(N)$ can be recursively expressed as:

$$M_k(N) = M_{k-1}(N) + 2S_2(N) + 3R(N)$$

with initial condition

$$M_2(N) = S_2(N)$$

that yields

$$M_k(N) = 2(k-2)S_2(N) + 3(k-2)R(N) + S_2(N)$$

$\blacksquare$

Theorem 1 *For any factor graph G, the time complexity of sorting N^r keys in PG_r is $O(r^2 S_2(N))$.*

Proof: By the algorithm of Section 3.2 the time taken to sort N^r keys in PG_r is the time taken to sort in a 2-dimensional subgraph and then merge blocks of N sorted sequences into increasing number of dimensions. The expression of this time is as follows:

$$S_r(N) = S_2(N) + M_3(N) + M_4(N) + \cdots + M_{r-1}(N) + M_r(N)$$

$$= (r-1)S_2(N) + (2S_2(N) + 4R(N))\sum_{i=3}^{r}(i-2)$$

$$= (r-1)^2 S_2(N) + 1.5(r-1)(r-2)R(N).$$

Since $S_2(N)$ is never smaller than $R(N)$, the time obtained is $S_r(N) = O(r^2 S_2(N))$. $\blacksquare$

The following corollary presents the asymptotic complexity of the algorithm and one of the main results of this paper.

Corollary 1 *If G is a connected graph, the time complexity of sorting N^r keys in PG_r is at most $O(r^2 N)$.*

Proof: The basic observation is that, if G is a connected graph, it is always possible to obtain an algorithm for PG_2 with complexity $S_2(N) = O(N)$. To do so we simply emulate the 2-dimensional grid in PG_2 with constant dilation and congestion [2]. Then, the $O(N)$ algorithm presented by Schnorr and Shamir [9] for sorting N^2 keys on two-dimensional $N \times N$ grid can be emulated by PG_2 with complexity $O(N)$, leading to $S_2(N) = O(N)$. Hence, any arbitrary N^r-node r-dimensional product network can sort with complexity $O(r^2 N)$. $\blacksquare$

5 Application to Specific Networks

In this section we obtain the time complexity of sorting for several product networks in the literature. To do so, we obtain upper bounds for the value of $S_2(N)$ for each network. Using this value in Theorem 1 will yield the desired running time.

Grid: Schnorr and Shamir [9] have shown that it is possible to sort N^2 keys in a N^2-node 2-dimensional grid in $O(N)$ time steps. This value of $S_2(N)$ implies that our algorithm will take $O(r^2N)$ time steps to sort N^r keys in a N^r-node r-dimensional grid. If the number of dimensions r is bounded, this expression simplifies to $O(N)$. This algorithm is asymptotically optimal when r is fixed since the diameter of the grid with bounded number of dimensions is $O(N)$, and a value may need to travel as far as the diameter of the network.

Mesh connected trees (MCT): This network was introduced in [3] and extensively studied in [2]. It is obtained as the product of complete binary trees. Due to Corollary 1 we can sort in the N^r-node r-dimensional mesh connected trees in $O(r^2N)$ time steps. If r is bounded, we again have $O(N)$ as the running time. This running time is optimal when r is fixed, because the bisection width of r-dimensional MCT is $O(N^{r-1})$ as shown in [2], and in the worst case we may need to move $O(N^r)$ values across the bisection of the network.

Hypercube: From the above analysis given for grids and the fact that the hypercube is a special case of grid with $N = 2$ fixed, it follows that the time to sort in the hypercube with our algorithm is $O(r^2)$. This running time is same as the running time of the well known Batcher's algorithm for hypercubes. In fact, Batcher's algorithm is a special case of the proposed algorithm.

Petersen Cube: Petersen cube is the r-dimensional product of Petersen graph. Product graphs obtained from the Petersen graph are studied in [?]. Like the hypercube, the product of Petersen graph has fixed N, and therefore the only way the graph grows is by increasing the number of dimensions. Since the Petersen graph is hamiltonian, its two-dimensional product contains the 10×10 grid as a subgraph. Thus, we can use any grid algorithm for sorting 100 keys on the two-dimensional product of Petersen graph in constant time. Consequently the r-dimensional product of Petersen graph can sort 10^r keys in $O(r^2)$ time.

Product of de Bruijn and shuffle-exchange networks: To sort in their two-dimensional instances we can use the embeddings of their factor networks presented in [3] which have small constant dilation and congestion. In particular, N^2-node shuffle-exchange network can be embedded in the two dimensional $N \times N$ product of shuffle-exchange networks with dilation 4 and congestion 2. Also N^2-node de Bruijn network can be embedded in the two dimensional $N \times N$ product of de Bruijn networks with dilation 2 and congestion 2. Sorting $n = N^2$ keys in shuffle-exchange or de Bruijn networks requires in $O(\log^2 n)$ time by the Batcher's algorithm. Thus, we can sort on the two dimensional product of shuffle-exchange or de Bruijn network by emulation of N^2-node shuffle-exchange or de Bruijn network in $S_2(N) = O(\log^2 N^2) = O(\log^2 N)$ time steps. Using this in Theorem 1, our algorithm will take $O(r^2 \log^2 N)$ time steps to sort N^r keys. Again, if r is bounded the expression simplifies to $O(\log^2 N)$.

References

[1] K. Batcher, "Sorting Networks and their Applications," in *Proceedings of the AFIPS Spring Joint Computing Conference*, vol. 32, pp. 307–314, 1968.

[2] K. Efe and A. Fernández, "Computational Properties of Mesh Connected Trees: Versatile Architecture for Parallel Computation," in *Proc. 1994 International Conference on Parallel Processing*, vol. I, (St. Charles, IL), pp. 72–76, CRC Press Inc., Aug. 1994.

[3] K. Efe and A. Fernández, "Products of Networks with Logarithmic Diameter and Fixed Degree," *IEEE Transactions on Parallel and Distributed Systems*, 1994. Accepted for publication. Also available as Tech. Rep. 93-8-1, CACS, University of SW. Louisiana, Lafayette, LA, Feb. 1993.

[4] A. Fernández, N. Eleser, and K. Efe, "Generalized Algorithm for Parallel Sorting on Product Networks," Tech. Rep. 95-1-1, CACS, University of SW. Louisiana, Lafayette, LA, Sept. 1995.

[5] D. Nassimi and S. Sahni, "Bitonic Sort on a Mesh-Connected Parallel Computer," *IEEE Transactions on Computers*, vol. C-27, pp. 2–7, Jan. 1979.

[6] D. Nath, S. N. Maheshwari, and P. C. P. Bhatt, "Efficient VLSI Networks for Parallel Processing Based on Orthogonal Trees," *IEEE Trans. Computers*, vol. C-32, pp. 569–581, June 1983.

[7] S. R. Öhring and S. K. Das, "The Folded Petersen Cube Networks: New Competitors for the Hypercube," in *Proceedings of the 5th IEEE Symposium on Parallel and Distributed Computing*, pp. 582–589, Dec. 1993.

[8] F. Preparata and J. Vuillemin, "The Cube-Connected Cycles: A Versatile Network for Parallel Computation," *Communications ACM*, vol. 24, pp. 300–309, May 1981.

[9] C. P. Schnorr and A. Shamir, "An Optimal Sorting Algorithm for Mesh Connected Computers," in *Proceedings of the 18th Annual ACM Symposium on Theory of Computing*, (Berkeley, CA), pp. 255–263, May 1986.

[10] H. Stone, "Parallel Processing with the Perfect Shuffle," *IEEE Transactions on Computers*, vol. C-20, pp. 153–161, Feb. 1971.

[11] C. D. Thompson and H. T. Kung, "Sorting on a Mesh-Connected Parallel Computer," *Communications ACM*, vol. 20, pp. 263–271, Apr. 1977.

PREFETCHING AND I/O PARALLELISM IN MULTIPLE DISK SYSTEMS

Keok-Kee Lee[1]
ekklee@ntuvax.ntu.ac.sg

Peter Varman[1,2]
pjv@rice.edu

[1]School of Applied Science
Nanyang Technological University
Singapore 2263

[2]Department of ECE
Rice University
Houston,Texas 77005

Abstract -- *The problem of improving the performance of external merging in a parallel I/O system using read-ahead prefetching and disk scheduling is studied. A Markov model is developed and numerically solved for a two-disk two-run system. Larger configurations are studied using simulation. Demand-Priority disk scheduling is shown to improve I/O performance.*

1. Introduction

We study the effects of read-ahead prefetching and disk scheduling on parallel I/O performance [1-8]. In read-ahead prefetching an I/O fetch for a block is made before it is actually required by the computation, and input proceeds concurrently with the CPU processing of the set of in-memory blocks. In a *single-disk* system this increases the overlap between CPU and I/O. As the CPU speed increases, this performance advantage diminishes, and disappears when the computation becomes I/O bound. In a *multiple-disk* I/O system, read-ahead prefetching can improve performance by overlapping operations of multiple disks, providing parallelism in the reads. A model of read-ahead prefetching and its effect on I/O parallelism is presented in the following sections.

2. System Model

The system consists of D independent disks and a disk cache. The data consists of R sorted *runs*, distributed evenly among the disks. A run consists of several data *blocks* placed completely on a single disk. The computation merges the R runs together. Within a run the blocks are consumed in sequential order. The leading block of a run is chosen with probability $1/R$ as the next block to be depleted [5,6,9]. The CPU depletes this block and requests the next block from this run. If it is present in the cache the merge continues immediately; else the CPU is blocked until the I/O read containing the requested block is completed.

I/O from the disks is done in units of a *chain*. A chain is a sequence of n consecutive blocks of a run; n is called the *chain length*. I/O is *asynchronous, i.e.,* the CPU initiates the I/O but does not wait for it to complete. T-block *read ahead* is used for each run. T reflects how much earlier an I/O request is made for a block before it is needed. When the number of cached blocks of a run falls below T, an I/O request for the next chain of that run is made. The merge process proceeds concurrently with the input operations until some run has no blocks left in the cache. Then the merge halts until the I/O request containing the desired block is completed. Table 1 describes the behavior precisely. Since I/O is asynchronous, procedure SignalIODone() is invoked by a completing I/O. This flags the availability of the blocks in the cache, and wakes up the merge process if it is waiting for the I/O.

Consider an example of D=R=T=2 and n=3. Let the runs on disks 1 and 2 be A and B respectively. Initially the cache has blocks A_0 - A_2 and B_0 - B_2. The system behavior depends on the block-depletion order. In the sequence below, A_i (B_i) refers to the depletion of block A_i (B_i), #j indicates an I/O request to disk j, and ∇ indicates that the merge waits for I/O at that point.

$$A_0, \#_1, A_1, B_0, \#_2, A_2, \nabla B_1, B_2, B_3, \#_2 B_4, B_5 \nabla$$

After A_0 is depleted, A has 2 cached blocks; since T= 2, a prefetch request is made to disk 1 for blocks A_3-A_5 (the I/O at time p in Fig. 1(a)). Next blocks A_1 and B_0 are depleted; when B_0 is depleted an I/O request for B_3-B_5 is made to disk 2. At this time (q), both disks are busy; when A_2 is depleted, the CPU suspends for A_3 to be read. When the merge resumes at r, assume that blocks B_1-B_5 are depleted. After B_3 is depleted (time t), a prefetch request for B_6-B_8 is made. When B_5 is depleted the merge waits till the I/O completes at time u. Note that at most one disk is busy between times r and u. The average I/O parallelism of the system depends on the relative times that both disks or just one of the disks is busy. In the next section we develop a Markov model for this system and compute the average I/O parallelism for various values of n and T.

2.1 <u>Markov Model</u> Two idealizations are made in the rest of the paper. An infinite speed CPU is assumed. Secondly, all I/Os that are initiated together are assumed to finish together. Consequently, the merge proceeds in *rounds*. Each round consists of a *computation phase* and an *input phase*. In the computation phase, blocks are depleted until some run has no cached blocks left. This run is the *demand run*. During this phase, depletion may also result in other runs having T or less cached blocks. These runs are referred to as *prefetch runs*. A chain of the demand and each prefetch run are fetched in the input phase of the round. Note that the computation phase takes zero time.

Fig.1(b) shows the idealized behavior of the system as the CPU speed increases. CPU operations are denoted by a zero-width dark box. I/O requests initiated at times p and q in Fig. 1(a), now occur at the same time w in Fig. 1(b). The times at which these requests complete (r and s) is x, at which time the depletion of blocks B_1 through B_5 begins. Only one I/O (to disk 2) is initiated during this computation phase at time y (t of Fig. 1(a)), and the round ends at time z.

Definition: **1.** A state of the Markov chain is the integer pair (n_1, n_2) where n_i, $i \in \{1,2\}$ is the number of cached blocks of run i at the start of the computation phase of a round. **2.** A *1-transition* is an I/O transition in which blocks are fetched from only one run, and a *2-transition* is one where blocks are fetched from both runs. **3.** $^nC_k = n!/k!(n-k)!$ is the number of ways of choosing k out of n items.

Lemma 1: A state (n_1, n_2) of the Markov chain satisfies (i) $n_i = n$, $min(n, T) + 1 \leq n_j \leq n+T$, $j \neq i \in \{1,2\}$ (ii) The number of states is $2 \ max\{n, T\} - 1$.

Lemma 2: The probabilties of transition between states in the Markov chain are stated in the tables 2(a)-2(g). There are no 1-transitions from (w,n) to (n,x) for $n \leq T$.

The Markov chain obtained is *ergodic*, and hence unique steady state probabilities exist. The average I/O parallelism is obtained by taking the weighted sum of the I/O parallelism of transitions weighted with the steady-state probability of taking that transition. The above system was solved numerically to obtain the average I/O parallelism and plotted in Figure 3.

2.2 Multiple Runs Per Disk

When there are more runs than disks, each disk may receive several I/O requests in a round. The order in which the requests are serviced affects the performance. We assume a First Come First Served (FCFS) disk-servicing policy. Consider the case when D=2, R=4, n=3, T=1. Let A, B (respectively C, D) be the runs on disk 1 (2). Initially, the cache is loaded with the first three blocks of each run. Assume the following block-depletion sequence. The subscripts on #

indicate the disk number and the run for which I/O is requested.

$$A_0, B_0, D_0, D_1, \#_{2D}, A_1, \#_{1A}, B_1, \#_{1B}, B_2, \nabla B_3, C_0, C_1, \#_{2C}$$
$$C_2 \nabla$$

```
Merge() {
Load cache with first n blocks of each of the R runs.
NumBlocksInCache[i] = n, ∀i=1 ... R.
NumBlocksOutstanding[i] = 0, ∀i=1 ... R.
while (TRUE) {
for each i = 1 ... R {
if (    (NumBlocksInCache[i]    ≤    T) and
NumBlocksOutstanding[i] == 0)) {
        Initiate IORequest for next n blocks of run i;
        NumBlocksOutstanding[i] += n; } }
Choose a run r, with probability 1/R to deplete next.
NumBlocksInCache[r] = NumBlocksInCache[r] - 1;
if (NumBlocksInCache[r] == 0)
    Wait for Asynchronous I/O for run r to complete. }}

SignalIODone() {
Assume run r has completed I/O.
NumBlocksOutstanding[r] = 0;
NumBlocksInCache[r] += n;
if (Waiting for I/O for run r)  WakeupMerge();}
```

Table 1: Program describing System Operation

When block D_1 has been depleted a prefetch request for D_3-D_5 is made to disk 2. Similarly prefetch requests for chains of A and B are also made to disk 1 in that order. When B_2 is depleted there are two I/O requests at disk 1 (runs A,B) and one I/O request at disk 2 (run D). B is the demand run; A and D are prefetch runs. With the FCFS policy, the disks will first serve runs A and D. Only then will B be serviced. The first I/O uses both disks, but the second uses only disk 1. The merge can proceed only after the second I/O is completed. After the depletion of C_2, the CPU again suspends and another I/O transfer is done from disk 2 to fetch blocks C_3-C_5. Figs. 2(a) and (b) show the behavior of the system for the above depletion sequence. The first I/O has a parallelism of 2, but the other two use just 1 disk each time. The poor parallelism is due to the FCFS disk scheduling policy; B is the demand run but is preceded by prefetch requests earlier in the queue.

Now assume that the scheduling at each disk is such that the I/O request of the demand run is served first. We will refer to this as *Demand Priority Scheduling*. In this case (Fig. 2(c)) after the depletion of block B_2, disk 1 will serve run B, and disk 2 will serve run D. After this I/O, the merge can resume. After depleting C_2, the CPU again suspends, and disks 1 and 2 service runs A and C respectively. The disks are now fully utilized, on both I/Os. Note that for the infinite-speed CPU model, all I/Os are only initiated at the end of the computation phase, so all requests at the disk can be ordered to give priority to the demand run. For a finite-speed CPU only queued requests will be prioritized.

	x = n	x ≤ w, x ≠ n	x > w
T+1 ≤ w ≤ n-	$1/2^w$	$^{n+w-x-1}C_{w-x}/2^{n+w-x}$	0
n ≤ w ≤ n+T	$(^{w-1}C_{w-n}+1)/2^w$	$^{n+w-x-1}C_{w-x}/2^{n+w-x}$	0

Table 2(a): 1-transition from state (w, n) to (x, n) n > T

	x > w	T+1 ≤ x ≤ w	x ≤ T
T+1 ≤ w ≤ n+T	0	$^{n+w-x-1}C_{w-x}/2^{n+w-x}$	0
n+1 ≤ w ≤ T	0	0	0

Table 2(b): 1-transition from state (w, n) to (x, n) n ≤ T

	n < x ≤ n+T	T < x ≤ n-1
T+1 ≤ w ≤ n+	0	$^{n+w-x-1}C_{w-x}/2^{n+w-x}$

Table2(c): 1-transition from state(w, n) to (n, x), n > T

	T < x ≤ n	n < x ≤ n+T
T+1 ≤ w ≤ n+T	0	$^{2n+w-x-1}C_{n-1}/2^{2n+w-x}$

Table2(d): 2-transition from state (w, n) to (x, n), n > T

	n+1 ≤ x ≤ Min(n+T, n+w)
n+1 ≤ w ≤ n+T	$^{2n+w-x-1}C_{n-1}/2^{2n+w-x}$

Table2(e): 2-transition from state (w, n) to (x, n), n≤T

	T < x ≤ n	n < x ≤ n+T
T+1 ≤ w ≤ n+T	0	$^{2n+w-x-1}C_{w-1}/2^{2n+w-x}$

Table 2(f): 2-transition from state (w, n) to (n, x), n > T

	n +1 ≤ x ≤ 2n
n+1 ≤ w ≤ n+T	$^{2n+w-x-1}C_{w-1}/2^{2n+w-x}$

Table 2(g): 2-transition from state (w, n) to (n, x), n ≤ T

3. Simulation Results

In the first simulation, D=R=2. For each n and T, we simulate a merge which consumes five million blocks of data. The results are plotted in Fig. 3 together with the analytical values obtained in Section 2.1. The correspondence between the two sets of results is excellent. This validates the simulation model.

Fig. 4 shows the average I/O parallelism against n for different T. For a given chain length, increasing T (that is

prefetching earlier) increases the disk parallelism; but increasing n for a fixed value of T *reduces* the parallelism. Note that the (worst-case) cache memory needed for a run is n+T blocks. The reduction in the parallelism noted above is contrary to our expectations that increasing cache increases I/O parallelism. For any value of trigger, the maximum parallelism is obtained with the smallest value of n = 1. There is a tradeoff in choosing n. Increasing n reduces the parallelism, but may also reduce the average block-access time, by amortizing seek and rotational latencies over all n blocks in the chain. A study to determine optimal choice of n and T to minimize total I/O time will be pursued in the future. Finally we compare the two disk scheduling policies. Simulations were done for D=5, R=50 with different combinations of *n* and T. Fig. 5 shows clearly the benefits of Demand Priority scheduling, which consistently outperformed FCFS.

4. Summary

The performance of external merging in a parallel I/O system using T-block read-ahead prefetching and appropriate disk scheduling was studied. Two parameters, the chain length (number of consecutive blocks fetched in a single I/O) and the trigger T (the number of blocks of a run in cache below which a prefetch is triggered) are used. A random block-depletion model is used to simulate the computation. The results indicate that the highest I/O parallelism is obtained using a chain size of 1 block, and as large a trigger as possible subject to the available cache size. Increasing the chain length for a fixed trigger degrades I/O parallelism, an anomolous behavior with respect to cache. Demand Priority scheduling improves the performance over FCFS.

REFERENCES

1. M. Y. Kim, "Synchronized disk interleaving", *IEEE Trans. Computers*, C-35(11), (1986), p. 978-988.
2. M. Livny, S. Khoshafian and H. Boral, "Multi-Disk management algorithms", in *Proc. ACM SIGMETRICS Conf.*, (1987), p. 69-77.
3. A.L.N.Reddy and P. Banerjee, "An evaluation of multiple-disk I/O systems", *IEEE Trans. Computers*, C-38(12), 1989, p. 1680-1690.
4. D. F. Kotz and C. S. Ellis, "Prefetching in file systems for MIMD multiprocessors", *IEEE Tran. Par. & Dist. Sys.*,1(2), 1990, p. 218-230.
5. J. B. Sinclair, J. Tang and P. Varman, "Impact of Data Placement on Parallel I/O Systems", *Proc. Intl. Conf. Parallel Processing*, 1993.
6. V. Pai , A. Schaffer, P. Varman, "Markov analysis of multiple-disk prefetching strategies for external merging", *Theoret. Comp. Sci.*, (128) 1994, 211-239.
7. J. S. Vitter and E.A.H. Shriver, "Optimal Disk I/O with Parallel Block Transfer", in *Proc. 1990 ACM STOC*, 1990, pp. 159-169..
8. L.Q. Zheng and P. A. Larson, *Speeding up External Mergesort*, C.S. Dept., University of Waterloo, TR-CS-92-40, August 1992.
9. S. C. Kwan and J. L. Baer, "The I/O performance of multiway mergesort and tagsort", *IEEE Trans. Comp.*, C-34(4), 1985, 383-387.

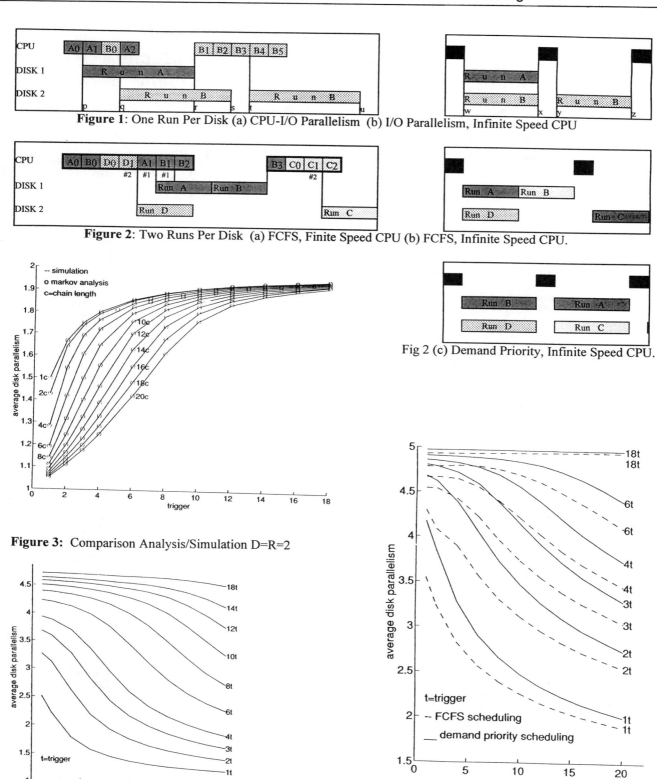

Figure 1: One Run Per Disk (a) CPU-I/O Parallelism (b) I/O Parallelism, Infinite Speed CPU

Figure 2: Two Runs Per Disk (a) FCFS, Finite Speed CPU (b) FCFS, Infinite Speed CPU.

Fig 2 (c) Demand Priority, Infinite Speed CPU.

Figure 3: Comparison Analysis/Simulation D=R=2

Figure 4: I/O Parallelism for D=R=5

Figure 5: FCFS vs Demand Priority Policy, D=5 R=50

Conflict-Free Path Access of Trees in Parallel Memory Systems With Application to Distributed Heap Implementation *

Sajal K. Das and Falguni Sarkar
Dept. of Computer Science
University of North Texas
Denton, TX   76203
E-mail: {das,sarkar}@cs.unt.edu

M. Cristina Pinotti
Istituto di Elaborazione della Informazione
CNR, Pisa
Italy
E-mail: pinotti@iei.pi.cnr.it

Abstract

Mapping the nodes of a data structure to a given set of memory modules in a multiprocessor system, such that different sets of nodes (called templates) can be accessed in parallel and without memory conflicts, is an important problem. The most commonly used templates of a tree are the subtrees of a given height, the nodes in any given level, and the paths from the root to the leaves. For conflict-free path access in a binary tree (or heap), we present for the first time two elegant algorithms. The load distribution of tree nodes among memory modules is well balanced except for one module which stores only the root. We also discuss how to apply these results for optimal implementation of a distributed heap on the hypercube architecture.

1 Introduction

A shared memory multiprocessor system consists of several processors and memory modules. In order to allow concurrent access of shared data, different portions or *templates* of the data structure should be stored in different memory modules. Another goal is to minimize the number of memory modules required to facilitate such a scheme.

The *data distribution* problem deals with the mapping mechanism of the nodes of a data structure to a given set of memory modules so as to allow *conflict-free* access to all instances of the templates of interest. In this paper, we consider tree structures which have wide applications such as in logic programming, divide-and-conquer computation, branch-and-bound algorithms, image processing, and so on. The most commonly used templates of a tree are subtrees of a given height, nodes at any level, or paths from the root to the leaves.

Access of subtrees in parallel has been studied by many researchers. Among them, the simple recursive solution due to Das and Sarkar [1] uses an optimal number of memory modules.

In this paper, we study conflict-free access of path templates for trees. In a top-down (or bottom-up) manipulation of an insert or delete operation in a heap, an element is tricked down (or bubbled up) along a given path from the root to a leaf. A par-

allel heap implementation is expected to support several operations (in batches) concurrently. Even for efficient implementation of a single insert/delete operation in parallel, it is desirable that the nodes (i.e. elements) lying on a path be stored in different memory modules so that they can be accessed and manipulated simultaneously by multiple processors. This motivates us to assign memory modules to the nodes of a heap so that any path from the root to any of the leaves can be accessed conflict-free.

One obvious solution to this problem is to allocate one memory module to every level of the tree (or heap). But this distributes the load in a very uneven fashion. In this paper, we present two algorithms for conflict-free path access in a binary heap and at the same time keep the load on different memory modules well balanced, except for one module which is assigned to the root. We then define a new type of data structure, called *slope-trees*, for efficiently implementing distributed heaps on the hypercube architecture.

2 Conflict-free paths in a heap

Suppose we are given M memory modules and a data structure, specified by a graph $G = < V, E >$, where V is the set of nodes in the data structure and E is the set of edges. Two nodes are connected by an edge if there is an instance of a template containing both of them. The conflict-free distribution of nodes into memory modules is like assigning *colors* from the set $\{0, 1, 2, \ldots, M-1\}$ to the nodes of V, such that all the nodes in any subgraph (or template) of interest are assigned different colors. Let $\mathcal{C}(x)$ denote the memory module or color assigned to the node x.

We present two recursive and load balanced algorithms such that any path of length h in a binary tree T_h of height h can be accessed conflict-free, using $h+1$ memory modules.

2.1 A recursive heap coloring algorithm

The root is assumed to be at level 0 in this algorithm, which is recursive in nature and requires no new color assignment on one of the subtrees as the heap grows. Given a coloring of a binary tree T_h, the tree T_{h+1} is colored by decomposing it into three parts – the root of T_{h+1}, the left subtree LT_h, and the right subtree RT_h. The color assignment of LT_h is the same as that of T_h using colors $\{0, 1, \ldots, h\}$. A new color $(h+1)$ is assigned to the root of T_{h+1}. To color RT_h,

*This work is partially supported by Texas Advanced Technology Program grant under Award No. TATP-003594031.

we first count how many times each color has been used in LT_h, and then assign the colors $\{0, 1, \ldots, h\}$ exactly the same way as in LT_h except that the color used least frequently in LT_h is replaced with the color most frequently used, and so on.

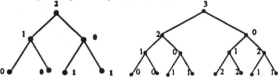

Figure 1: a) Coloring of T_2. b) Coloring of T_3.

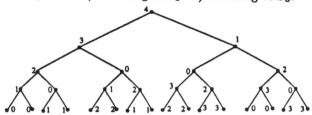

Figure 2: Coloring a heap T_4 using the coloring of T_3.

Figure 1(a) illustrates a coloring of T_2. The color 2 is used only once for the root, while the colors 0 and 1 are used three times each. To color T_3 in Figure 1(b), the coloring of the left subtree LT_2 is copied from Figure 1(a). The root is assigned a new color 3. The coloring of the subtree RT_2 is obtained from LT_2 such that color 0 is replaced with 2 and vice versa. This coloring in Figure 1(b) has been used in Figure 2 to color T_4, in which color 4 is used only for the root, and colors $\{0, 1, 2, 3\}$ are used 8, 7, 8, 7 times respectively. So keeping aside the new color, 0 and 2 are the most frequently used colors while 1 and 3 are the least frequently used. Our algorithm for coloring a binary heap T_{h+1} is formally described below.

Procedure Heap-Coloring-Recursive
1. Decompose T_{h+1} into the root, left subtree LT_h, and right subtree RT_h.
2. Color LT_h using colors from $S = \{0, 1, \ldots, h\}$.
3. Assign the new color $(h + 1)$ to the root.
4. Let f_i = frequency of color i in LT_h, for $0 \le i \le h$.
5. Sort f_i's in the increasing order. Let $f_{(i)}$ be the i-th element in this sorted sequence and $\mathcal{C}(i)$ be the corresponding color.
6. Define $\Phi : S \to S$, such that $\Phi(\mathcal{C}(i)) = \mathcal{C}(h - i)$.
7. Color RT_h with the help of LT_h but using color $\Phi(\mathcal{C}(i))$ wherever color $\mathcal{C}(i)$ was used.

It is easy to prove that the procedure Heap-Coloring-Recursive assigns a different color to every node on any path from the root to a leaf.

Let us analyze the load of each memory module, by finding bounds on the maximum and minimum loads among the modules. Let us define the sequence $\{f_{h,j} \mid 0 \le j \le h\}$ such that $f_{h,j}$ denotes the frequency of color j used for a complete binary tree T_h of height h. The minimum and maximum values in these sequences for binary trees of height $h \le 30$ are shown in Table 1. Let $x_h = \max_{0 \le j \le h}\{f_{h,j}\}$ and $y_h = \min_{0 \le j \le h}\{f_{h,j}\}$.

We state the following two lemmas and refer to [2] for their proofs.

Lemma 1: $y_h = x_{h-1} + 1$ for $h \ge 2$.

Lemma 2: The ratio $\frac{x_h}{y_h} \le 2$

Table 1: **Maximum and Minimum of $\{f_{h,i}\}$'s.**

h	x_h	y_h	h	x_h	y_h
3	6	4	17	17800	9661
4	8	7	18	32844	17801
5	15	9	19	61866	32845
6	24	16	20	117520	61867
7	46	25	21	222729	117521
8	78	47	22	422998	222730
9	130	79	23	820008	422999
10	237	131	24	1562374	820009
11	446	238	25	2972179	1562375
12	786	447	26	5665146	2972180
13	1442	787	27	10931673	5665147
14	2653	1443	28	21084763	10931674
15	5202	2654	29	40426936	21084764
16	9660	5203	30	78013980	40426937

Table 1 also implies that $\frac{x_h}{y_h} \ge \frac{3}{2}$ for $h \ge 5$, which is yet to be proved analytically for all values of h. Therefore, the load distribution among various memory modules is almost balanced, within a small constant factor.

Applying Lemmas 1 and 2, we can show that $x_h - y_h \le x_{h-1}$ and $y_h \le 2y_{h-1} - 1$. Furthermore, the sequence of elements $\{f_{h,j}\}$, for $0 \le j \le h$, exhibits several interesting properties. It is symmetric and hence for every even number h, the value x_h appears at least twice in the sequence $\{f_{h,j}\}$. This property leads to the proof that $\frac{x_{h+1}}{y_{h+1}} \ge \frac{3}{2}$ when h is even.

2.2 A better algorithm for heap coloring

The previous algorithm is recursive in the sense that the color of the existing nodes remains unchanged as the tree grows. However, it suffers from one major drawback, namely it is difficult to identify explicitly the color assignment to an arbitrary node. This may be an important requirement for heap management in parallel. Whenever a node is inserted or accessed, we need to know which memory module is to be accessed. In the following, we present an algorithm to determine the color of a node quickly. We also show that the load is more uniformly distributed among the modules.

Recall that a binary tree T_h of height h requires at least $(h + 1)$ colors to make every path conflict-free. For the sake of convenience, we count the levels of T_h from -1 to $h - 1$ and use the set of colors $\{-1, 0, 1, \ldots, h - 1\}$. As before, we fix one color, say -1 for the root. For simplicity, we describe the algorithm only for trees with even height $h = 2q$ for some q. The case for trees with odd heights can be handled with minor modification.

We partition T_h horizontally into two parts. The upper part, denoted as U, comprises the nodes of the tree between levels 0 (the root is at level -1) and $q - 1$. The bottom part, denoted as D, consists of all the nodes between levels q and $2q - 1$. The tree (except the root) is again partitioned vertically into two parts, each being a subtree T_{h-1}. The left (resp.

right) part, denoted as L (resp. R), is the subtree rooted at the left (resp. right) child of the root of T_h. Thus the original tree T_h (except the root) is divided into four parts: UL, UR, DL and DR as illustrated in Figure 4. Now colors 0 through $q-1$ are used for UL, using one color for each level of this subtree. Similarly, colors q through $2q-1$ are used for UR. Next, colors q through $2q-1$ are used for DL while colors 0 through $q-1$ are used for DR.

We describe how to color the part DR, since coloring of DL is done in a similar manner. Let us view DR as a $q \times q$ matrix M, each of whose elements is associated with a cluster of nodes. All the nodes in a level $l \geq q$ are grouped into q clusters which correspond to the elements in the $(l-q)$-th row of the matrix M. The first row of clusters is formed such that $\lfloor \frac{2^l}{q} \rfloor$ consecutive nodes from the q-th level belong to every cluster except the last cluster, which contains all nodes in level q that are not covered by the first $(q-1)$ clusters in the first row. The (i,j)-th cluster, where $1 \leq i < q$ and $0 \leq j < q$, is formed by taking the children of all nodes in the $(i-1,j)$-th cluster. Hence the maximum number of nodes in a cluster in the ith row can be $2^{i-q}\lfloor \frac{2^q}{q} \rfloor + (q-1)$. Next a *Latin square*[1] Q of size $q \times q$ is constructed as follows. The first row of Q is the set elements $\{0, 1, \ldots, q-1\}$. Every other row is formed by shifting the previous row cyclically one position to the right, i.e. Q is also a *circulant matrix*. All nodes in the cluster corresponding to the matrix element $M[i,j]$ are assigned the color $Q[i,j]$, the (i,j)-th element of Q.

This algorithm is formally presented below. Figure 3 illustrates coloring of a heap T_4 of height four.

Procedure Heap-Coloring-Better

1. Assign color -1 to the root.
2. Identify each node of the tree with an ordered pair (i,j) where i is the level of the node, and j is the index of the node within level i from left to right, $0 \leq j \leq 2^{i+1} - 1$,
3. (Coloring of UL) Assign color i to every node (i,j) such that $0 \leq i \leq q-1$ and $0 \leq j \leq 2^i - 1$.
4. (Coloring of UR) Assign color $q+i$ to every node (i,j) such that $0 \leq i \leq q-1$ and $2^i \leq j \leq 2^{i+1} - 1$.
5. Create a $q \times q$ Latin square Q with symbols from the set $\{0, 1, \ldots, q-1\}$. The (i,j)-th element of Q is defined as $Q[i,j] = (j-i) \bmod q$.
6. (Coloring of DL) For every node (i,j), where $q \leq i \leq 2q-1$ and $0 \leq j < 2^{i-q}\lfloor \frac{2^q}{q} \rfloor(q-1)$, decompose j as $j = l \cdot 2^{i-q}\lfloor \frac{2^q}{q} \rfloor + m$, for $0 \leq m \leq 2^{i-q}\lfloor \frac{2^q}{q} \rfloor - 1$. Assign color $Q[i-q+1, l] + q$ to the node (i,j). For every node (i,j) such that $i > (q-1)$ and $2^{i-q}\lfloor \frac{2^q}{q} \rfloor(q-1) \leq j \leq 2^i - 1$, assign color $Q[i-q+1, q] + q$. This step decomposes the left half of the i-th level of the heap into q clusters. Each of the first $(q-1)$ clusters contains $2^{i-q}\lfloor \frac{2^q}{q} \rfloor$ nodes while the remaining $2^i - (q-1)2^{i-q}\lfloor \frac{2^q}{q} \rfloor$

---

[1] A $q \times q$ matrix Q is called a *Latin square* [4] if every integer from the set $\{0, 1, 2, \ldots, q-1\}$ occurs exactly once in each row and each column of Q.

nodes belong to the q-th cluster.

7. (Coloring of DR) For every node (i,j), such that $q \leq i \leq 2q-1$ and $2^i \leq j < 2^{i-q}\lfloor \frac{2^q}{q} \rfloor(2q-1)i$, decompose $j = 2^i + l2^{i-q}\lfloor \frac{2^q}{q} \rfloor + m$ for $0 \leq m \leq 2^{i-q}\lfloor \frac{2^q}{q} \rfloor - 1$. Assign color $Q[i-q+1, l]$ to the node (i,j). For every node (i,j) such that $i > (q-1)$ and $2^i + 2^{i-q}\lfloor \frac{2^q}{q} \rfloor(q-1) \leq j \leq 2^{i+1} - 1$, assign color $Q[i-q+1, q]$.

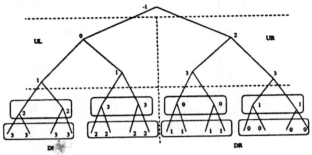

Figure 3: Illustration of Heap-Coloring-Better

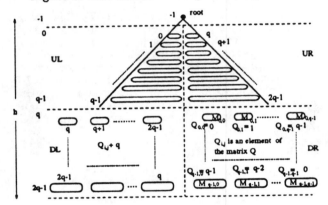

Figure 4: Sketch of algorithm Heap-Coloring-Better

It can be proved that procedure Heap-Coloring-Better assigns a different color to every node on any path from the root to a leaf in the heap. For proofs of the following two lemmas, refer to [2].

Lemma 3: For a binary heap T_h of height $h = 2q$, the algorithm Heap-Coloring-Better assigns the color $\mathcal{C}(i,j)$ to the node (i,j), where $\mathcal{C}(i,j) =$

$$
\begin{cases}
-1 & \text{if } i = -1 \\
i & \text{if } 0 \leq i < q \ \& \ 0 \leq j \leq 2^i - 1 \\
i + q & \text{if } 0 \leq i < q \ \& \ 2^i \leq j < 2^{i+1} \\
(\lfloor \frac{j}{2^i/q} \rfloor - (i-q)) \bmod q + q & \text{if } i \geq q \ \& \ 0 \leq j \leq 2^i - 1 \\
(\lfloor \frac{j-2^i}{2^i/q} \rfloor - (i-q)) \bmod q & \text{if } i \geq q \ \& \ 2^i \leq j \leq 2^{i+1} - 1.
\end{cases}
$$

Lemma 4: Let $f_{h,i}$ for $0 \leq i \leq h-1$, denote the frequency of color i occurring in a binary heap T_h of height $h = 2q$ for $q \geq 1$. Let $x_h = \max\{f_{h,i} \mid 0 \leq i \leq h-1\}$ and $y_h = \min\{f_{h,i} \mid 0 \leq i \leq h-1\}$. Then the ratio $\frac{x_h}{y_h}$ is asymptotically bounded by $1 + \frac{q}{\alpha}$, where $\alpha = \lfloor \frac{2^q}{q} \rfloor$.

3 Heap Implementation on Hypercube

Priority queues are very useful data structures for solving numerous problems based on the principle of selecting the smallest (or largest) element from a totally ordered set. Not much attention has been paid for implementing parallel priority queues on the distributed memory architectures. We have recently introduced a new data structure, called *slope-trees* [3], for efficiently storing a heap or priority queue on the hypercube architecture in a load balanced manner, having no additional communication overhead. This data structure facilitates the design of work-optimal parallel algorithms for performing *insert* and *deletemin* operations on a heap in the hypercube machine. The detailed scheme for mapping slope-trees into hypercubes and the heap manipulation algorithms are omitted due to space limitation. Interested readers may refer to Das, Pinotti and Sarkar [3].

In a distributed environment, heap properties suggest a communication pattern between a node with its parent and sibling nodes. So we look for a mapping of the heap-nodes to the hypercube processors which will ensure that a node X is mapped to a processor which is adjacent with the two processors in which the parent and sibling nodes of X are mapped. This is not achievable by embedding a binary tree into the hypercube while keeping the load on different processors balanced. This motivates us to define a new data structure, called slope-tree, defined below.

Definition: A *slope-tree*, $ST(l,h)$, of base l and height h is a tree with the following properties: (i) There are $h+1$ levels in the tree, where the root at level 0 has only one child at level 1; (ii) There are l nodes in the base level l; (iii) Every level k, for $1 \leq k \leq l-1$, contains exactly one node. The left child of this node is at level $k+1$ (when $k = l-1$, it is the leftmost node at level l), and the right child is the $(l-k)$-th node at level l, counting from left starting at 0; and (iv) Each node r, where $0 \leq r \leq l-1$, at the base level l is the root of a binary tree B_r of height at most $h-l+1$ and at least $h-l$. If the level $h-l+1$ of B_r for $0 \leq r \leq (l-1)$ is not completely full, then all the B_k's, with $r+1 \leq k \leq l-1$, are complete binary trees of height $h-l$, i.e. the last level of these trees is not filled. Note that a slope-tree is filled level by level from left to right.

Figure 5 shows a slope-tree, $ST(3,6)$ of base $l = 3$, height $h = 6$ and consisting of 45 nodes.

To map a slope-heap, $ST(l,h)$ into a hypercube with l nodes, we construct a Hamiltonian path, H, of length l using binary reflected gray codes. The top l levels of the slope-tree are embedded using H. The base level is embedded horizontally using H, starting from the left. All successive levels are mapped into the hypercube by circularly rotating H one step towards left. Our first implementation of priority queue on the *single-port* hypercube model achieves an optimal speed-up for inserting or deleting $b = \Omega(\log \log n)$ items in a b-bandwidth[2] slope-heap, using $O(\log n)$ processors, where n is the total number of items in the heap. In a second implementation, the insertion

---

[2] A b-bandwidth slope-heap contains b items in each node.

or deletion of b items is accomplished in $\mathcal{O}(\frac{h \log^2 b}{l})$ time using $\mathcal{O}(\frac{bl}{\log b})$ processors, where h is the height and l is the number of nodes in the base level of the slope-tree. Thus a 'nearly' optimal speed-up of $\mathcal{O}(\frac{lb}{\log^2 b})$ is achieved on the single port hypercube model. However, this approach yields an optimal speed-up when implemented on the (more powerful) *pipelined hypercube* model, requiring $\mathcal{O}(\frac{bl}{\log b})$ processors and $\mathcal{O}(\frac{h \log b}{l})$ time.

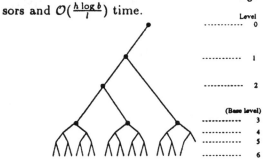

Figure 5: A slope-tree $ST(3,6)$ of base 3 & height 6.

4 Conclusions

We have proposed two solutions to the conflict-free data access for binary trees where the templates of interests are paths from the root to the leaves. (The second algorithm has also been generalized for any complete k-ary trees [2].) The number of nodes mapped onto different memory modules (i.e., colors) is almost balanced. It would be interesting to investigate the same problem for arbitrary trees, directed acyclic graphs or other kinds of special graphs.

The first approach for conflict-free path access, can be used as a general strategy for any recursive data structure (e.g. binomial tree) that is composed of two smaller copies. The ratio of maximum to minimum load will remain less than two in all such cases.

We have recently applied the concept of conflict-free path access to develop efficient parallel algorithms for heaps on the distributed memory hypercube machines, which achieves optimal speed-up [3]. In this regard we have introduced a new data structure, called slope-trees, to map a heap onto the hypercube processors in a load balanced manner.

References

[1] S. K. Das and F. Sarkar, "Conflict-Free Data Access of Arrays and Trees in Parallel Memory Systems", *Proc. of the sixth IEEE Symp. on Parallel and Dist. Proc.*, Dallas, Oct 1994, pp. 377-384.

[2] S. K. Das, F. Sarkar, and M.C. Pinotti, *Conflict-Free Access of Trees in Parallel Memory Systems and Its Generalization*, Tech Rep. CRPDC-94-21, Univ. of North Texas, Denton, Nov. 1994.

[3] S. K. Das, M.C. Pinotti, and F. Sarkar, *Optimal Parallel Priority Queues in Distributed Memory Hypercubes*, Tech Rep. CRPDC-94-23, Univ. of North Texas, Denton, Dec. 1994.

[4] J. Denes, and A. D. Keedwell, *Latin Squares and Their Applications*, Academic Press, NY, 1974.

SOLVING A TWO-DIMENSIONAL KNAPSACK PROBLEM ON A MESH WITH MULTIPLE BUSES

Darrell R. Ulm and Johnnie W. Baker
Department of Mathematics and Computer Science
Kent State University
Kent, OH   44242
dulm@mcs.kent.edu   jbaker@mcs.kent.edu

Abstract – This paper describes a parallelization of the sequential dynamic programming method for solving a 2D knapsack problem where multiples of n rectangular objects are optimally packed into a knapsack of size $L \times W$ and are only obtainable with guillotine-type (side to side) cuts. The parallel algorithm is described and analyzed for the mesh with multiple buses, and a parallel method that finds the solution set is included. This algorithm runs in $O(n+L+W)$ time for $O(n)$ objects and uses $O(LW)$ processors for a 2D knapsack problem with a capacity of $L \times W$. This result matches the time and cost obtainable on the CRCW PRAM implementation for the same problem.

INTRODUCTION

Knapsack problems are usually stated as finding a subset from the given set of objects such that the summed object profit is maximized while not exceeding the knapsack size or violating any constraints. Such problems often appear in computer science and operations research, e.g. in cutting stock applications. This problem is related to the well studied 0-1 knapsack problem [1] which has been solved somewhat efficiently with linear systolic arrays [2].

Many 2D knapsack algorithms considering a variety of problem constraints have been discovered [3]-[4]. The problem examined herein is in the class NP, but it can be solved sequentially in $O(LW(n+L+W))$ [4], where n is the number of objects, and L and W are the dimensions of the knapsack. This running time is not polynomial in terms of the input size since the the knapsack capacity is encoded in $\log_2(L)+\log_2(W)$ bits. Such a time bound is called pseudo-polynomial [1]. With this in mind an effective parallelization is sought to speed up the so-called inefficient sequential solution.

Recently, the use of buses either alone or alongside existing networks has been considered. The 2D knapsack problem is one such problem that greatly benefits from the use of multiple buses. The idea behind adding buses to existing networks is to reduce the communication diameter and increase the speed of certain operations [5]. There is a practical limit to the number of buses we can possibly add, yet this model of computation is buildable [6], and lately there has been much interest in developing algorithms for this architecture.

Mesh with Multiple Buses

A mesh with multiple buses is an architecture where each row and column of a $M \times N$ mesh oriented group of homogeneous processors has a bus. The normal north, south, east and west neighbor links also exist. Each processor in row i and column j is addressed by Cartesian coordinate (i, j). The upper left corner is labeled $(0, 0)$ while the lower rightmost is $(M-1, N-1)$. Each processor has $O(\log_2 MN)$ registers and in unit time may communicate a bit to a neighbor, broadcast a bit to a bus, read a bit from a bus, or perform a simple arithmetic or boolean operation. This machine operates in SIMD mode where instructions are broadcast in $O(1)$ time by one processor writing a command to its vertical bus, and then all processors on that bus broadcasting to their rows. The exclusive communication model allows multiple processors to read a bus concurrently with one processor writing. The common model permits any number of processors on a bus to concurrently write one bit as long as all processors send the same value [7]-[8]. It is assumed that the time to broadcast a bit on a bus is $O(1)$, a claim based on current research with actual hardware [9]-[10].

Maximum of n Values on a Bus.   Herein, a method is described to find the maximum of integer values stored one per processor connected to a single bus. This can be accomplished by using a simple bit-polling routine iterating k times for k-bit integer words. At each iteration, starting with the most significant bit, all processors with a 1 broadcast that value to the bus. If there were not any processors with a 1, then it continues to the next iteration. However, if at least one processor sent a 1, every sending processor remains active, recording the bit, while the other processors go inactive. After the last iteration, the remaining active processors contain the maximum value, and it may be broadcast bit-by-bit to all processors on the bus. On the mesh with multiple broadcast this procedure may execute on all rows or columns in parallel. [7].

2D KNAPSACK PROBLEM

The 2D knapsack problem involves cutting a rectangle into predefined sizes of smaller pieces, where each piece has an associated profit value, and the solution set is the list of cut pieces that adds up to the maximum profit obtainable [4]. This cutting problem only allows side-to-side or *guillotine* cuts of the knapsack, or in other words, all cuts must be made perpendicular from one edge of a rectangle to the other. Also, multiples of any one object are permitted in the solution. These are reasonable assumptions based on how real materials are actually cut, and furthermore these restrictions greatly simplify the complexity of the inherent problem where otherwise there would be no known pseudo-polynomial solution [11]. Algorithms to solve this problem include tree search or dynamic programming methods [12]. This paper employs the later technique.

The 2D knapsack problem seeks to fill an area of dimensions (L, W) with n rectangles of size (l_i, w_i) where $i = 1, 2, \ldots, n$. The profits are non-negative values, $\Pi_1, \Pi_2, \ldots, \Pi_n$, associated with each rectangle. Using these parameters, the maximum profit of $\Pi_1 z_1 + \Pi_2 z_2 + \ldots + \Pi_n z_n$ is to be computed where z_i is defined as a non-negative integer such that the knapsack is partitioned into z_i multiples of rectangle i, having the size (l_i, w_i) [4]. Objects may either have a fixed orientation or be allowed to rotate 90°, but either way the complexity of the algorithm is the same. An additional n objects of dimensions (w_i, l_i), corresponding to each unrotated object (l_i, w_i) with profit Π_i, are added when the problem calls for rotations.

The knapsack function $F(x, y)$, derived from dynamic programming techniques, is computed such that for any location (x, y), $F(x, y)$ is the largest profit obtainable and must satisfy the following inequalities corresponding to the guillotine cutting restrictions [4].

$$0 \leq x \leq L \quad , \quad 0 \leq y \leq W,$$
$$F(x, y) \geq 0,$$
$$F(x_1 + x_2, y) \geq F(x_1, y) + F(x_2, y),$$
$$F(x, y_1 + y_2) \geq F(x, y_1) + F(x, y_2),$$
$$F(l_i, w_i) \geq \Pi_i \quad (i = 1, \ldots, n)$$

Sequential Dynamic Programming Algorithm.
A solution to the problem without an exponential running time is obtained using dynamic programming. Each rectangular object will be considered in turn, and a table of the best cutting profits for each location in the knapsack will be kept. Then the entire table is scanned, and lengths and widths are summed to create a configuration of rectangles that produces the optimal profit. The following relations are used to derive the sequential algorithm shown below [11].

$$F_0(x, y) = \max\{0, \Pi_j | l_j \leq x \wedge w_j \leq y\}.$$
$$F_k(x, y) = \max\{F_{k-1}(x, y),$$
$$F_{k-1}(x_1, y) + F_{k-1}(x_2, y),$$
$$F_{k-1}(x, y_1) + F_{k-1}(x, y_2)\}$$
$$0 < x_1 \leq x_2 \quad , \quad x_1 + x_2 \leq x,$$
$$0 < y_1 \leq y_2 \quad , \quad y_1 + y_2 \leq y,$$

```
—SEQUENTIAL ALGORITHM—
for i ← 0 to L      -STEP 1-
  for j ← 0 to W
    F_{i,j} ← 0
for k ← 0 to n       -STEP 2-
  for i ← 0 to L
    for j ← 0 to W
      if (l_k ≤ i and w_k ≤ j and Π_k > F_{i,j})
        F_{i,j} ← Π_k
for i ← 0 to L      -STEP 3-
  for j ← 0 to W
    for k ← 0 to ⌊i/2⌋
      sum ← F_{k,j} + F_{i-k,j}
      if (sum > F_{i,j}) F_{i,j} ← sum
    for k ← 0 to ⌊i/2⌋
      sum ← F_{i,k} + F_{i,j-k}
      if (sum > F_{i,j}) F_{i,j} ← sum
```

In step 1 of this algorithm all locations in the knapsack are set to 0. In step 2 each object is considered, placing the highest profit values in all locations where an object fits. Step 3 scans the 2D table from the lowest row to the highest, summing all possible combinations of vertical and horizontal cuts at each location and retaining the two objects whose sum of profits is the largest. The code in above shows how $F(x, y)$ values are computed iteratively [11]. Since only guillotine cuts are used, the partial rectangular solution, at location (i, j), is cut into two pieces with a cut parallel to the x axis at some value of y. Because of symmetry, only x-cuts from $x = 1$ to $i/2$, and y-cuts from $y = 1$ to $j/2$ are considered for any given (i, j). The sequential algorithm has time complexity $O(LW(n+L+W))$, where n is the number of objects, and L, W are the dimensions of the knapsack[11].

THE PARALLEL ALGORITHM

The serial algorithm can be directly parallelized for the mesh with multiple buses architecture with $L \times W$ processors. Each processor is designated as $P_{i,j}$ and computes $F_{i,j}$. Optimal speedup is easily obtainable using a $L \times W$ processor CRCW PRAM machine, as the execution time is $O(n + L + W)$ [13].

The parallelization given below is accomplished by first noting that any at location in the array, $F_{a,b}$ is dependent on other $F_{i,j}$ entries at lower subscripted positions. In fact, the only values needed are those in row a at positions $j \leq b$ and those in column b at positions $i \leq a$. This data dependency prevents a row or column of F values from being computed in parallel, but a diagonal set of F values can be calculated concurrently. Since the number of diagonals traversed is $O(L + W)$, $O(\max\{L, W\})$ iterations are needed to scan all diagonals.

Let $P_{i,j}$ be a processor located inside the diagonal region, $diag/2 \leq i + j \leq diag$, where $diag$ is an integer denoting the current diagonal. Then the F value at this processor contributes to calculating $F_{diag-j,j}$ and $F_{i,diag-i}$. The parallel variables $V_{i,j}$ and $H_{i,j}$ hold shifted $F_{i,j}$ values that are added to the current location for vertical and horizontal directions respectively. In each iteration these values shift to the next location, and a new value is added to the shifting parallel variables $V_{i,j}$ and $H_{i,j}$. These shifting operations use N, S, E, W connections such that one shift is completed in unit time. After the sums are collected in participating processors, the maximum of all sums is computed from variables $sumV_{i,j}$ and $sumH_{i,j}$ for each row and column, and the result is stored in the processor on that row or column's diagonal. This is done using the previously described algorithm. A column finds the maximum of $sumV_{diag-j,j}$ through $sumV_{(diag-j)/2,j}$ storing the result in $F_{diag-j,j}$. Likewise, a row finds the maximum of $sumH_{i,diag-i}$ through $sumH_{i,(diag-i)/2}$ placing the result in $F_{i,diag-i}$. All operations inside this while loop only take $O(1)$ time.

—PARALLEL ALGORITHM—
—STEP 1—
forall processors $P_{i,j}$ do in parallel
    $F_{i,j} \leftarrow 0$, $V_{i,j} \leftarrow 0$, $H_{i,j} \leftarrow 0$
—STEP 2—
for $k \leftarrow 1$ to n
    forall processors $P_{i,j}$ do in parallel
        if $(i \geq l_k$ and $j \geq w_k)$
            $F_{i,j} \leftarrow max\{F_{i,j}, \Pi_k\}$
$diag \leftarrow 2$
—STEP 3—
while $(diag \leq L + W)$ do
    forall processors $P_{i,j}$ do in parallel
        if $(diag \leq i + j$ and $i > 0$ and $j > 0)$
            $V_{i,j} \leftarrow V_{i-1,j}$
            $H_{i,j} \leftarrow H_{i,j-1}$
        if $(i * 2 + j = diag)$ $V_{i,j} \leftarrow F_{i,j}$
        if $(j * 2 + i = diag)$ $H_{i,j} \leftarrow F_{i,j}$
        if $(diag/2 \leq i + j \leq diag)$
            $sumV_{i,j} \leftarrow F_{i,j} + V_{i,j}$
            $sumH_{i,j} \leftarrow F_{i,j} + H_{i,j}$
            $F_{diag-j,j} \leftarrow BUSMAX(sumV_{diag-j,j},$
               $sumV_{diag-j-1,j}, ..., sumV_{(diag-j)/2,j})$
            $F_{i,diag-i} \leftarrow BUSMAX(sumH_{i,diag-i},$
               $sumH_{i,diag-i-1}, ..., sumH_{i,(diag-i)/2})$
    $diag \leftarrow diag + 1$

Time Complexity Analysis. The initialization phase (1) takes unit time since broadcasting is done in $O(1)$. Step 2 sequentially loops through n objects in $O(n)$ time. In step 3, the diagonalization phase, we iterate through $O(L + W)$ diagonals. The inner part of the loop shifts $V_{i,j}$ and $H_{i,j}$ values using N, S, E, W connections in $O(1)$ time. Then in unit time $F_{i,j}$ values are appended to the shifting variables $V_{i,j}$, $H_{i,j}$, and a summation is taken of these values and the current data at $F_{i,j}$. The parallel maximums along all active processors on each row and column are computed using the previously mentioned routine. Because the BUSMAX function loops though $k = log_2 D$ bits where D is an integer of maximum size, and there are $O(L + W)$ diagonals to traverse, the total complexity of the while loop is $O((L + W)log_2 D)$. The final complexity for the 2D knapsack algorithm is thus $O((n + L + W)log_2 D)$ if the bit complexity is considered in all bus activity, such as broadcasting. Considering that only integers are transmitted, and the word size complexity is usually ignored in sequential algorithms, our complexity is $O(n + W + L)$.

OBTAINING THE SOLUTION CUTTING PATTERNS

An outline of the algorithm to find the optimal solution patterns is parallelized as follows, and the algorithm itself is given below. The algorithm assumes the values of parallel variables X and Y are assigned within the preceding dynamic programming algorithm. Here, $X_{i,j}$ is the x-cut value that was used to define the maximum profit $F_{i,j}$. If $X_{i,j} = 0$, then an x-cut was not used. Similarly, $Y_{i,j}$ stores the y-cut value used to calculate $F_{i,j}$. If the value is zero, then a y-cut was not used. If both $X_{i,j}$ and $Y_{i,j}$ are 0, then the (i, j) rectangle was not subdivided and contains only the single original object. This backtracking algorithm only transmits bits, so the final time complexity to backtrack has nothing to do with the machine word size.

In step 1 all locations are flagged as not being in the solution, and after step 2, the only active processor is the upper rightmost cell (L, W). The main loop, step 3, executes $O(L + W)$ times, and inside the main loop all operations take unit time. In each iteration the $newactive_{i,j}$ parallel variable is initialized to 0, then for all active cells on the diagonal, $X_{i,j}$ and $Y_{i,j}$ cut-values are checked to see if there exist two pieces that were used in the summation of $F(i, j)$. If both $X_{i,j}$ and $Y_{i,j}$ are 0, then no cut was made at that location. Otherwise the processors containing the two pieces that were added into that location, are flagged as active. This is done by two column broadcasts when $X_{i,j} > 0$ and then two row broadcasts when $Y_{i,j} > 0$. The diagonal backtracking insures that there are no write conflicts. When a processor is active and on the current diagonal, but the $X_{i,j}$ and $Y_{i,j}$ values are 0, then the rectangle was used but never cut and therefore must be an original object. This case flags such locations to be added into the solution set.

The initialization phases, (1) and (2), take $O(1)$ time. Step 3 sequentially loops $O(L+W)$ times where all commands inside the loop require $O(1)$ time. Bits are the only values broadcast so the backtracking algorithm runs in $O(L+W)$.

—SOLUTION RETRIEVAL—
—STEP 1—
forall processors $P_{i,j}$ do in parallel
  $act_{i,j} \leftarrow 0,\ insolut_{i,j} \leftarrow 0$
—STEP 2—
$act_{L,W} \leftarrow 1$
$diag \leftarrow L + W$
—STEP 3—
while $(diag \geq 2)$ do
  forall processors $P_{i,j}$ do in parallel
    $newact_{i,j} \leftarrow 0$
    if $(act_{i,j} = 1$ and $i + j = diag)$
      if $X_{i,j} > 0$
        $newact_{i-X_{i,j},j} \leftarrow 1$
        $newact_{X_{i,j},j} \leftarrow 1$
      elseif $Y_{i,j} > 0$
        $newact_{i,j-Y_{i,j}} \leftarrow 1$
        $newact_{i,Y_{i,j}} \leftarrow 1$
      else
        $insolut_{i,j} \leftarrow 1,$
    $act_{i,j} \leftarrow newact_{i,j}$
  $diag \leftarrow diag - 1$

SUMMARY

In this paper the 2-dimensional knapsack problem is addressed for a mesh with multiple buses with $L \times W$ processors. Using buses to broadcast data and to find the maximum value from all processors on a bus, the CRCW PRAM algorithm time of $O(n + L + W)$ can be matched. The sequential algorithm executes in $O(LW(n + L + W))$, and the previous best parallel algorithm runs in $O(n\gamma + (LW)log_2(max(L, W)))$ where $\gamma \leq log_2(max(L, W))$ on a hypercube with the same number of processors. A solution retrieval technique for this architecture is presented with a time complexity of $O(L + W)$ insuring that the overall algorithm performs in $O((n + L + W)log_2 D)$ if bit complexity is considered and $O(n + L + W)$ if bit complexity is ignored.

REFERENCES

[1] S. Martello and P. Toth, *Knapsack Problems*, John Wiley & Sons, New York, (1991), pp. 2-7.

[2] V. Aleksandrov and R. Andonov, "A Systolic Linear Array for the Knapsack Problem," *Parallel and Distributed Processing*, 17, (1991), pp. 285-299.

[3] J.C. Herz, "Recursive Computational Procedure for Two-dimensional Stock Cutting," *IBM Journal of Research and Development*, 16, (1967), pp. 462-469.

[4] P.C. Gilmore and R.E. Gomory, "Multistage Cutting Stock Problems of Two or More Dimensions," *Operations Research*, 13, (1965), pp. 94-120.

[5] J. Rothstein, "Bus Atomata, Brains, and Mental Models," *IEEE Trans. Systems Man Cybernetics (SMC-18)*, (2)4, (1988), pp. 522-531.

[6] M. Maresca and H. Li, "Connection Autonomy and SIMD Computers: a VLSI Implementation," *Journal of Parallel and Distributed Computing*, 7, (1989), pp. 302-320.

[7] R. Miller, V.K. Prasanna-Kumar, D.I. Reisis and Q.F. Stout, "Parallel Computations on Reconfigurable Meshes," *IEEE Transactions on Computers*, (42)6, (1993), pp. 678-696.

[8] K. Nakano, T. Masuzawa and N. Tokura, "A Sub-Logarithmic Time Sorting Algorithm on a Reconfigurable Array," *IEICE Transactions*, E-(74), (November, 1991), pp. 3894-3901.

[9] Q.F. Stout, "Meshes with Multiple Busses," *Proc. 27th IEEE Symposium on the Foundations of Computer Science*, (1986), pp. 264-273.

[10] K.P. Kumar and C.S. Raghavendra, "Array Processor with Multiple Broadcasting," *Proceedings of the 12th Int. Symp. Comput. Architecture*, (June, 1985), pp. 2-10.

[11] P.C. Gilmore and R.E. Gomory, "The Theory and Computation of Knapsack Functions," *Operations Research*, 15, (1966), pp. 1045-1174.

[12] N. Christofides and C. Whitlock, "An Algorithm for Two-Dimensional Cutting Problems," *Operations Research*, 26(1), (1977), pp. 30-45.

[13] D. Ulm and P.Y. Wang, "Solving a Two-Dimensional Knapsack Problem on SIMD Computers," *International Conference on Parallel Processing*, volume 3, (1992), pp. 181-184.

A Framework for Solving Geometric Problems on Enhanced Meshes*

V. Bokka, H. Gurla, S. Olariu, J. L. Schwing, L. Wilson

Department of Computer Science, Old Dominion University, Norfolk, VA 23529-0162

Abstract

In this work we show that a number of seemingly unrelated geometric problems can be solved simply and elegantly by formulating them as instances of a general computational paradigm called the Multiple Query problem (MQ, for short). An instance of the MQ problem consists of a collection $A = \{a_1, a_2, \ldots, a_n\}$ of *items*, a collection $Q = \{q_1, q_2, \ldots, q_m\}$ $(1 \le m \le n)$ of *queries*, a decision problem $\phi : Q \times A \rightarrow \{\text{"yes", "no"}\}$, and an associative and commutative function f operating on subsets of A. For every query q_i, let S_i be the set of items a_j in A for which $\phi(q_i, a_j) = \text{"yes"}$. The *solution* of q_i is defined to be $f(S_i)$. The goal is to solve all queries in Q. As it turns out [4], if A and Q are stored one item and at most one query per processor on a mesh with multiple broadcasting (MMB, for short) of size $\sqrt{n} \times \sqrt{n}$, then any algorithm that solves the MQ problem must take at least $\Omega(m^{\frac{1}{3}}n^{\frac{1}{6}})$ time in the worst case. Using the MQ paradigm we provide time-optimal solutions to a variety of problems in computer graphics, image processing, and robotics.

1 Introduction

The maturation of computer science has exacerbated the need to consolidate isolated algorithms and techniques into general computational paradigms. The benefits come in two flavors: first, it becomes clear that problems previously treated in isolation from one another belong to the same class, in the sense that they can be solved similarly, by formulating them as instances of a more general paradigm; second, once established, the paradigm is likely to become a powerful tool that can be used to solve a host of other related problems. The main contribution of this work is to present a unifying algorithmic framework for solving the following geometric problems:

1. Given a set A of non-intersecting line segments and a set B of points in the plane, for each point in B determine the closest segment intersected by a vertical ray originating at it.
2. Given a set A of arbitrary line segments and a set B of parallel lines, for every line in B determine the number of line segments in A it intersects.
3. Given point-sets A and B in the plane, for each point in B determine a point in A closest to it.
4. Given point-sets A and B in the plane, determine all the points in B that are interior to the convex hull of A. Somewhat surprisingly, we solve this problem without computing the convex hull of A.

*Work supported, in part, by NSF grant CCR-9407180

5. Given point-sets A and B, determine whether the convex hull of A (resp. B) is contained in the convex hull of B (resp. A). We provide a solution to this problem without computing the respective convex hulls.
6. Given point-sets A and B of points in the plane, determine whether they are linearly separable. If so, construct a separating line for A and B. As before, we provide a solution to this problem even if the convex hulls of A and B are not known.
7. Given separable point-sets A and B, compute the common tangents of their convex hulls; as before, we provide a solution to this problem even if the convex hulls of A and B are not known.

The algorithmic framework is provided by a generic solution to the following problem termed the *Multiple Query* problem (MQ, for short). A generic instance of the MQ problem has the following parameters:

- an arbitrary set $A = \{a_1, a_2, \ldots, a_n\}$ of items;
- an arbitrary set $Q = \{q_1, q_2, \ldots, q_m\}$ of queries;
- a decision problem $\phi : Q \times A \rightarrow \{\text{"yes", "no"}\}$;
- an associative and commutative function f operating on subsets of A.

For every query q_i in Q, let $S_i = \{a_j \in A \mid \phi(q_i, a_j) = \text{"yes"}\}$. The *solution* of q_i is $f(S_i)$. The goal is to find the solution of every query in Q.

An MMB of size $\sqrt{n} \times \sqrt{n}$ is a mesh-connected SIMD machine enhanced by the addition of row and column buses as illustrated in Figure 1. For details see [6, 8]. The MMB underlies the DAP family of computers [11] and has received a great deal of attention in the literature [1, 2, 3, 5, 6, 7, 8, 9, 10].

The authors have shown recently that if A and Q are stored one item and at most one query per processor on an MMB of size $\sqrt{n} \times \sqrt{n}$, then any algorithm that solves a non-trivial instance of the MQ problem must take at least $\Omega(m^{\frac{1}{3}}n^{\frac{1}{6}})$ time in the worst case. Our main contribution is to outline time-optimal solutions to the seven problems above on the MMB.

2 The Generic MQ Algorithm

Consider a generic instance of the MQ problem. The set A is stored, one item per processor, in an MMB $\mathcal{R}$ of size $\sqrt{n} \times \sqrt{n}$. The set Q is stored in the first $\frac{m}{\sqrt{n}}$ columns of $\mathcal{R}$, one query per processor. To make the notation less cumbersome, we write $s = m^{\frac{1}{3}}n^{\frac{1}{6}}$. In this notation, we view $\mathcal{R}$ as consisting of submeshes $R_{i,j}$ $(1 \le i, j \le \frac{\sqrt{n}}{s})$, of size $s \times s$, with $R_{i,j}$

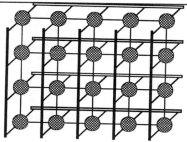

Figure 1: An MMB of size 4×5

involving processors $P(r,c)$ with $1 + (i-1)s \leq r \leq is$, $1 + (j-1)s \leq c \leq js$. Occasionally, we shall view $\mathcal{R}$ as consisting of submeshes $S_1, S_2, \ldots, S_{\frac{\sqrt{n}}{s}}$ of size $s \times \sqrt{n}$, with S_i $(1 \leq i \leq \frac{\sqrt{n}}{s})$ involving the submeshes $R_{i,1}, R_{i,2}, \ldots, R_{i,\frac{\sqrt{n}}{s}}$. Every S_i is called a *slice* of $\mathcal{R}$.

The algorithm consists of three distinct stages that we now outline.

Stage 1. The goal is to replicate the set Q in each submesh $R_{i,j}$. It is important to note that, at the end of the Stage 1, having replicated the queries, we have, in fact, partitioned the original instance of the MQ problem into several instances, each local to a $R_{i,j}$. Every local instance involves the subset of A stored by the processors in $R_{i,j}$, the set Q of queries, a decision problem, and a function f;

Stage 2. The goal of this stage is to solve, in each $R_{i,j}$, the local instance of the MQ problem; this will be done, in parallel, for all the submeshes $R_{i,j}$;

Stage 3. The goal of this stage is to combine the solutions obtained in Stage 2 into the solution of the original MQ problem.

We now look at each stage in detail.

Stage 1. The goal is to replicate the set Q into the leftmost $\frac{m}{s}$ columns of each $R_{i,j}$. To achieve this, we move the queries in each column k $(1 \leq k \leq \frac{m}{\sqrt{n}})$ of $\mathcal{R}$ into columns $(k-1)\frac{\sqrt{n}}{s} + 1$ through $k\frac{\sqrt{n}}{s}$ of each $R_{i,j}$. To begin, every processor $P(r,k)$ $(1 \leq r \leq \sqrt{n})$ broadcasts the query it holds horizontally to processor $P(r,r)$. In turn, processor $P(r,r)$ broadcasts the query received vertically to processors $P(ts + 1 + (r-1) \bmod s, r)$ $(0 \leq t \leq \frac{\sqrt{n}}{s} - 1)$.

As a result of this data movement, the queries in column k of $\mathcal{R}$ have been replicated in the diagonal processors of the submeshes in every slice. From now on, every slice is processed in parallel: the queries stored by the diagonal processors of $R_{i,1}$ are replicated, using the row buses in slice S_i, into the $(k-1)\frac{\sqrt{n}}{s} + 1$-th column of each $R_{i,j}$ in slice S_i. Next, the queries stored by the diagonal processors in $R_{i,2}$ are replicated, using the row buses in slice S_i, into the $(k-1)\frac{\sqrt{n}}{s} + 2$-th column of each $R_{i,j}$, and so on.

It is easy to see that the task of replicating the queries in one column of $\mathcal{R}$ takes $O(\frac{\sqrt{n}}{s})$ time. Therefore, as long as $m \geq \sqrt{n}$, the queries initially stored

in the leftmost $\frac{m}{\sqrt{n}}$ columns of $\mathcal{R}$ can be replicated in time $O(\frac{\sqrt{n}}{s} * \frac{m}{\sqrt{n}}) = O(\frac{m}{s}) = O(\frac{m^{\frac{2}{3}}}{n^{\frac{1}{6}}}) \subseteq O(m^{\frac{1}{3}} n^{\frac{1}{6}})$. In case $m < \sqrt{n}$ the queries are replicated in a way similar to the one described. The complexity of this data movement is, again, $O(\frac{m}{s}) \subseteq O(m^{\frac{1}{3}} n^{\frac{1}{6}})$ time. To summarize, we state the following result.

Lemma 2.1. The set Q of queries initially stored in the first $\frac{m}{\sqrt{n}}$ columns of $\mathcal{R}$ can be replicated into the first $\frac{m}{s}$ columns of each $R_{i,j}$ in $O(m^{\frac{1}{3}} n^{\frac{1}{6}})$ time. $\square$

Stage 2. The bus system is ignored and every $R_{i,j}$ will act as an unenhanced mesh. The way the local instance of the MQ problem is solved in each $R_{i,j}$ is application-dependent. In different applications, different strategies will be used. We assume that this stage takes $O(m^{\frac{1}{3}} n^{\frac{1}{6}})$ time.

Stage 3. At the end of Stage 2, every processor of each $R_{i,j}$ that stores a query q_u will store its local solution $f(S_u)$. We wish to combine these local solutions into the solution of q_u for the original instance of the MQ problem. Clearly, once the processing in Stage 3 is complete, the solution to the original instance of the MQ problem has been obtained.

The first goal of this stage is to arrange the ordered pairs $(q_u, f(S_u))$ in column-major order in the leftmost $\frac{m}{s}$ columns of every $R_{i,j}$, sorted by u, the index of their first component. By using the optimal sorting algorithm on the mesh this task can be performed in $O(m^{\frac{1}{3}} n^{\frac{1}{6}})$ time.

From now on, we distinguish two cases.

Case 1. $m \geq n^{\frac{1}{4}}$.

In this case, in every $R_{i,j}$ there is at least one full column of queries. The various slices of $\mathcal{R}$ are processed in parallel. For illustration, we show the processing in slice S_i. Let $(q_u, f(S_u))$ be a generic query-solution pair stored by a processor $P(r,c)$ in $R(i,1)$. By the data movement described in the preamble to this stage, a similar pair is stored by processors $P(r, c+ts)$ in $R_{i,t+1}$, for $1 \leq t \leq \frac{\sqrt{n}}{s} - 1$. In $\frac{\sqrt{n}}{s} - 1$ time units, sequentially, every processor $P(r, c+ts)$ broadcasts to $P(r,c)$ the second component of the pair $(q_u, f(S_u))$ it holds. Thus, in $O(\frac{\sqrt{n}}{s})$ time, $P(r,c)$ can accumulate the solutions of q_u in the whole slice S_i. Since, S_i has s buses, entire columns of queries can be processed in this way. Consequently, the process of accumulating the corresponding solutions for all the queries can be done, in each slice, in $O(\frac{\sqrt{n}}{s} * \frac{m}{s}) = O(\frac{m\sqrt{n}}{s^2}) = O(m^{\frac{1}{3}} n^{\frac{1}{6}})$ time.

Finally, after transposing the first $\frac{m}{s}$ columns into rows in each $R_{i,1}$ the above process can be repeated in the vertical submesh consisting of $R_{1,1}, R_{2,1}, \ldots, R_{\frac{\sqrt{n}}{s},1}$ thus accumulating for every query the corresponding solutions in $O(m^{\frac{1}{3}} n^{\frac{1}{6}})$ time.

Case 2. $m < n^{\frac{1}{4}}$.

In this case, the queries in each $R_{i,j}$ occupy only a

segment of the first column. We assume without loss of generality that for some positive integer c, $s = c * m$. Using local connections only, the m queries in $R_{i,j}$ $(1 \le j \le \frac{\sqrt{n}}{s})$ will be moved vertically, in lock step, into positions $[(j-1) \bmod c] * m + 1$ through $[(j-1) \bmod c] * m + m$ in the first column of $R_{i,j}$. This operation takes no more than $O(m^{\frac{1}{3}} n^{\frac{1}{6}})$ time.

We shall refer to consecutive groups of c of the $R_{i,j}$'s in slice S_i as a *run*. The motivation for this terminology comes from the observation that by the previous data movement, the queries in each run occupy distinct rows and so, using horizontal buses in S_i can be moved in parallel in $O(1)$ time. Specifically, we plan to move the queries in S_i into the columns of $R_{i,1}$. Since there are exactly $\frac{\sqrt{n}}{s} * \frac{m}{s} = \frac{m\sqrt{n}}{s^2} = m^{\frac{1}{3}} n^{\frac{1}{6}}$ runs, the operation of compacting these runs into $R_{i,1}$ takes $O(m^{\frac{1}{3}} n^{\frac{1}{6}})$ time.

Next, we sort the queries in each submesh $R_{i,1}$ $(1 \le i \le \frac{\sqrt{n}}{s})$ in row-major order by query index. This data movement guarantees that the solutions corresponding to the same query will occur next to one another. Proceeding row by row these solutions are accumulated and stored in the leftmost processor in each row. Note that no such processor can contain accumulated results pertaining to more than two distinct queries. Proceeding vertically, the final sums are accumulated for every distinct query in every submesh $R_{i,1}$. Now transposing columns into rows and shifting appropriately we create horizontal runs which will be compacted in $R_{1,1}$, where the queries are sorted again and, as before, the partial results are accumulated. The whole computation can be performed in $O(m^{\frac{1}{3}} n^{\frac{1}{6}})$ time. Thus, we have proved the following result.

Theorem 2.2. Provided that every local instance of the MQ problem in Stage 2 can be solved in $O(m^{\frac{1}{3}} n^{\frac{1}{6}})$ time, the original instance of the MQ problem involving a set A of n items and a set of m queries can be solved in $O(m^{\frac{1}{3}} n^{\frac{1}{6}})$ time on a mesh with multiple broadcasting of size $\sqrt{n} \times \sqrt{n}$. Furthermore, this is time-optimal. $\square$

3 Applications

The purpose of this section is to show that the MQ problem has many, and sometimes unexpected, applications to problems in image processing, pattern recognition, robotics, and computational geometry. Due to severe space limitations, we will only show how each particular problem can be stated as an instance of the MQ problem. For the full details the reader should refer to the journal version of this work.

3.1 The MULTI-LOCATION Problem

Let $A = \{a_1, a_2, \ldots, a_n\}$ and $Q = \{q_1, q_2, \ldots, q_m\}$ $(1 \le m \le n)$ be arbitrary sets of points in the plane. The *Multiple Point Location* problem, (MULTI-LOCATION, for short) asks to determine for

every subscript i $(1 \le i \le m)$ whether the query point q_i lies inside the convex hull $CH(A)$ of A. Our MULTI-LOCATION algorithm can be extended to solve the following related problems as well:

• SEPARABILITY: Determine whether the sets A and Q are linearly separable and if so, find a separating line;

• CONTAINMENT: Determine whether the convex hull of Q (resp. A) is contained in the convex hull of A (resp. Q);

• COMMON-TANGENTS: In case A and Q are separable, find their common supporting lines (i.e. tangents).

We note that if the convex hull of A is known, then the MULTI-LOCATION problem can be solved in $O(\sqrt{m})$ time by using the algorithm of Bhagavathi *et al.* [3]. However, just computing the convex hull of n points is known to take $\Omega(\sqrt{n})$ time on an MMB of size $\sqrt{n} \times \sqrt{n}$. The MULTI-LOCATION problem can be stated as an instance of the MQ problem with the following parameters:

• a set $A = \{a_1, a_2, \ldots, a_n\}$ of items which is precisely the given set of points in the plane;

• a set $Q = \{q_1, q_2, \ldots, q_m\}$ of queries consists of the m query-points;

• a decision problem $\phi : Q \times A \to \{\text{"yes"}, \text{"no"}\}$ such that $\phi(q_i, a_j) = \text{"yes"}$ if and only if the line collinear with the points q_i and a_j is a supporting line for $CH(A)$;

• for every i $(1 \le i \le m)$, let $S_i = \{a_j \in A \mid \phi(q_i, a_j) = \text{"yes"}\}$; we set $f(S_i) = S_i$, in other words, we are interested in the supporting lines, if they exist.

3.2 The CLOSEST-SEGMENT Problem

The *multiple closest segment* problem (CLOSEST-SEGMENT, for short), given a set A of non-intersecting line segments and a set Q of points in the plane, asks to determine for each point in Q, the closest segment in A (if any) intersected by a vertical ray emanating from it. It is well known that the CLOSEST-SEGMENT problem finds numerous applications ranging from visibility, to ray tracing, to robotics, to name just a very few. As it turns out, the CLOSEST-SEGMENT problem can be stated as a MQ problem. For definiteness, we write $A = \{a_1, a_2, \ldots, a_n\}$ and $Q = \{q_1, q_2, \ldots, q_m\}$. The corresponding instance of the MQ problem has the following parameters:

• the set $A = \{a_1, a_2, \ldots, a_n\}$ of segments;

• the set $Q = \{q_1, q_2, \ldots, q_m\}$ of query-points;

• a decision problem $\phi : Q \times A \to \{\text{"yes"}, \text{"no"}\}$ is such that $\phi(q_i, a_j) = \text{"yes"}$ whenever a_j is the first segment in A intersected by the vertical ray originating at q_j and going in the positive y-direction;

• a function f such that $f(S_i) = S_i$, where for every query q_i $(1 \le i \le m)$, we write $S_i = \{a_j \in A \mid \phi(q_i, a_j) = \text{"yes"}\}$.

3.3 The CLOSEST-POINT Problem

For two points p and q let $d(p, q)$ stand for the Euclidian distance between them. The *multiple closest point* problem, given sets A and Q of points in the plane, asks to determine for each point in Q, a point in A that is closest to it in the Euclidian distance sense. We shall refer to this problem as CLOSEST-POINT. CLOSEST-POINT is a fundamental problem that finds additional applications in geographic data processing, computer graphics, image processing, and computational geometry, to name just a few.

We now show that the CLOSEST-POINT problem can be stated as a MQ problem. For definiteness, we write $A = \{a_1, a_2, \ldots, a_n\}$ and $Q = \{q_1, q_2, \ldots, q_m\}$. To simplify the exposition, we shall assume that all the points in A are distinct. The corresponding instance of the MQ problem has the following parameters:

- the set $A = \{a_1, a_2, \ldots, a_n\}$ of points;
- the set $Q = \{q_1, q_2, \ldots, q_m\}$ of query-points;
- a decision problem $\phi : Q \times A \rightarrow \{\text{"yes", "no"}\}$ is such that $\phi(q_i, a_j) = \text{"yes"}$ whenever $d(q_i, a_j) = \min_{1 \leq k \leq n} d(q_i, a_k)$;
- for every i $(1 \leq i \leq m)$, let $S_i = \{a_j \in A \mid \phi(q_i, a_j) = \text{"yes"}\}$; we set $f(S_i) = \min\{j \mid a_j \in S_i\}$, in other words, the solution to query q_i is the point with the smallest subscript that is closest to q_i.

3.4 The MULTI-STABBING Problem

Let $A = \{a_1, a_2, \ldots, a_n\}$ be an arbitrary set of possibly intersecting line segments in the plane and let $Q = \{q_1, q_2, \ldots, q_m\}$ $(1 \leq m \leq n)$ be a set of parallel lines. The lines in Q will be referred to as *query* lines. The *Multiple Stabbing* problem, (MULTI-STABBING, for short) asks to determine for every query-line q_i, the number of segments in A it intersects. The MULTI-STABBING problem is a natural generalization of the stabbing line problem, finding applications to computer graphics, path planning, and computational geometry.

The line segments of A are assumed stored in arbitrary order, one segment per processor, in a mesh $\mathcal{R}$ with multiple broadcasting of size $\sqrt{n} \times \sqrt{n}$. The query-lines in Q are stored, one line per processor, in the first $\frac{m}{\sqrt{n}}$ columns of $\mathcal{R}$. We assume that all the query-lines are parallel to the x-axis and that the line segments are in general position, with no two endpoints sharing the same y-coordinate. Every line segment a_i is specified by its top and bottom endpoints, t_i and b_i, respectively.

The MULTI-STABBING problem can be stated as an instance of the MQ problem with the following parameters:

- the set $A = \{a_1, a_2, \ldots, a_n\}$ of items is precisely the given set of line segments;
- the set $Q = \{q_1, q_2, \ldots, q_m\}$ of queries consists of the m query-lines;
- a decision problem $\phi : Q \times A \rightarrow \{\text{"yes", "no"}\}$ such that $\phi(q_i, a_j) = \text{"yes"}$ if and only if the query-line q_i intersects segment a_j;

- for every i $(1 \leq i \leq m)$, let $S_i = \{a_j \in A \mid \phi(q_i, a_j) = \text{"yes"}\}$. We let $f(S_i) = \mid S_i \mid$, in other words, we are interested in the number of line segments "stabbed" by query-line q_i.

Thus, we have the following result.

Theorem 3.1. Arbitrary instances of MULTI-LOCATION, COMMON-TANGENTS, CLOSEST-SEGMENT, CONTAINMENT, MULTI-LOCATION, CLOSEST-POINT, MULTI-STABBING, and SEPARABILITY, problems involving a set of n "items" and m "queries" can be solved in $O(m^{\frac{1}{3}} n^{\frac{1}{6}})$ on a mesh with multiple broadcasting of size $\sqrt{n} \times \sqrt{n}$. Furthermore, this is time-optimal. $\square$

References

[1] D. Bhagavathi, V. Bokka, H. Gurla, S. Olariu, J. L. Schwing, and I. Stojmenovic, Time-optimal visibility-related problems on meshes with multiple broadcasting, *IEEETPDS*, 1995.

[2] D. Bhagavathi, P. J. Looges, S. Olariu, J. L. Schwing, and J. Zhang, A fast selection algorithm on meshes with multiple broadcasting, *IEEETPDS*, 5, (1994), 772–778.

[3] D. Bhagavathi, S. Olariu, W. Shen, and L. Wilson, A time-optimal multiple search algorithm on enhanced meshes, with applications, *JPDC*, 22, (1994), 113–120.

[4] V. Bokka, H. Gurla, S. Olariu, J. L. Schwing, Multi-querying, a new computational paradigm, Technical Report, Old Dominion University, 1994.

[5] Y. C. Chen, W. T. Chen, G. H. Chen and J. P. Shen, Designing efficient parallel algorithms on mesh connected computers with multiple broadcasting, *IEEETPDS*, 1, (1990) 241–246.

[6] V. K. Prasanna and C. S. Raghavendra, Array processor with multiple broadcasting, *JPDC*, 2 (1987) 173–190.

[7] V. K. Prasanna and D. I. Reisis, Image computations on meshes with multiple broadcast, *IEEE Trans. PAMI*, 11 (1989) 119–125.

[8] R. Lin, S. Olariu, and J. L. Schwing, An efficient VLSI architecture for digital geometry, *Proc. ASAP*, 1994, 392–403.

[9] S. Olariu, J. L. Schwing, and J. Zhang, Optimal convex hull algorithms on enhanced meshes, *BIT*, 33 (1993) 396–410.

[10] S. Olariu and I. Stojmenović, Time-optimal proximity problems on meshes with multiple broadcasting, *Proc. IPPS*, 1994, 94–101.

[11] D. Parkinson, D. J. Hunt, and K. S. MacQueen, The AMT DAP 500, 33rd *IEEE CSIC*, 1988, 196-199.

Binary-Exchange Algorithms on a Packed Exponential Connections Network*

Craig Wong, Catherine Stokley, Quynh-Anh Nguyen, Donna Quammen, and Pearl Y. Wang

Department of Computer Science

George Mason University

Fairfax, VA 22030–4444

{quammen,pwang}@cs.gmu.edu

Abstract

Packed Exponential Connections (PEC) is a uniformly augmented two-dimensional mesh network that attempts to solve the scalability and connectivity problems of very large parallel interconnection networks. This paper examines the implementation of a class of algorithms, referred to as binary-exchange algorithms, on a PEC network. We show that it is possible to perform Global Reduction, One-to-All Broadcast, one-dimensional Fast Fourier Transforms, and Bitonic Merges in $O(\sqrt[4]{N})$ on an N-processor PEC mesh. This compares favorably against standard two-dimensional mesh and hypercube solutions.

keywords: interconnection networks, recursive-doubling, binary-exchange, global reduction, Fast Fourier Transform, PEC networks

1   Binary-Exchange Algorithms

Many parallel algorithms use a common binary-exchange (or butterfly) scheme for the parallel processing of data. This scheme is also known as Ascend/Descend [6] and can be formulated as a recursive computation. For example, recursive doubling is a well-known binary-exchange computation which can be used to find the sum of n elements $\{a_0, a_1, \ldots, a_{n-1}\}$. This requires successive computations beginning with $a'_0 = a_0 + a_1$, $a'_2 = a_2 + a_3$, ..., $a'_{n-2} = a_{n-2} + a_{n-1}$, followed by $a''_0 = a'_0 + a'_2$, $a''_4 = a'_4 + a'_6$, and so forth. Similarly, the Cooley-Tukey one-dimensional Fast Fourier Transform algorithm uses a binary scheme for $k = \log_2 n - 1, \log_2 n - 2, \ldots, 0$ which calculates values of the form $a'_i = a_i + u \cdot a_{i+2^k}$ and $a'_{i+2^k} = a_i + v \cdot a_{i+2^k}$ where u and v are appropriate twiddle factors. Bitonic sorting also is built on top of this type of binary scheme.

A generic *binary-exchange* computation can be defined as one that repeatedly uses data values that are

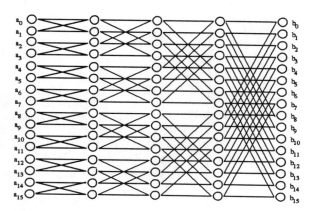

Figure 1: Data Movements in Binary-Based Computations

a power of two apart. Given an initial set of n data values, $\{a_0, a_1, \ldots, a_{n-1}\}$, we consider patterns of processing that involve pairs of the form $a'_i = a_i \otimes a_{i+2^k}$ and $a'_{i+2^k} = a_i \otimes a_{i+2^k}$ for $i = j..(j + 2^k - 1)$, $j = 0, 2^{k+1}, 2 \cdot 2^{k+1}, 3 \cdot 2^{k+1}, \ldots$ The operator $x \otimes y$ denotes an arithmetic, comparator, or set operation that is performed on x and y. The range for the variable k can be either from 0 up to $\log_2 n - 1$, or from $\log_2 n - 1$ down to 0.

The data movement patterns of binary-exchange schemes are illustrated in Figure 1 for $n = 16$. Note that the individual nodes in the diagram represent results of the computations and that the nodes in each column can be referenced as row-wise subscripted variables. For example, the nodes in column two could be labeled as $a'_0 = a_0 \otimes a_1$, $a'_1 = a_0 \otimes a_1$, $a'_2 = a_2 \otimes a_3$, $a'_3 = a_2 \otimes a_3$, etc. It is important to note that the operations performed by each node may be different from those performed by other nodes, and that the subscripted variable notation can denote a single value or a set of values.

The main difficulty in implementing this kind of algorithm on a parallel computer is the range of the communications paths required for completion of the algorithm. If you state that a_i and a_{i+1} are close to each other, then a_i and a_{i+2^k} would be considered far

*This research is partially sponsored by the NASA Goddard Space Flight Center, the CRA Distributed Mentorship Program, and the National Science Foundation

from each other. If the computer system has a shared single memory bank, then the time required to access either your nearest neighbor or a distant element is identical. However, on a distributed system this is frequently not the case. In a hypercube, one node has neighbors $a_{i+1}, a_{i+2^1}, \ldots, a_{i+2^k}$, where k equals $log_2 n$ and n is the total number of nodes, this kind of algorithm works well [6]. However, the wiring of a large degree hypercube is very difficult, and a complete hypercube is considered unusable for very large systems. In a mesh, however, the amount of non-local communication can make this pattern of communications inefficient. In a tree structure, contention at the root nodes may cause inefficiency. In this paper, we describe how this class of algorithms can be mapped to a *packed exponential connections* (PEC) network.

The PEC network tries to solve the problems of scale and connectivity by augmenting a two-dimensional mesh with additional longer connections [1, 7]. PEC was designed by W. Kirkman and was developed for use in the Digital Transform Machine [2], a fine-grain Configurable Hardware Computer. A prototype of this system has been built.

The distribution of PEC connections has been driven by an observed characteristic of applications: data-dependencies (and therefore data communication) tend to be localized. That is, within a parallel processing system, most data-transfers will be over short distances, with decreasing numbers over longer distances. However, binary-exchange algorithms have an equal number of long distance communications and short distance communications. Therefore, these applications may not be naturally mapped to a PEC. However, it will be shown that binary-exchange algorithms can be performed in $O(\sqrt[4]{N})$ on an N-processor PEC mesh. This compares favorably against standard two-dimensional mesh and hypercube solutions.

In Section 2, the PEC network is described. In Section 3, the steps describing the data movements of binary-exchange algorithms on a PEC network are presented. A proof of the implementation's correctness is also given. The effectiveness of this implementation is analyzed using two measures: the amount of information each node can access during and after a step in the computation; and, an upper bound on the amount of communications required for any entire binary-exchange computation.

Subsequent to this general development, we show in Section 4, how this technique can be used in several case studies involving All-to-One and One-to-All Broadcasting, one-dimensional Fast Fourier Transforms, and Bitonic Merges. Using the worst-case analyses developed in Section 3, an evaluation can be made which shows that these binary-exchange computations can be achieved efficiently on PEC networks.

2 PEC Network Description

A *packed exponential connections* (PEC) network is a $\sqrt{N} \times \sqrt{N}$ network in which each node has two sets of interconnections. Every node is connected to N,S,

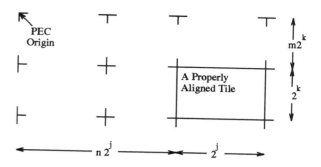

Figure 2: Properly aligned tiles.

E, W neighbors as well as to four other nodes a power of two away. These latter connections place each node in a second two-dimensional mesh of a different scale [1, 7]. This is frequently referred to as a 2D-PEC to distinguish it from a 1D-PEC described in [4, 5].

The nearest neighbor mesh is referred to as the $PEC(0)$ mesh and connects processors 2^0 apart. (This is the standard two-dimensional mesh network.) The four extra connections for each node which connect it to a second 2D mesh are indicated by a pre-assigned PEC value to the node. For example, a processor node labeled as a $PEC(1)$ will also be connected to a mesh of nodes that are a distance of 2^1 apart. Nodes labeled as $PEC(2)$ are connected to others a distance of 2^2 apart, etc. In general, nodes in a $PEC(k)$ mesh will connect the processors to neighbors 2^k away.

PEC has several properties which make it an attractive alternative to other networks. Many of the properties pertain to *processor farms*. A processor farm groups the nodes in one area of the processor array, and uses these processors together to implement a function, or a set of functions within the context of a larger application. If the farms are of a regular size (called *tiles*), then PEC can provide a predictable set of powers of two interconnections between them.

We define a *tile* to be a rectangular farm extending 2^j processors in one direction and 2^k in the other (j and k are positive integers). Such a tile is *properly aligned* if and only if its upper left processor is offset from the PEC origin by $n2^j$ in the first direction and $m2^k$ in the other (n and m are positive integers) as shown in Figure 2. A particularly useful characteristic of a properly aligned tile is called the *area constraint* of connections.

It can be proven that in the PEC network, every 2^j by 2^k properly aligned tile has exactly one connection of length 2^{j+k} and exactly one that is longer.

This implies that if it is desired to transfer data a distance of 2^{j+k}, such a connection is guaranteed to be within a properly aligned tile of size 2^j by 2^k. It should also be noted that if this tile is divided in half, each of the two resultant properly aligned tiles have one, and only one, connection of 2^{j+k-1}. As the farms become farther apart, the number of connections decrease. However, it is possible to assure that a connection of a certain power of two exists if the farm

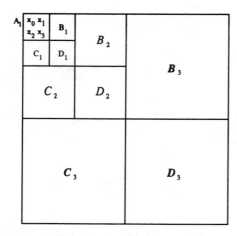

Figure 3: Mapping of data onto PEC nodes

is large enough.

PEC is scalable, has a simplistic wiring layout, a bi-section bandwidth of $\sqrt{N} \log \sqrt{N}$, where N is the total number of processors, and good routing characteristics [4, 5, 7]. Binary-exchange algorithms can be mapped naturally to a PEC network because of the network's recursive nature and the presence of the long distance connections. Subsets of information can be stored on PEC processor farms, and the communications links that are needed to process these subsets are guaranteed to exist on these farms. Our implementation of binary-exchange schemes uses the characteristics and constraints that define a PEC network.

3 Binary-Exchange Schemes on a PEC Network

A binary-exchange process that is to be applied to a data set $\{x_0, x_1, \dots, x_{N-1}\}$ can be performed on a $\sqrt{N} \times \sqrt{N}$ PEC network of N processors. In this discussion, we assume that each processor contains one of the N numbers to be processed. Also, the data that resides on the PEC processors are most naturally referenced using a quadrant ordering shown in Figure 3. Here, the values $\{x_i, x_{i+1}, x_{i+2}, x_{i+3}\}$ are positioned on PEC farms of size four which recursively form quadrant farms $\{A_j, B_j, C_j, D_j\}$ of size $2^j \times 2^j$.

A PEC implementation of a binary-exchange process will execute in $\log \sqrt{N}$ phases labeled $Phase(0)$, $Phase(1)$, ..., $Phase(\log_2 \frac{\sqrt{N}}{2})$, where each $Phase(i)$ consists of two processing steps that correspond to two steps in the binary-exchange computation. The two steps of each phase are applied to data subsets of size 2^{2i} residing on four PEC processor farms of size $2^i \times 2^i$. To perform the communications needed in $Phase(i)$, the $PEC(i)$ connections are used to connect nodes a distance of 2^i apart.

In the remainder of this section, we examine a PEC implementation for the binary-exchange computation

whose data movement patterns correspond to moving from left to right ($\rightarrow$) in Figure 1. Further, it is of interest to determine the amount and type of data that move between the PEC processors. The latter is used to prove the implementation correct, while both are needed for the case-studies discussed in Section 4.

To represent the patterns of data communicated in each phase of our implementation, we use the notation $I_1(i)$ and $I_2(i)$ to represent the *maximum* set of data values that could reside on $PEC(i)$ nodes at the end of the first and second steps of each $Phase(i)$. This gives an indication of what kind of information has been made available to the nodes during and after $Phase(i)$. The positional subscripts of the data communicated across $PEC(i)$ links will be represented as ordered pairs of subscripts and denoted as $T_1(i)$ and $T_2(i)$ for the two steps in $Phase(i)$, respectively. These subscripts correspond to the node positions represented in Figure 3.

3.1 Phase(0)

The data movement patterns in the first two steps of the binary-exchange computation can be implemented by using the $PEC(0)$ (i.e. nearest neighbor) connections. Groups of four single data values, (i.e. four PEC farms of size 1×1) are communicated to form pairs of data which reside on farms of size 1×2. These, in turn, communicate using $PEC(0)$ links to form data subsets residing on farms of size 2×2. These steps are accomplished by using nearest neighbor connections in the horizontal and vertical directions, and are respectively referred to as $Phase_1(0)$ and $Phase_2(0)$.

The horizontal and vertical communication steps of this operation are shown in Figure 4. In the horizontal communication step, pairs of horizontally adjacent numbers are bi-directionally communicated in parallel. As a result of this communication, horizontal pairs of processors now have collected a maximum of two data values. This can be denoted by $I_1^{(t)}(0)$ and $I_1^{(b)}(0)$ for each of the two processor farms of size 1×2. Here, (t) and (b) refer to top and bottom farms, respectively. As shown in Figure 4, processors at mesh coordinates $(0,0)$ and $(0,1)$ have now collected $I_1^{(t)}(0) = \{x_0, x_1\}$ and processors at $(1,0)$ and $(1,1)$ contain $I_1^{(b)}(0) = \{x_2, x_3\}$. Similarly, the pairs of data exchanged over the $PEC(i)$ links have subscripts $T_1(0) = <i, i+1>$, $i = 0, 2, 4, \dots$ and this corresponds to the type of data communication used in Step 0 of Figure 5.

Let the results of the computations performed at the nodes after the horizontal communications be denoted as a_i, where a_i can be single value or represent a set of values. In the subsequent vertical communication, the data on pairs of 1×2 farms are exchanged using the $PEC(0)$ communication links. Note that the data communicated have subscript pairings of $T_2(0) = <i, i+2>$ for $i = 0..1, 4..5, 8..9, \dots$. Hence, the horizontal and vertical communications in $Phase(0)$ correspond to Step 0 and Step 1 in

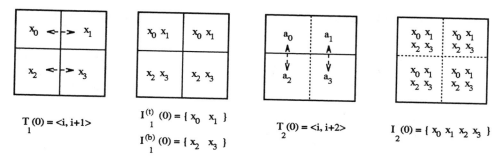

Figure 4: $Phase(0)$

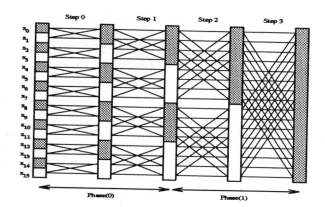

Figure 5: Data Movements and PEC Farm Groupings

the binary-exchange computation of Figure 5. In terms of the original data set, all processors on the 2×2 farms could now contain (maximally) $I_2(0) = \{x_i, x_{i+1}, x_{i+2}, x_{i+3}\}$ at the end of $Phase(0)$.

Since the communication operations in $Phase(0)$ are performed in parallel, the processors on all farms of size four could have computed individual results using the data on their farm. Each processor may have performed different computations with these data. Therefore, in preparation for $Phase(1)$, we *relabel* the computed results on all processors in Figure 6 as $\{a_0, a_1, \ldots a_N\}$.

3.2   Phase(1)

Data movements in $Phase(0)$ were accomplished using the connections provided by $PEC(0)$. Similarly, $Phase(1)$ uses the connections provided by $PEC(1)$ to perform the data movement patterns in Step 2 and Step 3 of the binary-exchange computation. As before, these communications links are first used horizontally in $Phase_1(1)$ to exchange data values that reside on pairs of two horizontally adjacent PEC farms of size $2^1 \times 2^1$.

The $PEC(1)$ connections that are used connect processors at coordinates $(0,1)$ to $(0,3)$, and $(1,0)$ to $(1,2)$ as shown in Figure 6. (Corresponding $PEC(1)$ connections are used in each pair of adjacent farms.)

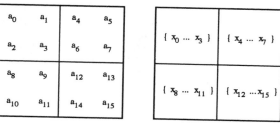

Initial Configuration

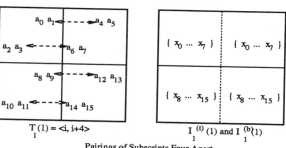

Pairings of Subscripts Four Apart

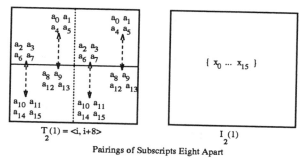

Pairings of Subscripts Eight Apart

Figure 6: $Phase(1)$

Since the farms are of size $2^1 \times 2^1$, the PEC area constraint property assures that there are two $PEC(1)$ connections in each farm. (Each link resides in a rectangular subfarm of the 2×2 farm.)

Assume that the per-node computations performed in $Phase(0)$ resulted in differing values on the four nodes of each farm, and that these are now to be combined with values on other farms. Then, the first task in $Phase(1)$ will be to collect the data on each subfarm onto the $PEC(1)$ nodes of each subfarm. After items a_0 and a_1 are collected onto the node at coordinates $(0,1)$, then these values can be exchanged with a_4 and a_5 (which were simultaneously collected) on node $(0,3)$ in the horizontally adjacent farm. Thus, $T_1(1) = <i, i+4>$ for $i = 0..3, 8..11, 16..19, \ldots$

After the horizontal communications, the data on each $PEC(1)$ node in the top $2^1 \times 2^2$ farms will be $I_1^{(t)}(1) = \{x_i, x_{i+1}, \ldots, x_{i+7}\}$, while the $PEC(1)$ nodes in the bottom farm have $I_1^{(b)}(1) = \{x_{i+8}, x_{i+9}, \ldots, x_{i+15}\}$ where i is a multiple of 16. After a vertical exchange of the information, all PEC(1) nodes maximally contain $I_2(1) = \{x_i, x_{i+1}, \ldots, x_{i+15}\}$ where i is a multiple of 16. The positional subscripts of the information exchanged in the vertical communications step are $T_2(1) = <i, i+8>$ for $i = 0..7, 16..23, 32..39, \ldots$

3.3 Phase(i)

$Phase(i)$ processes values on four farms of size $2^i \times 2^i$ using $PEC(i)$ communications links which connect nodes a distance of 2^i apart. On each farm of size $2^i \times 2^i$ there will be 2^i $PEC(i)$ links. This is because the farms are properly aligned and can be divided into 2^i subfarms of size $2^{\lfloor i/2 \rfloor} \times 2^{\lceil i/2 \rceil}$. (See [7] for more details.)

The data on horizontal pairs of $2^i \times 2^i$ farms is exchanged using these $PEC(i)$ links. If the computation necessitates using all newly computed results from $Phase(i-1)$, then this data must be collected from $PEC(i-1)$ nodes onto the $PEC(i)$ nodes first. After a horizontal exchange of data, a vertical exchange is performed using the same $PEC(i)$ links.

The following theorems quantify $I_1(i)$ and $I_2(i)$, which are the the maximum set of data values that could reside on $PEC(i)$ nodes at the end of the first and second steps of $Phase(i)$, as well as the ordered pairs of subscripts $T_1(i)$ and $T_2(i)$ for data communicated across $PEC(i)$ links.

Theorem 1 *After the horizontal communications in $Phase(i)$, $i \geq 0$, all $PEC(i)$ nodes in the top half of each $2^{i+1} \times 2^{i+1}$ farm can possibly have collected information from all the nodes of their $2^i \times 2^{i+1}$ subfarm, so that $I_1^{(t)}(i) = \{x_j, x_{j+1}, \ldots, x_{j+2^{2i+1}-1}\}$ where j is a multiple of $2^{2(i+1)}$. The $PEC(i)$ nodes in the bottom half will possibly have collected $I_1^{(b)}(i) = \{x_{j+2^{2i+1}}, x_{j+2^{2i+1}+1}, \ldots, x_{j+2^{2(i+1)}-1}\}$ where j is a multiple of $2^{2(i+1)}$. The positional subscripts of the*

data exchanged in this step are $T_1(i) = <j, j+2^{2i}>$ for $j = 0..2^{2i}-1, 2 \cdot 2^{2i}..3 \cdot 2^{2i}-1, 4 \cdot 2^{2i}..5 \cdot 2^{2i}-1, \ldots$

Theorem 2 *After the vertical communications, all $PEC(i)$ nodes can possibly have collected the data from all nodes on its $2^{i+1} \times 2^{i+1}$ farm, so $I_2(i) = \{x_j, x_{j+1}, \ldots, x_{j+2^{2(i+1)}-1}\}$ where j is a multiple of $2^{2(i+1)}$. The subscripts of information exchanged have the form $T_2(i) = <j, j+2^{2i+1}>$ for $j = 0..2^{2i+1}-1, 2 \cdot 2^{2i+1}..3 \cdot 2^{2i+1}-1, 4 \cdot 2^{2i+1}..5 \cdot 2^{2i+1}-1, \ldots$*

These claims can be established by using an inductive argument. Data from a pair of $PEC(i-1)$ nodes will be aggregated onto a $PEC(i)$ node. Utilizing the inductive assumption that each $PEC(i-1)$ node can have a copy of all the data on its respective $2^{\lfloor i/2 \rfloor} \times 2^{\lceil i/2 \rceil}$ subfarm of the $2^i \times 2^i$ farm, the aggregation of this data onto the $PEC(i)$ node will result in its maximally having a copy of all the data on its $2^i \times 2^i$ farm.

After the horizontal exchange of data between pairs of adjacent $2^i \times 2^i$ farms, each $PEC(i)$ node in a $2^i \times 2^{i+1}$ farm could have a copy of all the data on this farm. After a vertical exchange of this data, all 2^i $PEC(i)$ nodes of the $2^{(i+1)} \times 2^{(i+1)}$ farm could each have a copy of all $2^{2(i+1)}$ data values from the farm.

For $Phase(i)$, horizontal and vertical $PEC(i)$ links are used to communicate data between four farms of size $2^i \times 2^i$. The maximum amount of data possibly collected from the $PEC(i-1)$ nodes onto $PEC(i)$ nodes on all $2^i \times 2^i$ farms in $Phase(i)$ is $I_2(i-1) = \{x_j, x_{j+1}, \ldots, x_{j+2^{2i}-1}\}$ for $j = 0, 2^{2i}, 2 \cdot 2^{2i}, 3 \cdot 2^{2i}, \ldots$ Since the data residing on groups of four farms is ordered using the quadrant scheme, the distribution of data collected onto the $PEC(i)$ nodes will be identical on all four farms, although the subscript values will differ by 2^{2i}. Thus, the horizontal communications of data will result in each $PEC(i)$ node having exchanged data with subscripts $T_1(i) = <j, j+2^{2i}>$ for $j = 0..2^{2i}-1, 2 \cdot 2^{2i}..3 \cdot 2^{2i}-1, 4 \cdot 2^{2i}..5 \cdot 2^{2i}-1, \ldots$ A similar argument can be used to show that $T_2(i) = <j, j+2^{2i+1}>$ for $j = 0..2^{2i+1}-1, 2 \cdot 2^{2i+1}..3 \cdot 2^{2i+1}-1, 4 \cdot 2^{2i+1}..5 \cdot 2^{2i+1}-1, \ldots$

3.4 Analysis

The time complexity of the binary-exchange implementation on a PEC network can be measured in terms of the maximum per-node computations performed, and the number of communication steps that are used in each phase. Let $n = \log_2 \frac{\sqrt{N}}{2}$ denote the maximum phase performed.

To measure the amount of computation, we assume that a node's computations will combine any communicated values with values already stored on the processor. Let $U_1(i)$ represent the the actual data used in any per-node computations following the horizontal communications in $Phase(i)$, and $U_2(i)$ denote the data used in computations after a vertical communication. Then $0 \leq |U_j(i)| \leq |I_j(i)|$ for $j = 1, 2$.

Further, let $f(x)$ denote the amount of computation that is performed by any node in any $Phase(i)$, where x is the size of the data set involved in the computation. (We assume that all nodes perform the same computations on the data.) f and U will vary depending on the problem being solved. For example, in a global summation algorithm, individual nodes will combine one incoming data value with an already resident value. Then $|U_1(i)| = |U_2(i)| = 2$, and $f(x)$ for $x = 2$, is the time to add two numbers. Alternately, in bitonic merging, a list of 2^i values is sent across $PEC(i)$ links in $Phase(i)$ and compared pairwise with values already on the $PEC(i)$ nodes. Then $|U_1(i)| = |U_2(i)| = 2 \cdot 2^i$, and $f(x)$ for $x = 2 \cdot 2^i$ will be the time to do 2^i comparisons of pairs of values.

Using this notation, the total amount of computation time $C(N)$ can be bounded by

$$C(N) = \sum_{i=0}^{n} \left(f(|U_1(i)|) + f(|U_2(i)|) \right).$$

To measure the amount of communications, let $S(i)$ be the number of times data is sent across PEC links in $Phase(i)$ of the algorithm. Within each phase, the communications involve aggregating data from pairs of $PEC(i-1)$ nodes onto $PEC(i)$ nodes, and then sending this data across the links. The notation $S_{i-1,i}$ is used to denote the number of communications steps needed to collect sets of data residing on pairs of $PEC(i-1)$ nodes onto a $PEC(i)$ node. Similarly, let $V_1(i)$ and $V_2(i)$ be the amount of data to be sent across links in the horizontal and vertical steps of $Phase(i)$, respectively.

We assume that the number of values computed by a node in $Phase(i-1)$ is no more than $|V_2(i-1)|$ (i.e. the amount of data received by a node in the vertical communications of $Phase(i-1)$), and that only this computed data moves to $PEC(i)$ nodes in $Phase(i)$. If t_x denotes the maximum time needed to exchange two lists of size x across any PEC link, then the amount of communications in $Phase(i)$ can be expressed as $S(i) = S_{i-1,i} + t_{|V_1(i)|} + t_{|V_2(i)|}$.

A simple upper bound on $S_{i-1,i}$ can be obtained using mesh routing estimates. In the worst case, we can assume that the data items on two $PEC(i-1)$ nodes are sent to $PEC(i)$ nodes using only the nearest neighbor links of $PEC(0)$. Then

$$S_{i-1,i} \leq (2^{\lfloor i/2 \rfloor} + 2^{\lceil i/2 \rceil}) \cdot t_{|V_2(i-1)|}.$$

The total amount of communications required in any PEC-implemented binary-exchange algorithm will be bounded by

$$\begin{aligned} S(N) = \sum_{i=0}^{n} S(i) \quad &\leq \quad \sum_{i=0}^{n} [(2^{\lfloor i/2 \rfloor} + 2^{\lceil i/2 \rceil}) \cdot t_{|V_2(i-1)|} \\ &\quad + \; t_{|V_1(i)|} + t_{|V_2(i)|}]. \end{aligned}$$

The expressions $C(N)$ and $S(N)$ serve as a basis for measuring the amount of computation and communications time that are used for solving a particular binary-exchange problem on a PEC network. In practice, the actual amount of computation and communications time can only be determined for the particular problem being solved. The case studies in the Section 4 illustrate this further.

4 Case Studies

The approach described in this paper can be used to implement PEC algorithms for performing Global Reduction of N values residing on individual nodes into a single result. Similarly, a single value on one node can be broadcast to all other nodes, and algorithms such as one-dimensional Fast Fourier Transforms and bitonic merges can be implemented. In each case, the amount of computation and communications will differ due to the nature of the problem being solved. The remainder of this section summarizes the analysis results for implementations that solve these problems.

4.1 Global Reduction

For this problem, it is desired to apply a binary associative operator such as addition or maximum, to the individual data residing initially on N PEC nodes. As an example, we discuss the problem of finding the sum of all N items on a mesh.

Each horizontal and vertical communication of data in $Phase(0)$ involves sending one data value. Thus $|V_1(0)| = |V_2(0)| = 1$. The incoming data value is then added to the resident value so that two data items are involved in the per-node computation, i.e. $|U_1(0)| = |U_2(0)| = 2$ and $f(2)$ denotes the time to add two values. As a result of $Phase(0)$, each node in a 2×2 farm holds the sum of the four elements on the respective farm.

In each subsequent phase, only one data item is sent across PEC links in the horizontal and vertical communications steps so that $|V_1(i)| = |V_2(i)| = 1$. The positional subscripts of communicated data in each step will be $T_1(i)$ and $T_2(i)$, so $U_1(0) = \{x_j, x_{j+2^{2i}}\}$ and $U_2(0) = \{x_j, x_{j+2^{2i+1}}\}$. Each pair of values will be summed, and all $PEC(i)$ nodes on a resulting $2^{i+1} \times 2^{i+1}$ farm will contain the same summed value.

To prepare for $Phase(i)$, the single data values are sent to a $PEC(i)$ and so $S_{i-1,i} \leq (2^{\lfloor i/2 \rfloor} + 2^{\lceil i/2 \rceil}) \cdot t_1$. Hence the total amount of computation and communications will be

$$\begin{aligned} C(N) \quad &= \quad \sum_{i=0}^{n} 2f(2) \\ S(N) \quad &\leq \quad \sum_{i=0}^{n} \left[(2^{\lfloor i/2 \rfloor} + 2^{\lceil i/2 \rceil}) \cdot t_1 + 2t_1 \right]. \end{aligned}$$

Assuming that $f(2)$ and t_1 are constant, it can be shown that

$$C(N) = O(log_2 \frac{\sqrt{N}}{2})$$

$$S(N) = O(4\sqrt[4]{N} + 2\log_2 \frac{\sqrt{N}}{2} - 2).$$

At the end of these operations, the global sum of the initial data resides on all $PEC(n)$ nodes. If this value is to be distributed to all (or one particular node), then it can be sent from $PEC(n)$ nodes to $PEC(n-1)$ nodes, from $PEC(n-1)$ nodes to $PEC(n-2)$ nodes, etc., until it reaches all nodes. This requires an additional communications time of $\sum_{i=1}^{n} S_{i,i-1}$ where $S_{i,i-1}$ denotes the time to move data from a $PEC(i)$ node to two $PEC(i-1)$ nodes. Since $S_{i,i-1} = S_{i-1,i}$, the additional amount of communications will be $O(\sqrt[4]{N})$ using nearest neighbor routing estimates.

4.2 One-to-All Broadcasting

In one-to-all broadcasting, a single value resident on one of the PEC nodes is to be copied to all other nodes. One way to accomplish this is to send the value using $n+1$ phases as before. Without loss of generality, assume that the value in the top left-hand corner node of a PEC mesh is the value to be broadcast. The binary-exchange process proceeds as before, although $PEC(i)$ nodes on most farms will have no data to send or receive for several phase iterations. Thus, each communications operation involves at most one data value, and that value moves from $PEC(i-1)$ nodes to $PEC(i)$ nodes in preparation for $Phase(i)$. Thus, $|V_1(i)| = |V_2(i)| \leq 1$. The computation that is done at a node involves copying the incoming value to local memory, and is represented by $f(1)$. Hence, the amount of computation and communications are

$$C(N) = \sum_{i=0}^{n} 2f(1) = O(log_2 \frac{\sqrt{N}}{2})$$

$$S(N) \leq \sum_{i=0}^{n} [(2^{\lfloor i/2 \rfloor} + 2^{\lceil i/2 \rceil}) \cdot t_1 + 2t_1]$$

$$= O(4\sqrt[4]{N} + 2\log_2 \frac{\sqrt{N}}{2} - 2),$$

again assuming that $f(1)$ and t_1 are constant. Note that this provides a high upper bound for the diameter of a $\sqrt{N} \times \sqrt{N}$ PEC mesh as $O(\sqrt[4]{N})$.

4.3 Fast Fourier Transforms

In the one-dimensional Cooley-Tukey Fast Fourier Transform algorithm, repeated computations of the form $a'_i = a_i + u \cdot a_{i+2^k}$ and $a'_{i+2^k} = a_i + v \cdot a_{i+2^k}$ are performed on N input values for $k = \log_2 N - $

$1, \log_2 N - 2, \ldots, 0$. To implement these computations as a binary-exchange algorithm, each input data value a_i is assumed to initially reside on processor j of the quadrant ordering scheme, where j is the reversal of the binary representation of i. This address reversal technique is commonly used when implementing FFT's on a butterfly network [3].

By considering the input data to be distributed in this manner, each phase of the PEC binary-exchange implementation will correspond to pairs of decreasing k values in the Cooley-Tukey algorithm. For example, node computations after the nearest neighbor exchanges in $Phase(0)$ will correspond to iterations $k = \log_2 N - 1$ and $k = \log_2 N - 2$ in the FFT algorithm.

Unlike the two previous examples, the implementation of the Fast Fourier Transform algorithm involves moving differing amounts of data between nodes in each phase of the computation. In each $Phase(i)$, 2^i items are communicated by each $PEC(i)$ node to a corresponding $PEC(i)$ node in a horizontally adjacent farm. The positional subscripts of the communicated data have the form $T_1(i) = <j, j + 2^{2i}>$. Each $PEC(i)$ node will serially compute 2^i values $a'_j = a_j + u \cdot a_{j+2^{2i}}$ or $a'_{j+2^{2i}} = a_j + v \cdot a_{j+2^{2i}}$ for twiddle factors u and v. Hence, $|V_1(i)| = 2^i$ and $|U_1(i)| = 2 \cdot 2^i$.

Next, a vertical communication between $PEC(i)$ nodes exchanges the 2^i computed a' values so that each node computes new values using data having positional subscripts $T_2(i) = <j, j + 2^{2i+1}>$. $PEC(i)$ nodes on vertically adjacent farms will compute 2^i values $a''_j = a'_j + u' \cdot a'_{j+2^{2i+1}}$ or $a''_{j+2^{2i+1}} = a'_j + v' \cdot a'_{j+2^{2i+1}}$. This means $|V_2(i)| = 2^i$ and $|U_2(i)| = 2 \cdot 2^i$.

Thus,

$$C(N) = \sum_{i=0}^{n} 2f(2 \cdot 2^i)$$

$$S(N) \leq \sum_{i=0}^{n} \left[(2^{\lfloor i/2 \rfloor} + 2^{\lceil i/2 \rceil}) \cdot t_{2^{i-1}} + 2t_{2^i} \right].$$

The function $f(x)$ represents the time it takes to add x values two at a time, and if this is assumed to take cx time where c is a constant, then

$$C(N) = \sum_{i=0}^{n} 2c2^i = 2c(2^{n+1} - 1) = 2c\sqrt{N} - 2c$$

$$S(N) \leq \sum_{i=0}^{n} \left[(2^{\lfloor i/2 \rfloor} + 2^{\lceil i/2 \rceil}) \cdot t_{2^{i-1}} + 2t_{2^i} \right].$$

If $t_{2^{i-1}}$ is assumed to be constant (i.e., all communications of any amount of data take constant time), then as before

$$S(N) = O(4\sqrt[4]{N} + \log_2 \frac{\sqrt{N}}{2} - 2).$$

On the other hand, if the communication requires that $t_{2^{i-1}} = 2^{i-1}$, then

$$S(N) = O(\sqrt[4]{N^3}).$$

4.4 Bitonic Merge

The amount of computation and communications that are needed to accomplish bitonic merging of data will be similar to that of the Fast Fourier Transform. Each communications step in $Phase(i)$ involves 2^i data values. Each $PEC(i)$ node will do a pairwise comparison of the incoming data with its resident data. A series of bitonic merges can be used to perform sorting. Full details of the implementation are discussed in [7].

5 Summary

The analysis in the previous section has shown that binary-exchange or butterfly parallel algorithms can be implemented on a PEC network using a singular framework. This is also true for other networks. These algorithms can be implemented on a hypercube with communication costs of $O(log_2 N)$ and computation cost of $O(log_2 N)$, and implemented on a mesh with $O(\sqrt{N})$ communication and computation cost. We have shown that it is possible to perform Global Reduction, One-to-All Broadcasting, one-dimensional Fast Fourier Transforms, and Bitonic Merges in $O(\sqrt[4]{N})$ on an N-processor PEC mesh if communication costs are constant. Since PEC's connectivity falls between the connectivity of hypercubes and meshes, this ranking of costs would be expected. However, PEC was designed to have more close connections than long connections, and since, in binary-exchange computations, the number of long distance communications equals the number of short communications, this result was not obvious when we started.

It is important to note that if a PEC interconnection network is used in a massively parallel system with 2^{20} nodes (1,000,000 nodes), the highest PEC connection required by binary-exchange is $PEC(9)$. Thus, for binary-exchange implementations, the maximum number of items which would have to be moved across this connection is 2^9 or 512. This will happen only in the last phase of the implementation. The amount of data in this case may seem large, but is quite small if one considers that over a million items are being calculated. The maximum number of data items on any one node is 1024, well within the storage capacity on modern multiprocessors. However, we currently consider only communication buffer sizes of one or infinity. In future work, we will consider what would happen with a fixed buffer size.

For future work, we also intend to determine lower bounds on the communications costs, and to study the effects of a data set which is larger than the number of processors. It should be noted that the communications estimate used in the analysis was a very high upper bound. The work on 1D-PECs [4, 5, 7] would make these bounds much smaller. However, at this point we only wished to establish that all of the algorithms can be implemented within whatever bounds currently exist.

References

[1] W. W. Kirkman and D. Quammen. Packed Exponential Connections– A Hierarchy of 2-D Meshes. In *Proc. of the 5th International Parallel Processing Symposium*, pages 464–470. IEEE Computer Society, 1991.

[2] W. Worth Kirkman. The Digital Transform Machine. In *Proceedings of the Third Symposium on the Frontiers of Massively Parallel Computation*, pages 265–269. IEEE Computer Society, 1990.

[3] F. T. Leighton. *Introduction to Parallel Algorithms and Architectures: Arrays Trees Hypercubes*. Morgan Kaufmann, 1992.

[4] C.C. Lin and V. K. Prasanna. A Routing Algorithm for PEC Networks. In *Procedings of the Fourth Symposium on the Frontiers of Massively Parallel Computation*, pages 170–177. IEEE Computer Society, 1992.

[5] C.C. Lin and V. K. Prasanna. A Tight Bound on the Diameter of One-Dimensional PEC Networks. In *Proceedings of the IEEE 1992 Symposium on Parallel and Distributed Processing*, pages 368–375. IEEE Computer Society, 1992.

[6] F.P. Preparata and J. Vuillemin. The Cube-Connected Cycles: A Versatile Network for Parallel Computation. *Communications of the ACM*, 24(5):300–309, May 1981.

[7] D. Quammen and P. Y. Wang. Bitonic Sorting on 2D-PEC: An Algorithmic Study on a Hierarchy of Meshes Network. In *Proc. of the 8th International Parallel Processing Symposium*, pages 418–423. IEEE Computer Society, 1994.

DATA-PARALLEL PRIMITIVES FOR SPATIAL OPERATIONS*

Erik G. Hoel†
Computer Science Department
University of Maryland
College Park, Maryland 20742
hoel@cs.umd.edu

Hanan Samet
Computer Science Department
Center for Automation Research
Institute for Advanced Computer Sciences
University of Maryland
College Park, Maryland 20742
hjs@cs.umd.edu

Abstract − *Data-parallel primitives are presented for the R-tree and R^+-tree spatial data structures using the scan model. Algorithms are described for building these two data structures that make use of these primitives. The data-parallel algorithms are assumed to be main memory resident. The algorithms were implemented on a minimally configured Thinking Machines CM−5 with 32 processors containing 1GB of main memory.*

1 Introduction

Spatial data consists of points, lines, regions, rectangles, surfaces, volumes, and even data of higher dimension which includes time [18, 19]). The key issue is that the volume of the data is large. This has led to an interest in parallel processing of such data. There are two possible approaches termed image-space and object-space [8]. In particular, the object-space approach assigns one processor per spatial object (e.g., [2, 11, 12, 13]), while the image-space approach assigns one processor per region of space (e.g., [9]).

In this paper our focus is on object-space data-parallel representations of spatial data. The representations which we discuss sort the data with respect to the space that it occupies. The effect of the sort is to decompose the space from which the data is drawn into regions called *buckets*. Our presentation is for spatial data consisting of a collection of lines such as that found in road maps, utility maps, railway maps, etc. Of course, similar results could be obtained for other types of spatial data.

One approach known as an R-tree [10] buckets the data based on the concept of a minimum bounding (or enclosing) rectangle. In this case, lines are grouped (hopefully by proximity) into hierarchies, and then stored in another structure such as a B-tree [6]. All leaf nodes appear at the same level. Each entry in a leaf node is a 2-tuple of the form (R, O) such that R is the smallest rectangle enclosing the line segment O (where O points to the actual line segment). Each entry in a directory (non-leaf) node is a 2-tuple (R, P), where R is the minimal rectangle enclosing the rect-

angles in the child node pointed at by P. Each node in an R-tree of order (m, M), with the exception of the root, contains between $m \leq \lceil M/2 \rceil$ and M entries. The root node has at least 2 entries unless it itself is a leaf node. The drawback of the R-tree is that it does not result in a disjoint decomposition of space — that is, the bounding rectangles corresponding to different lines may overlap. Equivalently, a line may be spatially contained in several bounding rectangles, yet it is only associated with one bounding rectangle. This means that a spatial query may often require several bounding rectangles to be checked before ascertaining the presence or absence of a particular line.

The non-disjointness of the R-tree is overcome by a decomposition of space into disjoint cells. Each line is decomposed into disjoint sublines so that each of the sublines is associated with a different cell. There are a number of variants of this approach. They differ in the degree of regularity imposed by their underlying decomposition rules and by the way in which the cells are aggregated. The price paid for the disjointness is that in order to determine the area covered by a particular line, we have to retrieve all the cells that it occupies. The reason is that each line is decomposed into as many pieces (termed *q-edges*) as there are cells through which it passes. There are two principal methods: the R^+-tree [7] and variants of the PM quadtree [17, 20]. The principal difference between them is that the latter is based on a regular decomposition of space while the former is not. Here we study the former.

The R^+-tree is motivated by a desire to avoid overlap among the bounding rectangles. Each line segment is associated with all the bounding rectangles that it intersects. All bounding rectangles in the tree (with the exception of the bounding rectangles for the line segments at the leaf nodes) are non-overlapping. The result is that there may be several paths starting at the root to the same line segment only if the line segment is partitioned across multiple disjoint leaf nodes. This may lead to an increase in the height of the tree due to the possible multiple representation of a feature. However, retrieval time is usually sped up because of the non-overlapping nodes in the tree.

We use the scan model of parallel computation [5]. It is defined in terms of a collection of primitive operations that can operate on arbitrarily long vectors (single dimensional arrays) of data. Three

*This work was supported in part by the National Science Foundation under Grants IRI-92-16970 and HPCC/GCAG Grant BIR-93-18183.

types of primitives (elementwise, permutation, and scan) produce result vectors of equal length. A *scan* operation [21] takes an associative operator $\bigoplus$, a vector $[a_0, a_1, \cdots, a_{n-1}]$, and returns the vector $[a_0, (a_0 \bigoplus a_1), \cdots, (a_0 \bigoplus a_1 \bigoplus \cdots \bigoplus a_{n-1})]$. Blelloch [4] points out that the EREW PRAM model with the scan operations included as primitives is termed the scan model. The scan model considers all primitive operations (including scans) as taking unit time on a hypercube architecture. This allows sorting operations to be performed in $O(\log n)$ time.

In this paper our focus is on the primitives that are needed to efficiently construct these representations. Our goal is one of showing the reader how the analogs of relatively simple sequential operations can be implemented in a data-parallel environment. Our presentation assumes that the data-parallel algorithms are main memory resident. Our algorithms were implemented in C* on a minimally configured Thinking Machines CM–5 with 32 processors containing 1GB of main memory (the algorithms have also been run on a 16K processor CM–2).

The rest of this paper is organized as follows. We first discuss the data-parallel primitives that are used to construct the data structures. This is followed by the algorithms that are used to construct the two structures using the specified primitives. Finally, we provided some concluding remarks.

2 Spatial Primitive Operations

In this section we describe the primitive operations that are needed to construct an R-tree and an R+-tree. Several of the lower-level primitives have been described elsewhere (i.e., [14, 16]).

2.1 Cloning

Cloning (also termed *generalize* [16]) is the process of replicating an arbitrary collection of elements within a linear processor ordering. Figure 1 shows an example cloning operation. Cloning may be accomplished using an exclusive upward addition scan operation, an elementwise addition, and a permutation operator.

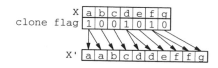

Figure 1: Example of a cloning operation.

Figure 2 details the various operations necessary to complete the cloning operation. The label **clone flag** indicates which elements of **x** must be cloned (i.e., **a**, **d**, and **g**). The basic technique is to calculate the offset necessary that each existing element must move toward the right in the linear ordering in order to make room for the new cloned elements.

This may be accomplished by employing an upward exclusive scan which sums the clone flags, as denoted by **up-scan(CF,+,ex)**. After the offset has been determined, an elementwise addition on the offset value (**F1**) and the position index (**P**) determines the new position for each element in the ordering (**ew(+,P,F1)**). Next, use a simple permutation operation to reposition the elements (**permute(X,F2)**). Cloning is completed when each of the cloning elements copies itself into the next element in the linear ordering (denoted by the small curved arrows in the figure).

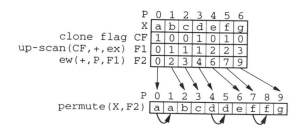

Figure 2: The mechanics of the cloning operation.

2.2 Unshuffling

Unshuffling physically separates two arbitrary, mutually exclusive and collectively exhaustive subsets of an original group. When applied without monotonic mappings, it has also been termed *packing* [15] or *splitting* [3]. Unshuffling can be accomplished using two inclusive scans (one upward and one downward), two elementwise operations (an addition and a subtraction), and a permutation operator. An example unshuffling operation is shown in Figure 3.

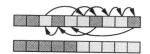

Figure 3: Example of an unshuffling operation.

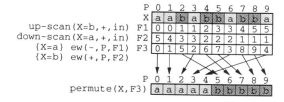

Figure 4: Example of the unshuffle operation.

Figure 4 illustrates the mechanics of the unshuffle operation for the data of Figure 3. The two different items to be unshuffled are **a** and **b**. Assume that the **a**'s are to be repositioned toward the left, and the **b**'s toward the right in our linear ordering. The basic technique is, for each element of the two groups, to calculate the number of elements from the other group that are positioned between itself and its desired position at either the left end or the right end. An upward inclusive scan (up-scan(X=b,+,in)) counts the number of b's between each a and the left end of the ordering. Similarly, a downward inclusive scan (down-scan(X=a,+,in)) counts the number of a's between each individual b and the right end of the linear ordering. Next, use two elementwise operations to calculate the new position index for each element of the linear ordering. For each a element, an elementwise subtraction of the calculated number of interposed b's (F1) from the original position index P determines the new position index (ew(-,P,F1)). Similarly, for each b element, an elementwise addition of the calculated number of interposed a's (F2) and the original position index P determines their new position indices (ew(+,P,F2)). Finally, given the new position indices in F3, permutation operation (permute(x,F3)) repositions each element into the proper position in the linear ordering.

2.3 Duplicate Deletion

Duplicate deletion (also termed *concentrate* [16]) is the process of removing duplicate entries from a sorted linear processor ordering. Figure 5 is an example of duplicate deletion (with the duplicate elements shaded). Duplicate deletion is accomplished using an upward exclusive scan operation, followed by a elementwise subtraction and finally a permutation operation.

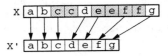

Figure 5: Example of a duplicate deletion operation.

Assume that the elements in the linear ordering are sorted by identifier. The basic technique is to count the number of duplicates between each element and the left side of the ordering. Next, move each element to the left by this number of positions. Consider Figure 6 where the elements are sorted and the duplicate items are marked (duplicate flag). An upward exclusive scan operation (up-scan(DF,+,ex)) sums the number of elements in the linear ordering that are to be deleted. Employ an elementwise operation (ew(-,P,F1)) to subtract the number of interposed items to be deleted (F1) from the element's position index P. Use this value as the new position index in a simple permutation operation (permute(X,F2)) to complete duplicate deletion.

2.4 Node Capacity Check

For spatial decompositions such as the R-tree whose node splitting rule focuses solely on the number of items in a node, a node capacity check can be used

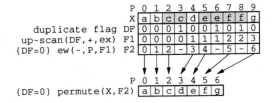

Figure 6: Example of the duplicate deletion operation.

to determine if a node is overflowing and needs to be split. This is done using a downward inclusive addition scan operation, followed by an elementwise write (or read) operation. In Figure 7, the downward scan is shown for an example dataset. Following the determination of the node counts, nodes whose bucket capacity is exceeded may be marked for subdivision.

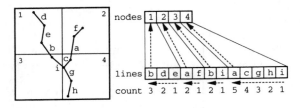

Figure 7: Example of the use of a downward inclusive segmented scan operation in a node capacity check.

2.5 Determining Legal Node Splits

A legal node split in an R-tree is defined as one where for a given minimal occupancy level of m/M (m is the minimal node capacity, and M is the maximal node capacity), each of the two resulting nodes contains at least m nodes. An example is provided in Figure 8 where there are two segment groups, containing data values a – j and k – t, respectively.

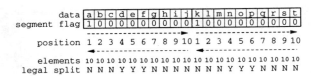

Figure 8: Example legal splits where the minimal occupancy level is 30%.

An upward inclusive segmented scan sums the number of elements in each segment group (i.e., the line labeled position). Next, a downward inclusive scan operation using the copy operator distributes the count of the number of elements across the segment group. Finally, each element in parallel determines whether or not it corresponds to a legal split by checking that its position is at least m and not more than elements - m (i.e., the line labeled legal split). In Figure 8, assuming a a minimum occupancy level of 30%, the legal splits correspond to lines c − e and j − l.

2.6 Selecting an R-tree Node Split

How to split an overflowing node is the subject of much research on sequential R-trees (i.e., the R*-tree [1]). For the data-parallel R-tree, we have developed two node splitting algorithms, each of varying computational complexity. In the first and simplest algorithm, the splitting axis (i.e., x or y axes) and the coordinate value are determined by finding the mean values along each axis of the midpoints of all bounding boxes in the line processor set in parallel via a sequence of scan operations. For each axis and segment group, the midpoints of the bounding boxes are first summed using a downward inclusive segmented scan operation (with the addition operator). The first node in the segment group then divides the sum by the number of bounding boxes in the segment group. This value is then broadcast [14] to all other nodes in the segment group with an upward segmented scan (using the copy operator). Each node then determines if it lies in the left or right resulting bounding boxes. Finally, use a small sequence of upward and downward inclusive scan operations (using either a min or max operator, depending upon the nature of the scan) to determine the physical extents of the two bounding boxes.

The split axis and coordinate value are chosen from the two possible splits (i.e., the means along the x and the y axes) so as to minimize the amount of area common to the two resulting nodes. This operation is $O(1)$ at each stage of the building operation as a constant number of scans dominates the computation.

The second node splitting algorithm first sorts all lines in the segment according to the left edge of their bounding boxes. A sequence of upward scan operations are used to determine the extents of the bounding box formed by all lines preceding a line in the sorted segment. A similar sequence of downward scans determines the bounding box for all following lines in the segment. For all legal splits, calculate the amount of bounding box overlap, and select the split corresponding to the minimal amount of overlap as the x-axis candidate. An analogous procedure is employed for the y-axis in obtaining the y-axis candidate split coordinate value. Next, select the candidate split coordinate value corresponding to the minimal bounding box overlap. In the event of a tie, use another metric such as choosing the split with the minimal bounding box perimeter lengths. This splitting algorithm is of complexity $O(\log n)$ at each stage of the building operation as we employ two $O(\log n)$ sorts and a constant number of scan operations.

Consider the example shown in Figure 9 consisting of four bounding boxes labeled A–D where the nodes have been sorted according to their left x-coordinate values. In this example, we are only considering the x-coordinate values of the bounding boxes though incorporation of y-coordinate values is straightforward.

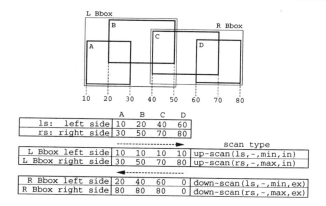

		A	B	C	D	scan type
ls:	left side	10	20	40	60	
rs:	right side	30	50	70	80	
L Bbox left side		10	10	10	10	up-scan(ls,-,min,in)
L Bbox right side		30	50	70	80	up-scan(rs,-,max,in)
R Bbox left side		20	40	60	0	down-scan(ls,-,min,ex)
R Bbox right side		80	80	80	0	down-scan(rs,-,max,ex)

Figure 9: Example showing the determination of the x-coordinate values for the left and right bounding boxes.

In the figure, the left and right coordinate values of the four nodes are indicated on the lines labeled ls:left side and rs:right side, respectively. For example, node B has left and right x-coordinate values 20 and 50 respectively, while node C has left and right x-coordinate values 40 and 70. Assume that a node is grouped with all nodes on its left when forming the bounding boxes (i.e., node C is grouped with nodes A and B when forming node C's left and right bounding boxes). As shown in Figure 9, an upward minimum inclusive scan on the left coordinate value determines the left x-coordinate value for the bounding box on a nodes left side (L Bbox left side). Similarly, an upward maximum inclusive scan on the right x-coordinate values establishes the right x-coordinate value for the bounding box on a nodes left side (L Bbox right side). Thus, for node B, we see that the left and right coordinate values for the bounding box to its left (i.e., the bounding box containing nodes A and B, labeled L Bbox in Figure 9), have x-coordinate values 10 and 50 (i.e., the lines labeled L Bbox left side and L Bbox right side). Analogous downward min/max exclusive scans determine the left and right x-coordinate values of the bounding box to the right of each node. Observe that the left and right x-coordinate values for the bounding box to the right of node B (i.e., a bounding box containing nodes C and D, labeled R Bbox in Figure 9) have values 40 and 80, respectively.

3 Construction Algorithms

In this section we show how to build an R-tree and an R+-tree. The algorithms are brief and make use of the primitives described in the previous section.

3.1 R-tree Construction

The data-parallel construction algorithm differs from the sequential algorithm as instead of inserting line

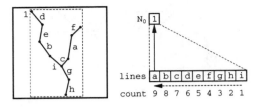

Figure 10: Initial configuration of a data-parallel R-tree.

segments sequentially into the data structure, all line segments are inserted simultaneously. The data-parallel algorithm proceeds as follows. Initially, one processor is assigned to each line of the data set, and one processor to the resultant data-parallel R-tree (e.g., Figure 10 which assumes an order $(1, 3)$ R-tree). In the figure, the label N_0 denotes the R-tree node processor set, with the associated square region containing the identifier of the R-tree node associated with the R-tree node processor. The term *segment* refers to the collection of line processors associated with a particular R-tree node processor. Within the line processor set, the nine square regions contain the line identifiers. A downward scan operation is performed on the line processor set to determine the number of lines associated with the single R-tree node processor (the **count** field beneath the line processor set in Figure 10). The number of lines in the segment is then passed by the first line in the linear ordering to the single R-tree node processor (i.e., the arrow from line **a** to node 1 in Figure 10). If the number of lines in the segment exceeds the node capacity M, then split the root node into two leaf nodes and a root node (as is similarly done with the sequential R-tree). Insert the two new leaf nodes into the R-tree node processor set, with the former root node/processor updated to reflect the two new children.

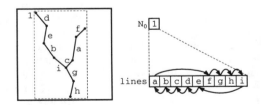

Figure 11: Un-shuffle operation.

Use the second of the two R-tree node splitting algorithms (detailed in Section 2.6) to select the splitting axis and coordinate value. Next, apply an un-shuffle operation to concentrate those line processors together into two new segments, each of which corresponds to one of the two R-tree leaf node processors (see Figure 11). For example, all lines which have a midpoint that is less than the split coordinate value are monotonically shifted toward the left, while those whose midpoint is greater than the split coordinate value are monotonically shifted toward the right among the line

processors. Figure 12 is the result of the un-shuffle operation on the lines in Figure 11. Note that the root node is associated with two segments in the line processor set **A** (i.e., (a,b,e,h) and (c,d,f,g,i)), and must itself be subdivided in an analogous manner.

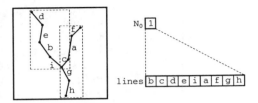

Figure 12: Result of the un-shuffle operation.

At this stage after the first root node split and line redistribution, we have two segments in the line processor set, and two different R-tree processor sets N_0 and N_1 (each set corresponding to a node at a different height in the tree), as shown in Figure 13.

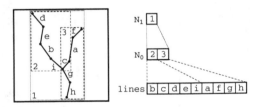

Figure 13: Completion of root node split operation.

The algorithm now proceeds iteratively, with each segment in the collection of line processors determining the number of lines it contains, and transmitting the count to the associated node processor (see Figure 14). If the number of lines in the segment exceeds the node capacity M, then the segment (and corresponding node processor) must subdivide. Note that this subdivision process may result in processors that correspond to internal nodes in the tree splitting themselves (with these splits possibly propagating upward through the tree).

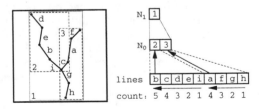

Figure 14: Broadcasting line counts to the associated nodes.

The process terminates when all nodes in the node processor set have at most M child processors (either internal nodes or line processors) as shown in Figure 15. The root node corresponds to the single processor in set N_2, the leaf nodes are contained in processor set N_0, and all lines are grouped in segments of length less than or equal to 3 in the line processor set (recall that we are dealing with an order $(1,3)$ R-tree in our example).

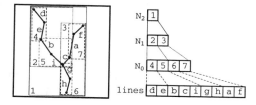

Figure 15: Final data-parallel R-tree.

Given n lines, the data-parallel R-tree building operation takes $O(\log^2 n)$ time, where each of the $O(\log n)$ stages requires $O(\log n)$ computations (a constant number of scans, clonings, and two sorts).

3.2 R$^+$-tree Construction

The R$^+$-tree construction algorithm is similar to that of the R-tree with a few additional modifications. In order to facilitate the R$^+$-tree's disjoint decomposition, the method for handling splitting nodes must be modified. The R$^+$-tree node splitting algorithm first sorts all lines in the node according to the left edge of their bounding boxes as is similarly done with the R-tree. Note that this splitting process is described for selecting a possible x-axis split; an analogous procedure will be followed for selecting a possible y-axis split. Next, for each legal node split, the coordinate value of the left edge is broadcast to each of the lines in the node being split.

In an iterative process which depends on the number of legal splits in a node, each node that corresponds to a legal split in turn broadcasts the coordinate value of its left side. Each line in parallel clips itself against the split coordinate value. If the line intersects the split axis and coordinate such that it would lie in each of the two resulting nodes that correspond to the split, the line processor clips itself against each of the two possible resulting nodes. Each resulting line determines in which of the two new nodes it is contained. For example, in the case of an x-axis split, a line can lie in either the node which is comprised of all lines to the left of the split coordinate value, or the node which consists of all lines to the right of the split coordinate value. The definition of an R$^+$-tree requires that each node at a given level of the tree is disjoint from all other nodes. In order to ensure this disjoint decomposition, some lines will have to be split across multiple nodes in the final decomposition. Once each line determines the node in which it lies, use a sequence of scan operations to determine the bounding box that contains the lines in the two new nodes. Finally, compute the perimeter of the two resulting bounding boxes.

The splitting process continues for each of the legal node splits and split axes. Once all legal node splits

have been determined and the resulting node perimeters are computed, the split axis and coordinate value that correspond to the minimal perimeter of the two resulting nodes is selected as the final node split value. In the event of a tie, some other metric such as the split with the minimal bounding box areas may be employed. Any lines that intersect the chosen split axis and coordinate are cloned. After cloning any intersecting lines, an un-shuffle operation concentrates those line processors together into two new segment groups, each of which corresponds to one of the two new R$^+$-tree leaf node processors that result from the node split. Figure 16 is the result of the first round of node splitting. Note that the split of the root node results in a new root node and an additional level of R$^+$-tree node processors.

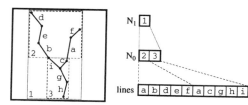

Figure 16: Result of first round of R$^+$-tree node splitting.

The insertion algorithm proceeds iteratively. Each node determines the number of lines it contains, and transmits the count to the associated R$^+$-tree node processor. If the number of lines in the node exceeds the node capacity M, then the node (and corresponding R$^+$-tree node processor) are split. The construction process is done when no nodes are overflowing.

R$^+$-tree Node Splitting Complexities

Note that the leaf node subdivision process may result in processors that correspond to internal nodes in the R$^+$-tree being forced to split when the number of their children (e.g., leaf nodes) exceeds the node capacity. These internal node splits may possibly propagate up to the root node of the tree (termed *upward splits*).

An additional complication in the node splitting process arises if splitting an internal node forces splitting some of the descendents (both nodes and lines) of the node being split. Unlike the R-tree which does not enforce a disjoint decomposition, an upward internal node split may result in selecting a split axis and a coordinate value that intersects the descendents of the splitting node. The disjoint decomposition requires that any intersecting descendents (nodes or lines) must also be split. Splitting the descendents of a node is termed a *downward split*. The R$^+$-tree construction process terminates when all nodes in the node processor set have at most M child processors (either internal R$^+$-tree nodes or line processors).

For example, consider Figure 17 where node **A** is the parent of nodes **B**, **C**, **D**, **E**, and **F**. Assume that the

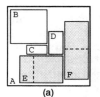 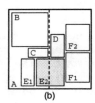 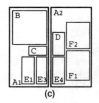

Figure 17: Example of upward and downward splits.

node capacity is five, and that nodes E and F in Figure 17a are being split. The result of their subdivisions into nodes E_1, E_2, F_1 and F_2 is shown in Figure 17b. Because parent node A now has seven children, the splitting of nodes E and F results in an upward split of A. Assuming that node A splits along the dashed line shown in Figure 17b, a downward split of child node E_2 is necessary to maintain the disjoint decomposition. Figure 17c is the result of parent node A's subdivision, and the downward split of node E_2.

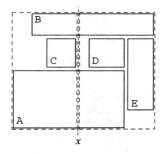

Figure 18: Example of the left and right bounding boxes corresponding to the split coordinate x associated with bounding box C when splitting an R^+-tree node.

In comparison to the the data-parallel R-tree, the data-parallel R^+-tree construction process suffers with respect to the number of scan operations necessary to choose the locally optimal node split. Recall that a sequence of ten or so scans is all that is necessary with the data-parallel R-tree in selecting the node split axis and coordinate. With the R^+-tree, due to its disjoint decomposition of space, the process requires considerably more scan operations. Because the R^+-tree decomposition is disjoint, it is sometimes necessary to split the children of internal nodes during the subdivision process. Thus it is necessary to have additional knowledge (as compared with the R-tree) of the contents of the children of a node (e.g., the spatial extents of the grandchildren) when splitting the node.

For example, consider Figure 18 where nodes A – E must be regrouped and the x-axis split corresponding to node C. Node C is grouped in its entirety with all portions of nodes A and B that fall to the left of C's rightmost x coordinate). If we were to apply a similar sequence of scan operations as was done with the data-parallel R-tree, each node could independently and in parallel determine the two associated bounding boxes that correspond to all portions of nodes to the left of split coordinate x (the left halves of nodes A and B, and all of node C), and all portions of nodes to the right of split coordinate x (the right halves of nodes A and B, and all of nodes D and E). For example, in Figure 18, the left bounding box corresponding to the split associated with node C consist of the left portions of nodes A and B, as well as node C in its entirety. Similarly, the right bounding box consists of the right portions of nodes A and B, and nodes D and E in their entirety. Note that this approach ignores the contents of all nodes when determining the left and right bounding boxes corresponding to a given split axis and coordinate.

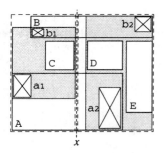

Figure 19: Figure 18 with the addition of two children each for nodes A and B (i.e., the crossed boxes a1, a2, b1, and b2). Shaded regions indicate the correct left and right bounding boxes corresponding to split coordinate x.

Employing he R-tree-like approach to forming the bounding boxes is flawed however. As we are working with a disjoint decomposition of space, it sometimes becomes necessary to consider the contents of the nodes when determining the left and right bounding boxes that would result from a candidate split. For example, consider Figure 19 where node A contains *only* two children (labeled a1 and a2), as does node B (similarly labeled b1 and b2). For our purposes, it is not necessary to show any other node's children. In Figure 19, the left bounding box corresponding to split coordinate x contains the region formed by the portions of nodes A and B that lay to the left of the split coordinate (regions a1 and b1), as well as node C in its entirety. The bounding box containing a1, b1), and C is shown as the left shaded region in Figure 19. Notice that this shaded region is considerably smaller than the left bounding box formed using the R-tree-like technique in Figure 18. Similarly, the actual right bounding box that corresponds to split coordinate x is shown as the right shaded region in Figure 19. As we cannot accurately determine the left and right bounding boxes corresponding to each candidate split, it becomes necessary to employ a much more computationally expensive technique. For each possible legal split, an independent sequence of scans is necessary For each possible split, each node determines the physical extent of its children in each of the two possible resulting subnodes. Once this is determined, scans are then used to determine the actual corresponding subnode bounding boxes across the segment of splitting nodes. Thus, rather than employing ten or so scans to determine the best node split (as is the case with the R-tree), a sequence of ten scans times the number of legal splits is required. This is often a considerably larger number of scans.

From this example, it is clear why the simple R-tree technique using a small sequence of scan operations is not applicable to the R$^+$-tree as knowledge of the contents of the nodes being regrouped and possibly split must also be considered when determining the locally optimal node split.

4   Conclusion

A number of data-parallel primitive operations used in building spatial data structures such as the R-tree and the R$^+$-tree were described as well as the algorithms. These primitives have been used in the implementation of other data-parallel spatial operations such as polygonization and spatial join [11, 12, 13]. It would be interesting to see whether these primitives are sufficient for other spatial operations and whether a minimal subset of operations can be defined. This is a subject for future research.

References

[1] N. Beckmann, H.-P. Kriegel, R. Schneider, and B. Seeger. The R*-tree: An efficient and robust access method for points and rectangles. In *Proc. of the 1990 ACM SIGMOD Intl. Conf. on Management of Data*, pages 322–331, Atlantic City, NJ, May 1990.

[2] T. Bestul. *Parallel Paradigms and Practices for Spatial Data*. PhD thesis, University of Maryland, College Park, MD, Apr. 1992.

[3] G. E. Blelloch. Scans as primitive parallel operations. *IEEE Trans. on Computers*, 38(11):1526–1538, Nov. 1989.

[4] G. E. Blelloch. *Vector Models for Data-Parallel Computing*. MIT Press, Cambridge, MA, 1990.

[5] G. E. Blelloch and J. J. Little. Parallel solutions to geometric problems on the scan model of computation. In *Proc. of the 1988 Intl. Conf. on Parallel Processing*, volume 3, pages 218–222, St. Charles, IL, Aug. 1988.

[6] D. Comer. The ubiquitous B-tree. *ACM Computing Surveys*, 11(2):121–137, June 1979.

[7] C. Faloutsos, T. Sellis, and N. Roussopoulos. Analysis of object oriented spatial access methods. In *Proc. of the 1987 ACM SIGMOD Intl. Conf. on Management of Data*, pages 426–439, San Francisco, May 1987.

[8] J. D. Foley, A. van Dam, S. K. Feiner, and J. F. Hughes. *Computer Graphics Principles and Practice*. Addison–Wesley, Reading, MA, second edition, 1990.

[9] W. R. Franklin and M. Kankanhalli. Parallel object-space hidden surface removal. *Computer Graphics*, 24(4):87–94, Aug. 1990.

[10] A. Guttman. R-trees: A dynamic index structure for spatial searching. In *Proc. of the 1984 ACM SIGMOD Intl. Conf. on Management of Data*, pages 47–57, Boston, June 1984.

[11] E. G. Hoel and H. Samet. Data-parallel R-tree algorithms. In *Proc. of the 1993 Intl. Conf. on Parallel Processing*, volume 3, pages 49–53, St. Charles, IL, Aug. 1993.

[12] E. G. Hoel and H. Samet. Data-parallel spatial join algorithms. In *Proc. of the 1994 Intl. Conf. on Parallel Processing*, volume 3, pages 227–234, St. Charles, IL, Aug. 1994.

[13] E. G. Hoel and H. Samet. Performance of data-parallel spatial operations. In *Proc. of the 20th Intl. Conf. on Very Large Data Bases*, pages 156–167, Santiago, Chile, Sept. 1994.

[14] Y. Hung and A. Rosenfeld. Parallel processing of linear quadtrees on a mesh-connected computer. *Journal of Parallel and Distributed Computing*, 7(1):1–27, Aug. 1989.

[15] C. P. Kruskal, L. Randolph, and M. Snir. The power of parallel prefix. *IEEE Trans. on Computers*, 34(10):965–968, Nov. 1985.

[16] D. Nassimi and S. Sahni. Data broadcasting in SIMD computers. *IEEE Trans. on Computers*, C-30(2):101–107, Feb. 1981.

[17] R. C. Nelson and H. Samet. A consistent hierarchical representation for vector data. *Computer Graphics*, 20(4):197–206, Aug. 1986.

[18] H. Samet. *Applications of Spatial Data Structures: Computer Graphics, Image Processing, and GIS*. Addison–Wesley, Reading, MA, 1990.

[19] H. Samet. *The Design and Analysis of Spatial Data Structures*. Addison–Wesley, Reading, MA, 1990.

[20] H. Samet and R. E. Webber. Storing a collection of polygons using quadtrees. *ACM Trans. on Graphics*, 4(3):182–222, July 1985.

[21] J. T. Schwartz. Ultracomputers. *ACM Transactions on Programming Languages and Systems*, 2(4):484–521, Oct. 1980.

A PARABOLIC LOAD BALANCING METHOD [1]

Alan Heirich & Stephen Taylor

Scalable Concurrent Programming Laboratory

California Institute of Technology

MS 256-80, Pasadena, CA 91125

email: steve@scp.caltech.edu

Abstract – *This paper presents a diffusive load balancing method for scalable multicomputers. In contrast to other schemes which are provably correct the method scales to large numbers of processors with no increase in run time. In contrast to other schemes which are scalable the method is provably correct and the paper analyzes the rate of convergence. To control aggregate cpu idle time it can be useful to balance the load to specifiable accuracy. The method achieves arbitrary accuracy by proper consideration of numerical error and stability.*

This paper presents the method, proves correctness, convergence and scalability, and simulates applications to generic problems in computational fluid dynamics (CFD). The applications reveal some useful properties. The method can preserve adjacency relationships among elements of an adapting computational domain. This makes it useful for partitioning unstructured computational grids in concurrent computations. The method can execute asynchronously to balance a subportion of a domain without affecting the rest of the domain.

Theory and experiment show the method is efficient on the scalable multicomputers of the present and coming years. The number of floating point operations required per processor to reduce a point disturbance by 90% is 168 on a system of 512 computers and 105 on a system of 1,000,000 computers. On a typical contemporary multicomputer [19] this requires 82.5 μs of wall-clock time.

INTRODUCTION

The scale of scientific applications and the computers on which they run is growing rapidly. Disciplines such as particle physics and computational fluid dynamics pose numerous problems which until recently have exceeded the memory and cpu capacities of existing computers. Moreover these disciplines appear ready to pose new problems which exceed the capacities of the largest computers that will be built in this decade. The *Grand Challenges* represent a set of problems solvable by computers with TeraFlops (10^9) performance. If present trends continue these will be supplanted within a decade by a set of challenges requiring PetaFlops (10^{12}) performance.

This increase in problem size is made possible by the dramatic reductions in size and cost of VLSI technology and parallel computing. Currently our research in computational fluid dynamics uses scalable multicomputers with roughly 500 processors [19, 23, 24]. Research efforts are underway to develop within three years time scalable multicomputers with hundreds of thousands of functional units [16, 21]. These trends suggest that limits to the growth of scientific computing are more likely to be determined in the next several years by software technology than by hardware limitations.

A potentially limiting software technology is the class of methods to solve the load balancing problem. Most numerical algorithms require frequent synchronization. If a load distribution on a multicomputer is uneven then some processors will sit idle while they wait for others to reach common synchronization points. The amount of potential work lost to idle time is proportional to the degree of imbalance that exists among the processor workloads. Since this loss also increases with processor count it can be valuable to control the accuracy of the resulting balance and to trade off the quality of the balance against the cost of rebalancing.

This paper presents a load balancing method for mesh connected scalable multicomputers with any number of processors. The method can balance the load to an arbitrary degree of accuracy. Theory and experiment show the method is inexpensive under realistic conditions. While it is effective on contemporary systems with under 1,000 processors it is specifically intended to scale to systems with tens and hundreds of thousands of processors. The total wall clock time for the method decreases as the processor count increases. The method was developed to solve the dynamic (run time) load balancing problem. It has also proven useful for cases of static (initial load time) balancing.

The method assumes the computation is sufficiently fine grained that work can be treated as a

[1] The research described in this report is sponsored primarily by the Advanced Research Projects Agency, ARPA Order number 8176, and monitored by the Office of Naval Research under contract number N00014-91-J-1986. The second author is also supported in part by President's Young Invetigator grant number ASC-9157650 from the National Science Foundation.

continuous quantity. This is reasonable in an application domain like computational fluid dynamics in which units of work represent grid points in a simulation. Each processor is typically responsible for thousands of grid points and can exchange individual points with other processors.

Cybenko [6] has published a method which is similar in several respects to this one. He proves asymptotic convergence of an iterative scheme on arbitrary interconnection topologies. Two other articles have appeared more recently on diffusive load balancing methods. Boillat [4] demonstrates polynomial convergence on arbitrary interconnection topologies using a Markov analysis. Horton objects to the polynomial convergence demonstrated in [4] and the lack of bounds on accuracy in the previous work. He applies a multigrid concept [11, 14] to accelerate convergence of a simple diffusive scheme. The method presented in this paper, while developed independently, resembles a special case of Cybenko's method restricted to mesh connected topologies. It differs from Cybenko's method with regard to issues of numerical stability. The objections in [11] lack rigor. This paper refutes them via formal demonstrations of convergence rates, accuracy, and scaling. These results show that convergence is rapid, accuracy is limited only by machine precision, and superlinear speedup can be achieved for cases of practical interest in CFD.

A PARABOLIC MODEL

A number of articles have proposed solutions to the load balancing problem in recent years [2, 5, 10, 12, 13, 15, 17, 18, 22]. Many of these solutions are reliable and efficient for computer systems with small numbers of processors. Unfortunately many of them do not scale well to systems with large numbers of processors. It is well known that in a scalable algorithm the amount of work performed in parallel grows more rapidly than the amount of work performed in the serial part of the calculation as the size of the problem increases [1, 9]. Scalable algorithms tend to be highly concurrent and this fact often makes it difficult to prove that they are correct. A load balancing method for scalable multicomputers should be scalable but should not sacrifice reliability.

It is worth noting that a class of random placement methods have been proposed for scalable multicomputers [2, 10]. These methods are scalable and are reliable under the assumption that disturbances occur frequently and have short lifespans. These assumptions do not hold in a domain like CFD where disturbances arise occasionally and are long lasting. As a result we are unable to take advantage of the rewards these methods offer.

A method is *reliable* if it can be shown to compute a correct solution within a predictable number of steps. The simplest reliable load balancing method collects load statistics from all processors, computes the average load, and broadcasts the average to all processors. Each processor then exchanges work with it's neighbors so that the new workloads equal this average. Unfortunately this simplest method is not *scalable* because it is inherently serial. The number of terms in the calculation of the average load is proportional to the number of processors in the computer system. Although this cost can be made logarithmic with an octree data structure there is a more severe cost associated with interprocessor communication. The current state of the art in mesh routing technology requires a nonconflicting communication path for each message. The opportunities for path conflicts known as "blocking events" increase factorially with the number of processors in the computer system. This simplest reliable method is not scalable because the time lost to blocking events can grow factorially with the size of the computer system.

A method is *scalable* if it can run efficiently on computer systems with very large numbers of processors. Due to the effects of Amdahl's law [1] most scalable methods are concurrent algorithms in which the computation is distributed among the processors in the computer system. What these methods gain in scalability they often lose in reliability because they lack formal proofs of correctness and convergence.

As one example of the problems which can arise in concurrent algorithms consider a simple concurrent method in which each processor adjusts it's load to equal the average of the loads at it's immediate neighbors. This method is distributed and scalable and is easily seen to be convergent. Unfortunately it is well known that it converges to solutions of the Laplace equation $\nabla^2 \Phi = 0$. This equation is known to admit sinusoidal solutions which are not equilibria. As a result this method, although scalable, is not reliable.

This paper leverages existing numerical and analytic techniques to derive a reliable and scalable load balancing method. The method is based on properties of the parabolic heat equation $u_t - \alpha \nabla^2 u = 0$. The heat equation describes a process in nature whereby thermal energy diffuses away from hot regions and into cold regions in a volume until the entire volume is of the same temperature. A literal interpretation of the terms in the heat equation reads that the rate of change in temperature u at each point in the volume is determined by the local curvature $\nabla^2 u$ times a diffusion rate α. This paper applies finite difference techniques to derive an unconditionally stable discrete form of this equation, and uses a scalable iterative method to invert the resulting coefficient matrix. The end result is a concurrent load balancing method which is scalable, reliable and efficient.

A PARABOLIC ALGORITHM

The algorithm consists of a simple arithmetic iteration which is performed concurrently by every processor in the multicomputer. Each step of the iteration requires 7 floating point operations at each processor. The iteration calculates an expected workload at each processor. Processors periodically exchange units of work with their immediate neighbors in order to make their actual workload equal to this expected workload. The algorithm contains parameters which control the accuracy of the resulting solution. In many applications an accuracy of 10% is sufficient. In these cases only 24 iterations are required to reduce a point disturbance by 90% regardless of the size of the multicomputer. This paper presents the algorithm for a three dimensional mesh. The reduction to two dimensions is described in the discussion section.

Initialization

Specify the accuracy α desired in the resulting equilibrium. For example, to balance to within 10% choose $\alpha = 0.1$. Determine the interval ν at which processors will exchange work with their immediate neighbors according to the formula

$$\nu = \left\lceil \frac{\ln \alpha}{\ln \frac{6\alpha}{1+6\alpha}} \right\rceil \geq 1 \qquad (1)$$

Note that in the interval $0 < \alpha < 1$ ν is less than or equal to 3:

$$\nu = \begin{cases} 2: & 0 < \alpha \leq 0.0445 \\ 3: & 0.0445 < \alpha < 0.622 \\ 2: & 0.622 \leq \alpha < 0.833 \\ 1: & 0.833 \leq \alpha \end{cases}$$

Execution

At every processor x, y, z adjust the workload $u_{x,y,z}$ as follows:

$$u_{x,y,z}^{(0)} = u_{x,y,z}$$

for m=1 to ν

$$u_{x,y,z}^{(m)} = \frac{u_{x,y,z}^{(0)}}{1+6\alpha} + \left(\frac{\alpha}{1+6\alpha}\right) \left(u_{x+1,y,z}^{(m-1)} + \right. \qquad (2)$$
$$u_{x-1,y,z}^{(m-1)} + u_{x,y+1,z}^{(m-1)} + u_{x,y-1,z}^{(m-1)} +$$
$$\left. u_{x,y,z+1}^{(m-1)} + u_{x,y,z-1}^{(m-1)} \right)$$

endfor

Exchange $\left(u_{x,y,z}^{(\nu)} - u'^{(\nu)}\right) \alpha$ units of work with every neighbor u'.

$$u_{x,y,z} = u_{x,y,z}^{(\nu)}$$

Repeat these steps until reaching equilibrium. Much of the following analysis will be concerned with formulating an exact statement of the number of repetitions required to reach equilibrium. The motivation behind this analysis is to support claims of correctness, convergence, and accuracy. The analysis develops a theory of convergence for any disturbance and applies this to the specific case of a point disturbance. Analysis appears to be less practical in many cases than conservative estimates derived from simulations. Following the analysis this paper presents simulations from cases of interest in CFD and distributed operating systems. The paper concludes with a summary and discussion.

RELIABILITY AND SCALABILITY

This section demonstrates reliability of the method by showing that for any initial disturbance every component of the disturbance vanishes at an exponential rate. It demonstrates scalabilty by defining a lower bound on this rate as a function of α, n. It applies the resulting theory to the case of a point disturbance and demonstrates that weakly superlinear speedup can be achieved under realistic assumptions. The appendix contains technical derivations in support of this analysis.

ACCURACY OF THE JACOBI ITERATION

Since the algorithm is intended to observe strict accuracy of $O(\alpha)$ it is important to verify that each stage of the algorithm observes this accuracy. The justification of accuracy for the finite difference equation is given in the appendix. The accuracy of the coefficient matrix inversion can be verified by analyzing the spectral radius of the Jacobi iteration.

From the Geršgorin disc theorem [7] the eigenvalues λ of (2) are bounded $|\lambda| \leq \frac{6\alpha}{1+6\alpha}$. Since the row and column sums are constant and the iteration matrix is nonnegative ([7], theorem 8.1.22) the spectral radius equals the row sum

$$\rho\left(D^{-1}T\right) = \frac{6\alpha}{1+6\alpha} \qquad (3)$$

Define the error in a current value $\vec{u}^{(m)}$ under the iteration (2) as $e(\vec{u}^{(m)}) = (\vec{u}^{(m)} - \vec{u}^*)$ where $\vec{u}^*$ is the fixed point of the Jacobi iteration. Then for $\nu > 0$

$$e(\vec{u}^{(\nu)}) = e((D^{-1}T)^{\nu}\vec{u}^{(0)}) = (D^{-1}T)^{\nu}e(\vec{u}^{(0)}) \qquad (4)$$

which converges when $\rho(D^{-1}T) < 1$ since $\rho((D^{-1}T)^\nu) = (\rho(D^{-1}T))^\nu$. In order for the algorithm to correctly reduce the error it is necessary to compute the desired load at each time step to an appropriate accuracy. In order to quantify the error define the infinity norm

$$\|e(\vec{u}^{(m)})\|_\infty = \max_{x,y,z}\left|e(u_{x,y,z}^{(m)})\right| = \max_{x,y,z}\left|u_{x,y,z}^{(m)} - u_{x,y,z}^*\right|$$

Using this norm define a necessary condition to improve the accuracy of the solution $\vec{u}$ by a factor α in ν steps to be $\|e(\vec{u}^{(\nu)})\|_\infty \le \alpha\|e(\vec{u}^{(0)})\|_\infty$. From (4) this is satisfied when $(\rho(D^{-1}T))^\nu \le \alpha$ and thus (1)

$$\nu = \left\lceil \frac{\ln\alpha}{\ln\rho(D^{-1}T)} \right\rceil = \left\lceil \frac{\ln\alpha}{\ln\frac{6\alpha}{1+6\alpha}} \right\rceil \qquad (5)$$

The method is unconditionally stable and the cost of this stability is small (in fact it is free for $0.833 \le \alpha < 1$). This suggests the possible use of large time steps to deal with worst case disturbances.

ELAPSED TIME FOR THE DIFFUSION

This section determines the number of artificial time steps τ required to reduce the load imbalance by a factor α. It does this by considering the eigenstructure of the finite difference equation (22) which is rearranged to express the change in load with each iteration

$$\begin{aligned} u_{x,y,z}(t+dt) - u_{x,y,z}(t) = \alpha\,[&u_{x+1,y,z}(t+dt)+ \\ &u_{x-1,y,z}(t+dt) \\ +&u_{x,y+1,z}(t+dt)+ \\ &u_{x,y-1,z}(t+dt) \\ +&u_{x,y,z+1}(t+dt)+ \\ &u_{x,y,z-1}(t+dt) \\ -&6u_{x,y,z}(t+dt)] \end{aligned}$$

or as a vector equation with matrix operator L

$$\vec{u}(t+dt) - \vec{u}(t) = \alpha L\vec{u}(t+dt) \qquad (6)$$

Any load distribution $\vec{u}(t)$ can be written as a weighted superposition of eigenvectors $\vec{x}$ of L

$$\vec{u}(t) = \sum_{i,j,k} a_{i,j,k}(t)\vec{x}_{i,j,k}$$

Using this fact rewrite the vector equation (6) as

$$\sum_{i,j,k} a_{i,j,k}(t+dt)\vec{x}_{i,j,k} -$$

$$\sum_{i,j,k} a_{i,j,k}(t)\vec{x}_{i,j,k} = \alpha \sum_{i,j,k} L a_{i,j,k}(t+dt)\vec{x}_{i,j,k} \qquad (7)$$

Using the definition of $L\vec{x}_{i,j,k}$ and the eigenvalues of L

$$L\vec{x}_{i,j,k} = -\lambda_{i,j,k}\vec{x}_{i,j,k}$$

$$\begin{aligned} \lambda_{i,j,k} = 2\Big[&3 - \cos\left(2\pi i/n^{1/3}\right) - \\ &\cos\left(2\pi j/n^{1/3}\right) - \cos\left(2\pi k/n^{1/3}\right)\Big] \end{aligned} \qquad (8)$$

(7) can be further simplified to

$$\sum_{i,j,k} \left(a_{i,j,k}(t+dt)\vec{x}_{i,j,k}\left[1+\alpha\lambda_{i,j,k}\right] - a_{i,j,k}(t)\vec{x}_{i,j,k}\right) = 0$$

and by the completeness and orthonormality of the eigenvectors

$$a_{i,j,k}(t+dt)\left[1+\alpha\lambda_{i,j,k}\right] - a_{i,j,k}(t) = 0$$

$$a_{i,j,k}(dt) = \frac{a_{i,j,k}(0)}{1+\alpha\lambda_{i,j,k}} \qquad (9)$$

It is apparent from equation (9) that convergence of the individual components is dependent upon the eigenvalues $\lambda_{i,j,k}$. Reducing the amplitude of an arbitrary component i,j,k by α in τ steps of the method requires $[1+\alpha\lambda_{i,j,k}]^{-\tau} \le \alpha$. The worst case occurs for the smallest positive eigenvalue $\lambda_{0,0,1} = (2 - 2\cos(2\pi/n^{1/3}))$ which corresponds to a smooth sinusoidal disturbance with period equal to the length of the computational grid. To reduce such a disturbance requires

$$\tau = \left\lceil \frac{\ln\alpha^{-1}}{\ln\left[1+\alpha\left(2-2\cos\frac{2\pi}{n^{1/3}}\right)\right]} \right\rceil \qquad (10)$$

Convergence of this slowest component approaches $\ln\alpha^{-1}$ for large n since

$$\lim_{n\to\infty} \ln\left[1+\alpha\left(2-2\cos\frac{2\pi}{n^{1/3}}\right)\right] = 1$$

Convergence of highest wavenumber component $\lambda_{(n^{1/3})/2-1,(n^{1/3})/2-1,(n^{1/3})/2-1}$ is rapid because

$$\tau = \left\lceil \frac{\ln\alpha^{-1}}{\ln\left[1+(6-\epsilon)\alpha\right]} \right\rceil \qquad (11)$$

ANALYSIS OF A POINT DISTURBANCE

This section considers a specific case of a point disturbance and derives an inequality relating τ, n and α. The purpose in doing this is to provide an exact prediction of convergence of a known case in order to demonstrate scaling properties. The procedure followed is to describe the initial amplitudes of the components of the disturbance and then solve an inequality which describes the magnitude of the disturbance over time. A periodic domain is assumed for the purpose of analysis. Simulations presented later in this paper verify that convergence is similar on aperiodic domains.

The following text uses the Poisson bracket $\langle \cdot, \cdot \rangle$ to represent the inner product operator. When discussing loads or eigenvectors it uses $\vec{u}[x, y, z]$ or $\vec{x}_{i,j,k}[x, y, z]$ to denote the vector element which corresponds to location x, y, z of the computational grid with the convention that $[0, 0, 0]$ is the first element of the vector. Then the initial disturbance confined to a particular processor x, y, z can be written as a superposition of eigenvectors of L

$$\vec{u}(0) = \sum_{l,m,n} a_{l,m,n}(0) \vec{x}_{l,m,n} \tag{12}$$

Assume the initial disturbance $u(0)$ to be zero at every element except $[x, y, z]$. Then

$$\langle \vec{x}_{i,j,k}, \vec{u}(0) \rangle = \vec{x}_{i,j,k}[x, y, z] \tag{13}$$

This is equal to the initial amplitude $a_{i,j,k}(0)$ of each eigenvector $\vec{x}_{i,j,k}$

$$
\begin{aligned}
\langle \vec{x}_{i,j,k}, \vec{u}(0) \rangle &= \left\langle \vec{x}_{i,j,k}, \sum_{l,m,n} a_{l,m,n}(0) \vec{x}_{l,m,n} \right\rangle \\
&= \sum_{l,m,n} \langle \vec{x}_{i,j,k}, \vec{x}_{l,m,n} \rangle a_{l,m,n}(0) \\
&= \sum_{l,m,n} a_{l,m,n}(0) \delta_{il} \delta_{jm} \delta_{kn} \\
&= a_{i,j,k}(0) \tag{14}
\end{aligned}
$$

The computational domain has periodic boundary conditions and as a result the origin of the coordinate system is arbitrary. Without loss of generality place the origin at the source of the disturbance and take $x = y = z = 0$. This has no effect on the eigenvectors $\vec{x}_{i,j,k}$ and from (13), (14)

$$a_{i,j,k}(0) = \vec{x}_{i,j,k}[0, 0, 0] \tag{15}$$

Placing the origin at the source of the disturbance is particularly convenient when we consider the first element of the eigenvectors $\vec{x}_{i,j,k}[0, 0, 0]$. L has $(n^{1/3})/2$ distinct eigenvalues $\lambda_{i,j,k}$ each of algebraic multiplicity two. Each of these eigenvalues has geometric multiplicity eight, ie. has eight linearly independent associated eigenvectors of unit length

$$\vec{x}_{i,j,k}[x, y, z] = c_{i,j,k} F_1 \left(2\pi x i / n^{1/3} \right)$$
$$F_2 \left(2\pi y j / n^{1/3} \right) F_3 \left(2\pi z k / n^{1/3} \right) \tag{16}$$

where each F_i is either sin or cos. By choosing $x = y = z = 0$ this expression (16) is zero except

for the single eigenvector for which $F_1(x) = F_2(x) = F_3(x) = \cos(x)$. Without loss of generality restrict further consideration to initial disturbances of the form

$$u[0, 0, 0](0) = \sum_{i,j,k} c_{i,j,k} \vec{x}_{i,j,k}[0, 0, 0] = \sum_{i,j,k} c_{i,j,k}^2 \tag{17}$$

From (9) define the time dependent disturbance at any location x', y', z'

$$
\begin{aligned}
\vec{u}&[x', y', z'](\tau dt) \\
&= \sum_{i,j,k} c_{i,j,k} \left[1 + \alpha \lambda_{i,j,k} \right]^{-\tau} \vec{x}_{i,j,k}[x', y', z'] \\
&= \sum_{i,j,k} c_{i,j,k}^2 \left[1 + \alpha \lambda_{i,j,k} \right]^{-\tau} \cos \left(2\pi x' i / n^{1/3} \right) \\
&\qquad \cos \left(2\pi y' j / n^{1/3} \right) \cos \left(2\pi z' k / n^{1/3} \right) \tag{18}
\end{aligned}
$$

The appendix demonstrates that $c_{i,j,k} = (8/n)^{1/2}$ and thus the disturbance is a summation of equally weighted eigenvectors. From (17) and (18) the time dependent disturbance at $0, 0, 0$ is therefore

$$
\begin{aligned}
\vec{u}[0, 0, 0](\tau dt) = \frac{8}{n} \sum_{i,j,k} &\left[1 + \alpha 2 \left(3 - \cos \frac{2\pi i}{n^{1/3}} - \right.\right. \\
&\left.\left. \cos \frac{2\pi j}{n^{1/3}} - \cos \frac{2\pi k}{n^{1/3}} \right) \right]^{-\tau} \tag{19}
\end{aligned}
$$

Solving $\vec{u}[0, 0, 0](\tau dt) \leq \alpha$ yields

$$
\begin{aligned}
\frac{8}{n} \sum_{i,j,k} &\left[1 + \alpha 2 \left(3 - \cos \frac{2\pi i}{n^{1/3}} - \right.\right. \\
&\left.\left. \cos \frac{2\pi j}{n^{1/3}} - \cos \frac{2\pi k}{n^{1/3}} \right) \right]^{-\tau} \leq \alpha \tag{20}
\end{aligned}
$$

where i, j, k are indexed from 0 to $\left(n^{1/3} \right)/2 - 1$ and the case $i = j = k = 0$ is omitted.

Table 1 and figure 1 present solutions of the inequality (20). These are exact predictions of the number τ of exchange steps which must occur to reduce a point disturbance by a factor α on a periodic domain. Figures 2 and 3 are simulated results for two CFD cases. The first case of partitioning an unstructured grid is a point disturbance. In the simulation τ is observed to match the theoretical prediction exactly. The second case of rebalancing after a grid adaptation demonstrates the value of estimating τ from simulations. The initial disturbance is not a point and the simulation is observed to require 170 exchange steps before the worst case discrepancy drops to 10% of it's original value.

SIMULATIONS

This section presents simulations of three cases in which the method is useful. The first case is a bow shock resulting from a million grid point CFD calculation [23]. The disturbance dissipates rapidly and is nearly gone after 10 exchange steps. The second case computes an initial load distribution for a million point unstructured grid problem on a 512 node multicomputer. The simulation suggests the method may be highly competitive with Lanczos based approaches presented recently in [3, 20]. The third case simulates a multicomputer operating system under conditions of random load injection. This case demonstrates that the method can effectively balance large disturbances which occur frequently and randomly.

Parameters for the simulations are based on two scalable multicomputers. The first is a 512 node J-machine [19] which is in use at Caltech for research in CFD and scalable concurrent computing. The second is a hypothetical 1,000,000 node J-machine. These two design points represent a continuum of scalable multicomputers. The method is practical at both ends of this continuum and presumably at all points in between.

Wall clock times are based on a hand coded implementation of the method in J-machine assembler and assumes 32 MHz processors. Each repetition of the method requires 110 instruction cycles in 3.4375 μs. All simulations are run with $\alpha = 0.1$ and $\nu = 3$ resulting in 3 iterations between each exchange of work.

Bow Shock Adaptation

In CFD calculations it is common to adapt a computational grid in response to properties of a developing solution. This simulation considers an adaptation that results from a bow shock in front of a Titan IV launch vehicle with two boosters (figure ??). The grid has been adapted by doubling the density of points in each area of the bow shock. As a result the initial disturbance shows locations in the multicomputer where the workload has increased by 100% due to the introduction of new points. After 10 exchanges of work the imbalance has already decreased dramatically. After 70 exchanges the disturbance is scarcely visible. This simulation assumes a 1,000,000 processor J-machine. This example illustrates the weak persistence of low spatial frequencies.

Partitioning an Unstructured Grid

In parallel CFD applications the static load balancing problem has been the subject of recent attention [3, 20]. In addition to finding an equitable distribution of work this problem must observe the additional constraint of preserving adjacency relationships among elements of an unstructured computational grid. The load balancing method presented in this paper can satisfy the adjacency constraint if at each exchange step it selects exchange candidates that observe the constraint.

This simulation models an initial point disturbance by assigning 1,000,000 points of an unstructured computational grid to a single host node of a 512 processor J-machine (see figure ??). It executes the algorithm while observing the adjacency constraint at each exchange step. The initial disturbance was reduced by 90% after 6 exchange steps in exact agreement with theory ($\tau(0.1, 512)$ in table 1). After 59 exchange steps the worst case discrepancy was 999 grid points. After 162 steps the worst case discrepancy was 200 grid points, 10% of the load average. A balance within 1 grid point was achieved on the 500th step.

Random Load Injection

To be useful in practical contexts the method must be able to rebalance disturbances faster than they arise. This simulation (figure ??) demonstrates the behavior of the method on a 1,000,000 processor J-machine under demanding conditions. An initially balanced distribution is disrupted repeatedly by large injections of work at random locations. Injection magnitudes are uniformly distributed between 0 and 60,000 times the initial load average. The simulation alternates repetitions of the algorithm with injections at randomly chosen locations. After 700 repetitions and injections the worst case discrepancy was 15,737 times the initial load average. This is less than the average injection magnitude of 30,000 at each repetition. This demonstrates the method was balancing the load faster than the injections were disrupting it. The last random injection occured on step 700. After 100 additional exchange steps without intervening injections the worst case discrepancy had reduced from 15,737 to 50 times the initial load average.

SUMMARY AND DISCUSSION

This paper has demonstrated that a diffusion based load balancing method is efficient, reliable, and scalable. It has shown through rigorous theory and empirical simulation that the method is inexpensive for important generic problems in CFD. This section discusses a few remaining points and indicates future directions for this research.

An important property of the method is it's ability to preserve adjacency relationships among elements of a computational domain. Preserving adjacency permits CFD calculations to minimize their communication costs. The method preserves adjacency if it does so at each exchange step.

To make this requirement concrete consider an unstructured computational grid for a CFD calculation. When the time comes for the load balancing method to select grid points to exchange with neighboring processors it selects points in such a way that average pairwise distance among all points is minimal. One way to do this is to assume that each processor represents a volume of the computational domain and to select for exchange those grid points which occupy the exterior of the volume. The selected points would transfer to adjacent volumes where their neighbors in the computational grid already reside. In order for this to be practical it must be inexpensive to identify these exterior points. In problems where the computational grid has been generated contiguously this is presumably a simple matter because adjacent points will have data structures that identify their neighbors. In more general problems where the data is not already ordered the use of priority queues appears promising due to their $O(n \log n)$ complexity.

It is worth noting that the method can be used to rebalance a local portion of a computational domain without interrupting the computation which is occurring on the rest of the domain. This can be useful in CFD problems where some portions of the domain converge more quickly than others and adaptation might occur locally and frequently.

This paper presented the method and analysis for a domain which has periodic boundary conditions and is logically spherical. In practice multicomputer meshes are rarely periodic. In our simulations we implemented aperiodic boundaries by imposing the Neumann condition $\partial u/\partial x = 0$ in each direction x, y, z. This requires a simple modification to the iteration (2) so that processors immediately outside the mesh appear to have the same workload as processors one step inside the mesh. For example, if the processors are indexed from 1 through n in the x dimension then $u_{0,y,z} = u_{2,y,z}$ and $u_{n+1,y,z} = u_{n-1,y,z}$.

The algorithm is presented for three dimensional scalable multicomputers. It reduces for two dimensional cases by redefining ν and the iteration (2) as follows:

$$\nu = \left\lceil \frac{\ln \alpha}{\ln \frac{4\alpha}{1+4\alpha}} \right\rceil \geq 1$$

$$u_{x,y}^{(m)} = \frac{u_{x,y}^{(0)}}{1 + 4\alpha} +$$

$$\left(\frac{\alpha}{1+4\alpha} \right) \left(u_{x+1,y}^{(m-1)} + u_{x-1,y}^{(m-1)} + u_{x,y+1}^{(m-1)} + u_{x,y-1}^{(m-1)} \right)$$

The worst case behavior of the method is determined by disturbances of low spatial frequency. This

is the basis of the objections in [11] and a conventional response would be to apply a multigrid method [11, 14]. Such a method can have logarithmic scaling in n and $O(n)$ convergence. The analysis presented in this paper demonstrates that wall clock times can actually decrease as n increases and this suggests considering other methods of treating the worst case disturbance. One such method would be to use very large time steps in order to accelerate convergence of the low frequency components. The unconditional stability of this method makes this an attractive option. Although this would increase the error in the high frequency components these components can be quickly corrected by local iterations. We are presently considering the costs associated with such iterations.

APPENDICES

This appendix presents technical justifications for statements relied upon in the analysis.

FINITE DIFFERENCE FORMULATION OF THE HEAT EQUATION

Consider the parabolic heat equation in three dimensions

$$u_t = \nabla^2 u = u_{xx} + u_{yy} + u_{zz} \tag{21}$$

Taylor expanding in t with all derivatives evaluated at (x, y, z, t)

$$u(x, y, z, t + dt) = u(x, y, z, t) + u_t dt + \mathrm{O}(dt^2)$$

$$u_t = \left(\frac{u(x, y, z, t + dt) - u(x, y, z, t)}{dt} \right) +$$

$$\mathrm{O}(dt)$$

Obtain the second order terms by expanding in spatial variables where omitted coordinates are interpreted as (x, y, z, t)

$$u(x + dx, \cdot, \cdot, \cdot) = u(x, \cdot, \cdot, \cdot) + u_x dx +$$

$$u_{xx} \frac{dx^2}{2} + u_{xxx} \frac{dx^3}{6} +$$

$$\mathrm{O}(dx^4)$$

$$u(x - dx, \cdot, \cdot, \cdot) = u(x, \cdot, \cdot, \cdot) - u_x dx +$$

$$u_{xx} \frac{dx^2}{2} - u_{xxx} \frac{dx^3}{6} +$$

$$\mathrm{O}(dx^4)$$

$$u(x + dx, \cdot, \cdot, \cdot) + u(x - dx, \cdot, \cdot, \cdot) =$$

$$2u(x, \cdot, \cdot, \cdot) + u_{xx} dx^2 +$$

$$\mathrm{O}(dx^4)$$

$$u_{xx} =$$

$$\left(\frac{u(x+dx,\cdot,\cdot,\cdot)+u(x-dx,\cdot,\cdot,\cdot)-2u(x,\cdot,\cdot,\cdot)}{dx^2}\right) +$$

$$O(dx^2)$$

Similar expansions in y, z show that the heat equation can be rewritten

$$\frac{u(\cdot,\cdot,\cdot,t+dt)-u(\cdot,\cdot,\cdot,t)}{dt} =$$

$$\left(\frac{u(x+dx,\cdot,\cdot,\cdot)+u(x-dx,\cdot,\cdot,\cdot)-2u(\cdot,\cdot,\cdot,\cdot)}{dx^2}\right) +$$

$$\left(\frac{u(\cdot,y+dy,\cdot,\cdot)+u(\cdot,y-dy,\cdot,\cdot)-2u(\cdot,\cdot,\cdot,\cdot)}{dy^2}\right) +$$

$$\left(\frac{u(\cdot,\cdot,z+dz,\cdot)+u(\cdot,\cdot,z-dz,\cdot)-2u(\cdot,\cdot,\cdot,\cdot)}{dz^2}\right) +$$

$$O(dt,dx^2,dy^2,dz^2)$$

From 21) the finite difference formulation is

$$\frac{u(x,y,z,t+dt)-u(x,y,z,t)}{dt} =$$

$$\frac{1}{dx^2}\left(u(x+dx,y,z,t)+u(x-dx,y,z,t)+\right.$$

$$u(x,y+dy,z,t)+u(x,y-dy,z,t)+$$

$$u(x,y,z+dz,t)+$$

$$\left. u(x,y,z-dz,t)-6u(x,y,z,t)\right)$$

Setting $\alpha = \frac{dt}{dx^2}$ and taking the spatial terms on the right at time $t+dt$ yields an unconditionally stable implicit scheme

$$u(x,y,z,t) = (1+6\alpha)u(x,y,z,t+dt)-$$

$$\alpha\left[u(x+dx,y,z,t+dt)+\right. \qquad (22)$$

$$u(x-dx,y,z,t+dt)+$$

$$u(x,y+dy,z,t+dt)+u(x,y-dy,z,t+dt)+$$

$$\left. u(x,y,z+dz,t+dt)+u(x,y,z-dz,t+dt)\right]$$

JACOBI ITERATION TO COMPUTE $A^{-1}u$

In order to compute solutions at successive time intervals dt we must invert the relationship $u^{(t)} = Au^{(t+dt)}$ by solving

$$u^{(t+dt)} = A^{-1}u^{(t)} \qquad (23)$$

From (22) it is apparent that A has diagonal terms $(1+6\alpha)$ and six offdiagonals α. Let $A = (D-T)$

where D is this diagonal. Then $(D-T)u^{(t+dt)} = u^{(t)}$ implies $u^{(t+dt)} = D^{-1}Tu^{(t+dt)} + D^{-1}u^{(t)}$. This relation is satisfied by fixed points of the Jacobi iteration

$$\left[u^{(t+dt)}\right]^{(m)} = D^{-1}T\left[u^{(t+dt)}\right]^{(m-1)} + D^{-1}u^{(t)} \qquad (24)$$

The matrix $D^{-1}T$ has a zero diagonal and six offdiagonal terms $\frac{\alpha}{1+6\alpha}$. D^{-1} is a diagonal matrix with terms $\frac{1}{1+6\alpha}$. The iteration (24) is the central loop of the method (2). Note that this iteration is everywhere convergent with spectral radius defined by (3).

UNIT IMPULSE DERIVATION

This section demonstrates that the eigenvector normalization constant $c_{i,j,k}$ is equal to $(8/n)^{1/2}$ for all eigenvectors $\vec{x}_{i,j,k}$. From (16) a necessary condition for a normalized eigenvector is

$$1 =$$

$$c_{i,j,k}^2 \sum_{x,y,z} \cos^2\left(2\pi\frac{xi}{n^{1/3}}\right)\cos^2\left(2\pi\frac{yj}{n^{1/3}}\right)\cos^2\left(2\pi\frac{zk}{n^{1/3}}\right)$$

$$= c_{i,j,k}^2 \frac{1}{8}\sum_{x,y,z}\left(1+\cos 4\pi\frac{xi}{n^{1/3}}\right)X\left(1+\cos 4\pi\frac{yj}{n^{1/3}}\right)X$$

$$\left(1+\cos 4\pi\frac{zk}{n^{1/3}}\right)$$

$$= c_{i,j,k}^2 \frac{1}{8}\sum_{x,y,z}\left\{\left(1+\cos 4\pi\frac{yj}{n^{1/3}}\right)X\left(1+\cos 4\pi\frac{zk}{n^{1/3}}\right)\right.$$

$$+\sum_x \cos\left(4\pi\frac{xi}{n^{1/3}}\right)\sum_{y,z}\cos\left(4\pi\frac{yj}{n^{1/3}}\right)(1+$$

$$\left. \cos 4\pi\frac{zk}{n^{1/3}}\right)\right\} \qquad (25)$$

Simplify the preceding expression by the following

Lemma 1 $\sum_x \cos\left(4\pi\frac{xi}{n^{1/3}}\right) = 0$

Proof:

$$\sum_x \cos\left(4\pi\frac{xi}{n^{1/3}}\right) = \sum_x Re\left(e^{i\frac{4\pi xi}{n^{1/3}}}\right)$$

$$= Re\sum_x e^{i\frac{4\pi xi}{n^{1/3}}}$$

$$= Re\sum_x \left(e^{i\frac{4\pi i}{n^{1/3}}}\right)^x$$

$$= Re\left[\frac{e^{i\frac{4\pi i}{n^{1/3}}}\left(1-\left(e^{i\frac{4\pi i}{n^{1/3}}}\right)^{n^{1/3}}\right)}{1-e^{i\frac{4\pi i}{n^{1/3}}}}\right]$$

$$= 0$$

Q.E.D.

Repeated application of lemma (1) to equation (25) yields

$$1 = c_{i,j,k}^2 \frac{1}{8} \sum_{x,y,z} \left(1 + \cos 4\pi \frac{yj}{n^{1/3}}\right)\left(1 + \cos 4\pi \frac{zk}{n^{1/3}}\right)$$

$$= c_{i,j,k}^2 \frac{1}{8}\left[\sum_{x,y,z}\left(1 + \cos 4\pi \frac{zk}{n^{1/3}}\right) + \right.$$

$$\left. \sum_{x,y,z}\left(\cos 4\pi \frac{jy}{n^{1/3}}\right)\left(1 + \cos 4\pi \frac{zk}{n^{1/3}}\right)\right]$$

$$= c_{i,j,k}^2 \frac{1}{8}\left[\sum_{x,y,z} 1 + \sum_{x,y,z}\left(\cos 4\pi \frac{zk}{n^{1/3}}\right)\right]$$

$$= c_{i,j,k}^2 \frac{1}{8}n \qquad (26)$$

From which we conclude

$$c_{i,j,k} = \left(\frac{8}{n}\right)^{1/2} \qquad \forall\, i,j,k$$

ACKNOWLEDGEMENTS

We thank Michael Noakes for a concurrent implementation of the iteration (2) in J-Machine assembler. We are grateful to Andrew Conley for helpful discussions and criticisms.

References

[1] Amdahl, G. Validity of the single processor approach to achieving large scale computing capabilities. *Proc. AFIPS Comput. Conf.* **30** (1967) 483–485.

[2] Athas, W. C. & Seitz, C. L. Multicomputers: message passing concurrent computers. *IEEE Comp.* **21** (1988) 9–24.

[3] Barnard, S. T. & Simon, H. D. A fast multilevel implementation of recursive spectral bisection for partitioning unstructured problems. To appear in *Concurrency: Pract. Exp.* (1993).

[4] Boillat, J. E. Load balancing and poisson equation in a graph. *Concurrency: Pract. Exp.* **2** (1990) 289–313.

[5] Brugé, F. & Fornili, S. L. A distributed dynamic load balancer and it's implementation on multitransputer systems for molecular dynamics simulation. *Comp. Phys. Comm.* **60** (1990) 39–45.

[6] Cybenko, G. Dynamic load balancing for distributed memory multiprocessors. *J. Parallel Distrib. Comput.* **7** (1989) 279–301.

[7] Horn, R. A. & Johnson, C. R. *Matrix Analysis.* Cambridge University Press, New York, NY, 1991.

[8] Golub, G. H. & Van Loan, C. F. *Matrix Computations* (2nd edition). The Johns Hopkins University Press, Baltimore, MD, 1989.

[9] Gustafson, J. L. Reevaluating Amdahl's law. *Comm. ACM* **31** (1988) 532–533.

[10] Hofstee, H. P., Lukkien, J. J. & van de Snepscheut, J. L. A. A distributed implementation of a task pool. *Research Directions in High Level Parallel Programming Languages,* Banatre, J. P. & Le Metayer, D. (eds.), Lecture Notes in Computer Science **574** (1992) 338-348, Springer: New York.

[11] Horton, G. A multi-level diffusion method for dynamic load balancing. *Parallel Comp.* **19** (1993) 209–218.

[12] Kumar, V., Ananth, G. Y. & Rao, V. N. Scalable load balancing techniques for parallel computers. Preprint 92-021. Army High Performance Computing Research Center, Minneapolis, MN, 1992.

[13] Lin, F. C. H. & Keller, R. M. The gradient model load balancing method. *IEEE Trans. Soft. Eng.* **SE-13** (1987) 32–38.

[14] McCormick, S. F., *Multigrid Methods.* Society for Industrial and Applied Mathematics, Philadelphia, PA, 1987.

[15] Marinescu, D. C. & Rice, J. R. Synchronization and load imbalance effects in distributed memory multi-processor systems. *Concurrency: Pract. Exp.* **3** (1991) 593–625.

[16] Keckler, S. W. & Dally, W. J. Processor coupling: integrating compile time and run time scheduling for parallelism. *Proc. ACM 19th Int'l. Symp. on Comp. Arch.*, Queensland, Australia (1992) 202-213.

[17] Ni, L. M., Xu, C. & Gendreau, T. B. A distributed drafting algorithm for load balancing. *IEEE Trans. Soft. Eng.* **SE-11** (1985).

[18] Nicol, D. M. & Saltz, J. H. Dynamic remapping of parallel computations with varying resource demands. *IEEE Trans. Comp.* **37** (1988).

[19] Noakes, M. & Dally, W. J. System design of the J-machine. *Proc. 6th MIT Conf. on Advanced Research in VLSI.* Dally, W. J. (Ed.). MIT Press, Cambridge, MA, 1990, pp. 179–194.

[20] Pothen, A., Simon, H. D. & Liou, K. Partitioning Sparse Matrices with Eigenvectors of Graphs. *SIAM J. Matrix Anal.* **11** (1990) 430–452.

[21] Seitz, C. L. Mosaic C: an experimental fine-grain multicomputer. *Proc. International Conference Celebrating the 25th Anniversary of INRIA, Paris, France, December 1992*, Springer-Verlag, New York, NY, 1992.

[22] Shen, S. Cooperative Distributed Dynamic Load Balancing. *Acta Informatica* **25** (1988) 663–676.

[23] Wang, J. C. T. & Taylor, S. A Concurrent Navier-Stokes Solver for Implicit Multibody Calculations. To appear in *Proc. Parallel-CFD '93*, Paris, France, June 1993.

[24] Rowell, J. Lessons Learned on the Delta. *High Perf. Comput. Rev.* **1** (1993) 21–24.

Table 1: Solutions to equation (20) for increasing processor count n and accuracy α. τ represents the number of exchange steps in the method. See figure 1.

$\tau(\alpha, n)$		n (total processors)						
		64	512	4,096	8,000	32K	256K	10^6
	0.1	7	6	6	5	5	5	5
α	0.01	152	213	229	173	157	145	141
	0.001	2,749	5,763	10,031	10,139	9,082	7,564	7,003

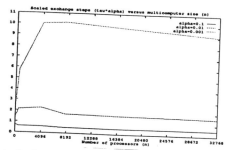

Figure 1: Scaled number of exchange steps $\tau\alpha$ to achieve accuracy α for various n, α following a point disturbance. See table 1 and equation (20). Each line is scaled by α. All lines are initially increasing for small n and asymptotically decreasing for larger n demonstrating weak superlinear speedup.

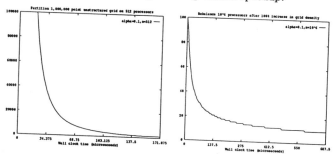

Figure 2: Time course of disturbances for simulated CFD cases. Left: largest discrepancy among 512 processors partitioning an unstructured computational grid. The initial disturbance of 1,000,000 points confined to a single processor is reduced by 90% after 6 exchanges (20.625 μs) in exact agreement with theory ($\tau(0.1, 512)$ in table 1). Right: largest discrepancy among 1,000,000 processors rebalancing a disturbance following a bow shock adaptation. $\alpha = 0.1, \nu = 3$ in both cases. Wall clock times assume a 32 MHz J-machine. The interval between exchange steps is 3.4375 μs.

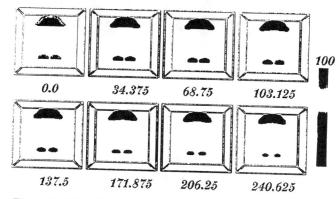

Figure 3: Disturbance following a bow shock adaptation of a computational grid on a million processor J-machine. First frame is the initial disturbance resulting from the adaptation. Subsequent frames are separated by 10 exchange steps. The disturbance is reduced dramatically by the second frame. After 70 exchange steps only weak low frequency components remain.

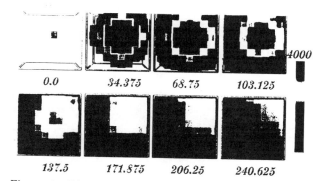

Figure 4: Disturbance during an initial load of a million point unstructured computational grid onto a 512 processor J-machine. The first frame represents the entire grid assigned to a host node on the multicomputer. This is a point disturbance and the resulting behavior is in exact agreement with the analysis presented earlier in this paper. Subsequent frames are separated by 10 exchange steps. After 70 exchange steps the workload is already roughly balanced. A balance within 1 grid point was achieved after 500 exchange steps.

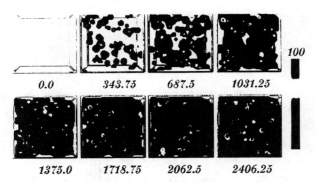

Figure 5: Disturbances resulting from rapid injection of large random loads on a million processor J-machine. After each exchange step a point disturbance is introduced at a randomly chosen processor. The average value of each point disturbance is 30,000 times the initial system load average. The interval between successive frames is 100 exchange steps. After 700 injections the worst case discrepancy was 15,737 times the initial load average. This demonstrates the algorithm was balancing the load faster than disturbances were created. After load injection ceased an additional 100 repetitions with no new disturbance reduced the worst case discrepancy from 15,737 to 50 times the initial load average.

PERFORMANCE GAINS FROM LEAVING IDLE PROCESSORS IN MULTIPROCESSOR SYSTEMS

E. Smirni*, E. Rosti†, G. Serazzi‡, L.W. Dowdy*, K.C. Sevcik§

esmirni@vuse.vanderbilt.edu
rose@dsi.unimi.it
serazzi@ipmel2.elet.polimi.it
dowdy@vuse.vanderbilt.edu
kcs@csri.toronto.edu

Abstract

When a parallel job arrives at a multiprocessor and there are idle processors, it is a common practice to assign all available processors to the job. In this paper the performance advantages of not allocating all available processors in multiprocessor systems are investigated. The class of adaptive space-sharing policies is considered. Analytical models of simple adaptive policies are examined. Complex policies are presented and investigated by means of experiments on a 512 node Intel Paragon. For all experiments, real parallel applications are used as the system workload. Results are reported for single and multiclass cases. Sensitivity analysis with respect to the workload interarrival time distribution is presented.

Performance gains derived from saving computational power (i.e., leaving idle processors while there are parallel jobs in the waiting queue) are identified as a function of the system size and workload characteristics. The more variable the workload characteristics are, the higher the performance improvements. The results show that leaving processors idle is beneficial in multiclass situations or when the job arrival rate is quite variable.

1 MOTIVATION

Multiprogramming is a common way of improving performance of multiprocessor systems. Since few applications can take full advantage of all system processors and since it is not uncommon that several parallel jobs are waiting to use the multiprocessor, performance can improve by allocating a subset of the processors to each of several jobs. Software related issues such as maximum application parallelism, diminishing return from assigning additional processors, and variable speedup characteristics of various jobs, motivate different processor allocation strategies for multiprogramming parallel systems.

Several processor allocation strategies have been presented since 1982 [11, 7, 15, 18, 12, 6, 2, 19, 3, 8, 20, 9, 14, 13, 16, 1]. The proposed strategies differ from each other in the way they allocate resources to the executing programs. Each strategy is designed to work well under certain conditions. Depending upon the system architecture, the memory organization (i.e., shared or distributed), the topology of the interconnection network, and the parallel application characteristics, preemptive or non-preemptive policies are more appropriate. Under preemptive policies, parallel programs can be suspended while in execution to allow for processor redistribution according to the system load. Time-sharing policies and dynamic space-sharing policies are considered preemptive. Non-preemptive [17], adaptive [13, 14], or run-to-completion policies [19, 1] do not allow processor reallocation while a job is in execution. Each job is assigned to a limited set of processors (i.e., a partition). The number of processors per partition is usually a function of the system load (i.e., the arrival rate of parallel jobs) and the speedup characteristics of the arriving jobs. The higher the workload arrival rate, the smaller the partition size should be to guarantee low average job response time and high system throughput [13, 14, 1].

The low implementation cost of non-preemptive policies makes them particularly appropriate for distributed memory message passing architectures [17]. However, non-preemptive policies are not as flexible as dynamic ones, which can adapt the number of allocated processors to each job according to system load fluctuations. Various solutions have been proposed to improve upon non-preemptive policy flexibility. A possible approach is to set aside some processors as a reserve for unpredictable changes in the workload intensity.

---

*Department of Computer Science - Vanderbilt University - Tennessee, USA.

†Dipartimento di Scienze dell'Informazione - Università di Milano, Italy.

‡Dipartimento di Elettronica e Informazione, Politecnico di Milano, Italy.

§Computer Systems Research Institute, University of Toronto, Canada.

This work was partially supported by Italian MURST 40% and 60% projects and by sub-contract 19X-SL131V from the Oak Ridge National Laboratory managed by Martin Marietta Energy Systems, Inc. for the U.S. Department of Energy under contract no. DE-AC05-84OR21400.

In this paper, the relative benefits of restricting the number of processors assigned to a job under specific conditions are examined. Policies that keep some of the available processors idle for later allocations are investigated. Such an approach is somewhat counter-intuitive, since idle processors do not contribute to maximum system utilization. Several cases where preserving idle processors is beneficial to performance, are identified. In the single processor case, there is no advantage of keeping the processor idle. When multiple processors are available, assigning an additional processor to a job may not improve performance significantly and it may be better to allocate the processor to a potential future arrival. Thus, it may be better to save some processors when future is uncertain. As uncertainty increases, it is more important to keep some processors unallocated. As expected, the relative performance gains depend upon the workload characteristics.

Analytical models of simple two-partition policies are presented and solved. The impact on performance is investigated. Factors such as offered system load, workload arrival process, workload speedup and composition (single class versus multiclass) are considered. The analysis of more complex policies is performed experimentally on a 512 node Intel Paragon multiprocessor. An LU decomposition kernel is used as the test workload for the experiments.

This paper is organized as follows. Section 2 presents the results of the analysis of simple policies. Section 3 reports the results of the experimental study of complex policies. Section 4 summarizes the findings and concludes the paper.

2 SIMPLE POLICIES

In this section, Markovian models of simple two-partition policies and performance analysis based on their solution are presented. Simple policy models can be solved analytically and provide useful insights for more complex systems. Initially, interarrival and execution times are assumed to be exponentially distributed.

2.1 Policy Description

The policies considered in this section assign either all processors or half of the processors to a job, based on information available at scheduling time. As a base case, a policy that never keeps processors idle if there are jobs waiting for execution is considered. Processor Saving (PS) policies which do not always allocate all available processors are derived from the base case. A systematic way of constructing PS policies is suggested.

Figure 1 depicts the Markov diagrams of the base case policy (top diagram) and two different PS policies (two lower diagrams). A state is described by the notation (q, r_p) where q is the number of jobs awaiting processor assignment, r is the number of executing jobs, and p indicates the partition assigned to the executing jobs. If the job has been assigned half the system, then p is h. If the

job is assigned the whole system, then p is w. State $(3, 1_w)$ indicates that there are 3 jobs waiting in queue, and 1 job is executing on the entire system. In state 0 the system is

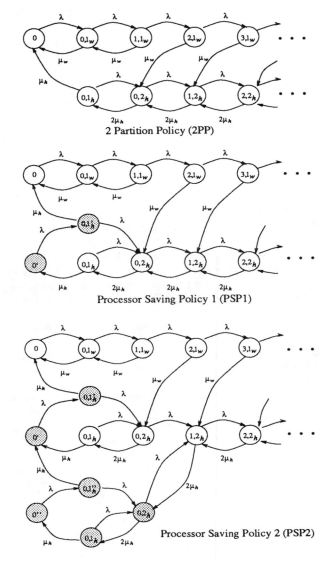

Figure 1: Markov diagrams of three allocation policies: 2PP, PSP1, and PSP2, which exhibit increasing amounts of processor saving.

completely idle. Shaded states indicate that a "processor saving" decision has been, or will be, made. λ and μ_i represent the job arrival rate and the job average execution rate on partition i, respectively. The base case 2-partition policy keeps all processors busy in presence of waiting jobs. The two PS policies are examples of possible strategies for saving processing power where in cases processors are kept idle, even though there are waiting jobs. In all cases, the policies are non-preemptive.

The base case 2-Partition Policy (2PP) (top diagram in

Figure 1) assigns the whole system to a single job if there is a single job requesting service in the system. Upon completion of a job that has been allocated the entire system, and if there are two or more jobs in the queue, the first two are scheduled, each being allocated half of the processors. With 2PP, at most two partitions are active, even when the waiting queue grows longer than 2.

The first Processor Saving policy PSP1 (middle diagram in Figure 1) is obtained by changing the scheduling decisions in the base case policy. When a job arrives at an empty system and the previously completed job had been allocated only half the processors (shaded $(0')$ state), the next arriving job is only allocated half the processors (shaded $(0, 1'_h)$ state). Thus the system "remembers" that both its partitions were busy in the recent past. It decides to "save" processing power since the past states suggest that new arrivals are highly probable. If another job arrives before the newly scheduled one completes, it will find an immediately available partition. If no arrivals occur, a light load condition is inferred and the policy returns to the 2PP behavior (state (0)).

By increasing the amount of memory introduced in the policy, the return to the base case behavior can be further delayed. The bottom diagram of Figure 1 describes another processor saving policy (PSP2) derived by adding 4 states to the PSP1 case. Under PSP2, the system remembers that both partitions were recently busy with one waiting job (state $(1, 2_h)$). Because of this recent past, the next two arrivals will be assigned only half the processors instead of all processors. The justification is that it is likely that several jobs may arrive in the near future and saving processors for such a case is beneficial.

The procedure followed for constructing PSP2 can be applied systematically to generate new policies with even more memory levels than PSP2 and PSP1. The more "levels" of memory are introduced, the more slowly the policy returns to the original 2PP behavior. Introducing additional states in order to remember the recent past behavior is equivalent to considering higher order Markov chains.

2.2 Analytical Results

The models presented in the previous section are solved analytically using the global balance equations technique and performance figures are derived. The results are reported in this section. The impact of workloads with different speedups and various intensities (i.e., arrival rates to the system) is investigated. The performance metric adopted is the response time ratio [1], defined as the ratio of the average response time under a given policy to the average response time under the 2-Partition Policy. By varying the job arrival rate λ, the workload intensity is also varied. The system is studied under the range of utilization levels.

Figure 2 plots the response time ratio against system utilization under policy PSP1 (solid line) and PSP2 (dashed line) with a flat speedup workload ($\mu_w = \mu_h$), a

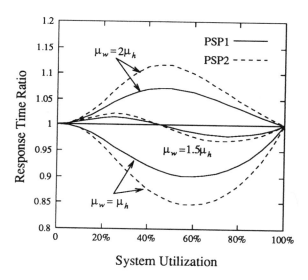

Figure 2: Response time ratio of the PS policies to the 2-Partition Policy (horizontal line) for three speedup workloads (linear, flat, and intermediate).

linear speedup workload ($\mu_w = 2\mu_h$), and with an intermediate speedup workload ($\mu_w = 1.5\mu_h$). Since the response time ratio (relative to the base case 2PP) is given, performance above the horizontal line at 1 indicates that it is better to always allocate all processors using 2PP. Similarly, performance below the horizontal line at 1 indicates that it is better to not allocate all processors and use a processor saving policy.

With a flat workload both PSP1 and PSP2 outperform 2PP. The maximum performance gain using PSP1 is 10% and the maximum performance gain using PSP2 is 15%. The PSP2 policy consistently outperforms PSP1 since it further delays the return to the 2PP type behavior. In the limit, with a flat workload, a fixed equipartitioning policy that assigns one processor to each job (i.e., an M/M/m server [5]), achieves the optimal performance. Such a policy exhibits a maximum performance gain over 2PP, since there is no advantage in assigning the entire system to an incoming job.

With a linear workload, there is no advantage in saving computational power. Performance loss results across the entire utilization range. This explains why PSP1 and PSP2 perform worse than the base 2PP policy in Figure 2.

As the speedup increases, spanning from flat to linear, the gain derived from PS strategies progressively reduces. Wave shaped response time ratios that intersect the 2PP line are obtained for workloads with intermediate speedups. The abscissa of the intersection depends on the workload speedup. The region to the left of the intersection point indicates the utilization interval where it is best to allocate all available processors. The region to the right of the intersection point indicates where it is best not to allocate all processors. As the system utilization increases, it is better to leave processors unallocated. As speedup increases, the intersection moves rightward, reaching 1 in the limit when the workload is nearly linear. Thus, as the speedup of the workload increases, it is better to allocate all processors as soon as possible. Nevertheless, since the difference between the total gains and losses for the two extreme speedups for a given PS policy is positive, if workload characteristics are not known, then using a PS policy improves performance more than it can hurt it. The potential maximum overall performance improvement due to a PS policy is greater than the potential maximum overall performance degradation.

As Figure 2 shows, for intermediate speedups, PSP1 performs better than PSP2 at low load, while their relative performance is reversed at higher load. At low load, keeping memory of long queues hurts performance since the system rarely returns to such states. A policy like 2PP that returns to larger partitions faster, following a temporary burst of job arrivals, performs better. On the contrary, at higher load, it is better not to allocate all processors after a temporary drop in job arrivals.

The results of the analysis of the simple policies considered in this section offer insight in the behavior of processor saving policies. The value of simple system modeling is the ability to isolate and focus on the peculiar aspects of the system under study abstracting the details which do not contribute to the observed phenomenon. When the analytical solution is computationally feasible, restricted models like the ones adopted here are useful for system analysis.

3 COMPLEX POLICIES

The policies presented in this section are more complex than those of Section 2. A wider variety of partition sizes is allowed. In the previous section, policies for which only two partitions are possible (whole and half) were examined analytically. In this section, more complex PS policies are considered for which analytical models are not feasible. Simulation and experimentation provide the alternative analysis techniques. An actual scheduler implementing the policies studied has been simulated on a real multiprocessor system. Actual applications were scheduled to arrive according to a Poisson arrival stream, which were then actually executed on the assigned partition.

The experimental platform is a 512 node Intel Paragon, a wormhole routed message passing multiprocessor system with distributed memory [4]. Results are presented for two system sizes, 32 and 64 processors. Experiments with 128, not reported here, confirm the findings of the smaller systems.

The implemented workload is an LU decomposition algorithm. It is executed on various matrix sizes, yielding different speedup curves. Four matrix sizes, namely 32×32, 64×64, 128×128, and 256×256, are considered. The corresponding speedup curves are reported in Figure 5.

A policy that never keeps processors idle in the presence of waiting jobs and a PS version derived from it have been implemented. The policies are described in the following section. The results of the base case experiment (i.e., with a Poisson arrival stream of single class jobs), are presented after the policy description. Sensitivity analysis with respect to the interarrival time distribution and experiments with multiclass workloads are also provided.

3.1 The Policies

The policies presented in this section allow for a wide variety of partition sizes. Partitions of various sizes provide greater flexibility to the workload fluctuations, thus improving the policy adaptability to unpredictable behavior. The goal is to mimic dynamic behavior using a non-preemptive policy thus avoiding the overhead of processor redistribution experienced by dynamic policies. With a wide variety of partition sizes, there are more opportunities for processor saving assignments, since the set of possible system states is larger. The pseudo-code algorithms of both the base case and the PS policies are given.

No Idle Processors Policy (NIP)
This policy is the base case of the experimental study and is analogous to the 2PP policy of Section 2. The NIP policy splits the entire processor set among the jobs waiting in queue for execution. Jobs are scheduled as long as there are free processors. When the number of free processors is smaller than the computed partition size, the job is scheduled on the remaining available processors. No information about the workload characteristics is needed.

Several policies have been presented in the literature that do not leave processors idle when there are pending requests [6, 3, 14, 1]. These policies divide the currently free processors among the waiting jobs, instead of the entire processor set. Such an approach is not adopted here in order to provide a uniform comparison with the PS policy considered. The partition size assigned to each job in this paper is computed as the total number of processors divided by the number of jobs in the queue. It has been demonstrated [13] that computing the partition size in this way results in better performance. The pseudo-code of the NIP algorithm is illustrated in Figure 3.

Processor Saving Adaptive Policy (PSA)
The PSA policy is based on a heuristic whose objective is to allocate an equal number of processors to all jobs in the system. Given the current state of the system, the new partition size is computed. If the number of free processors is smaller than the computed partition size, then no job is scheduled and the free processors are kept idle. In this case, a processor saving (i.e., non-work-conserving) decision is made, since some processors are kept idle while jobs are waiting for execution. The free processors will be assigned either: 1) if additional processors are released as jobs complete execution, or 2) if new jobs arrive, which force a smaller partition size. In the first case, a larger partition is granted at the expense of a longer waiting time.

```
job_in_queue := length(waiting_jobs_queue)
if (job_in_queue > 0) then {
    part_nb := min(system_size, job_in_queue)
    part_size := [system_size/part_nb + 0.5]
    while ((free_proc > 0) and
        (job_in_queue > 0)) do {
        if (part_size > free_proc) then
            part_size := free_proc
        partition := find(system, part_size)
        schedule(job, partition, part_size)
    }
}
```

Figure 3: No Idle Processors policy.

The potential gain or loss depends on the workload characteristics. If the waiting time for the partition size to become available is expected to be short and the waiting job can utilize the extra processors, then saving processors is beneficial. However, if the waiting time is long, or if the extra processors would not be efficiently utilized, or if additional jobs arrive which force a smaller target size, then it would be more beneficial to not keep available processors idle in the presence of waiting jobs. By adjusting the partition size, the PSA policy can vary the amount of processor saving that occurs. This is the key feature of the PSA policy. In the version analyzed here, memory of previous partition sizes is used. The motivation is that the partition size should change more slowly than the system state. A small partition size indicates many jobs in the system. If all jobs suddenly finish, it may not be wise to increase the partition size immediately in anticipation of another burst in job arrivals. Memory of the number of processors released by the last departing job (i.e., the size of the last released partition), is recorded in the variable last_released_part. A scheduling decision is made based upon this history variable and the current system state (i.e., the number of currently executing jobs (running_jobs), the number of free processors, and the number of partitions (part_nb)). No information about workload characteristics is needed. The pseudo-code of the algorithm is illustrated in Figure 4.

The PSA policy adapts to workload changes. Since no knowledge of the workload speedup characteristics is required to compute the partition size, the PSA policy is relatively robust. This policy has been introduced in [13], where the focus was on the policy robustness and the benefits of the processor saving aspects were not explored.

3.2 Experiments - Single Class Workload

In this section, experimental results on the Intel Paragon with various single class workloads for two system sizes, 32 and 64 processors, are presented. An LU decomposition algorithm is implemented and executed for four different matrix sizes. Each matrix size yields a dif-

```
job_in_queue := length(waiting_jobs_queue)
if (job_in_queue > 0) then {
    if (job_in_queue = 1) and
    ((running_jobs = 0 and
        last_released_part ≠ system_size) or
    (running_jobs = 1 and part_size = system_size))
    then {
        if (part_nb > 2) then
            part_nb := part_nb - 1
        else
            part_nb := 2
    }
    else {
        if (job_in_queue ≥ part_nb) then
            part_nb := min(system_size, job_in_queue)
        elseif ((part_nb - running_jobs) > 1) then
            part_nb := part_nb - 1
    }
    part_size := [system_size/part_nb + 0.5]
    while (free_proc > 0 and job_in_queue > 0) do {
        partition := find(system, part_size)
        schedule(job, partition, part_size)
    }
}
```

Figure 4: Processor Saving Adaptive policy.

ferent speedup curve (see Figure 5).

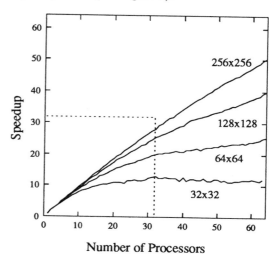

Figure 5: Speedup curves on the Intel Paragon for an LU decomposition application with various matrix sizes.

A Poisson stream of execution requests is submitted to the scheduler. The range of possible system utilizations is considered by changing the job arrival rate. Arriving requests join the waiting queue and the scheduler services them in FIFO order by allocating the computed number of processors (according to the policy) to the job at the head of the queue. Once a job is allocated a set of processors, the LU decomposition is executed. For every arrival rate considered, an arrival stream of four thousand jobs is submitted to the system and the average response time is measured.

The response time ratios of the PSA policy versus the NIP policy are plotted in Figure 6. The response time

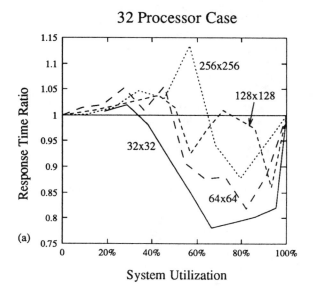

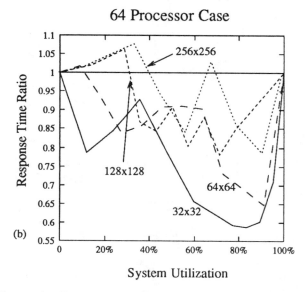

Figure 6: Response time ratios of the PSA policy to the NIP policy for the experiments on the Intel Paragon with 32 and 64 processors, for an LU decomposition application with various matrix sizes.

ratios follow the trend observed in the analytical results

obtained in Section 2.2. For a system size of 32 (see Figure 6(a)), the maximum gain is obtained for the lowest speedup (i.e., the 32×32 case). The intersection point of the reference horizontal line and the PSA policy curve occurs at low system utilization and around 35%. As the speedup increases, the intersection point moves to the right. For the 256 case the intersection is around 65% of the system utilization. Because of measurement instabilities, the curves are not as smooth as desired[1]. Regardless of the spikes in the curves, there is a clearly observed trend.

In the 64 processor case, the response time ratios of the PSA policy are reported against the system utilization in Figure 6(b). Note the different y-axis scales between Figures 6(a) and 6(b). The performance gain of the PSA policy with the larger system is more significant. Although the span of the speedup curves is more spread than in the 32 processor case (see Figure 5), for the same speedup type, the PSA policy performs better as the system size increases. As an example, consider the speedup of the 32×32 matrix for 32 processors and the speedup of the 64×64 matrix for 64 processors. Both speedups exhibit almost identical behavior, showing the same efficiency (40%) for the maximum number of processors (32 and 64, respectively). However, the performance gains observed in the larger system are more pronounced. The intersection point is closer to the lower end of the utilization range. As the system size increases, the impact of wasted power due to not allocating processors as soon as possible is less severe. This is due to the fact that more jobs are in execution in larger systems. In a 2-partition system, where at most two jobs are executing, idle processors hurt performance more severely since the individual job response time has higher variance. The basic observation is that processor saving policies become more advantageous: 1) as the speedup behavior of the workload decreases, 2) as the system utilization increases, and 3) as the system size increases.

3.3 Sensitivity Analysis

In this section, the sensitivity analysis with respect to the interarrival time distribution for the single class workload is considered. Hyperexponential and hypoexponential interarrival times are examined. Results are reported for a single matrix size, 64×64, on 32 processors only. The results for larger system sizes and other matrix sizes are similar. Sensitivity analysis with respect to the application execution time distribution was not possible since a real workload with negligible variance in the execution time was used. Sensitivity analysis with real workloads with variance in the execution time will be the subject of future investigation.

---

[1]Even though each experiment was run multiple times, the average response times produced by each run were very close. It is possible that the jagged behavior of the curves will be alleviated by using different seeds in the exponential interarrival times for the incoming jobs, as well as different parallel workloads.

With hypoexponentially distributed interarrival times (i.e., with coefficients of variation less than 1), the variance in the interarrival times is smaller than with exponentially distributed interarrival times. The opposite is true with hyperexponential distributions, with coefficients of variation greater than 1. The hypothesis is that as the workload becomes more unpredictable (i.e., as the CV increases), the better it is to save processors to compensate for sudden workload changes in the future. Thus, as the CV increases, the PSA policy is expected to outperform a policy that allocates all processors as soon as possible. Likewise, if the arrival process is more regular and has smaller variance, sudden changes in the arrival process are not likely to occur, and all processors can be kept busy without hurting performance.

scalability, and communication behavior. As in the previous section, the hypothesis is that as the workload becomes more unpredictable, meaning that more classes can be identified in the workload, the better it is to save some processors to respond to unpredictability.

The multiclass mixes considered are constructed from variable percentages of the single class components used in the single class analysis. Let the notation (W%, X%, Y%, Z%) represent a multiclass mix where W% of the arriving jobs perform LU decomposition on a matrix of size 32×32, X% of the arriving jobs solve a matrix of 64×64, Y% solve a matrix of size 128×128, and Z% solve a matrix of size 256×256. Since each matrix size leads to unique speedup characteristics (see Figure 5), the resulting workload exhibits a typical multiclass behavior.

Matrix Size 64x64

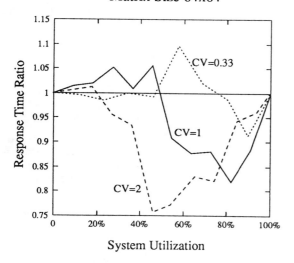

Multiclass Workload Mixes

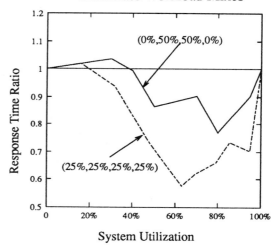

Figure 7: Response time ratios of the PSA policy to the NIP policy for the experiments on the Intel Paragon with 32 processors with hypoexponential and hyperexponential interarrival times as a function of the system utilization.

Figure 8: Response time ratios of the PSA policy to the NIP policy for the experiments on the Intel Paragon with 32 processors and two multiclass workload mixes.

Figure 7 shows the response time ratios of the PSA policy against system utilization for coefficients of variation equal to 2 (dashed line), 1 (the base case, solid line), and 0.33 (dotted line). The experiments confirm that as the coefficient of variation increases, the intersection point moves to the left and that the curves drop. This indicates that keeping processors idle is more beneficial as the arrival process becomes more unpredictable.

3.4 Experiments - Multiclass Workload

In this section, multiclass workloads are considered. Multiclass workloads often are more representative of actual workloads since applications running on parallel systems vary widely with respect to execution requirements,

In Figure 8, response time ratios for two workload mixes, namely (25%, 25%, 25%, 25%) and (0%, 50%, 50%, 0%) are reported. When the mix is balanced at (25%, 25%, 25%, 25%), the workload is the most heterogeneous. In this case, the intersection point is the leftmost and the benefit of saving processors is the highest. As the workload becomes more homogeneous and closer to single class, the intersection point moves to the right. These experiments support the hypothesis that if the workload components vary considerably, performance may improve by not allocating all processors as soon as possible.

4 CONCLUSIONS

In this paper, the advantage of not allocating processors to parallel applications, thus reserving some of the computational power for jobs yet to arrive, is investigated. Keeping some processors idle at the end of a scheduling round instead of allocating all processors whenever jobs are available to use them proves beneficial in certain cases:

- when the applications do not have good speedup characteristics,
- when the system utilization is medium to high,
- when the system size is large,
- when the job arrival rate is highly variable, or
- when the workload is multiclass.

The results of analytical models of 2-partition processor saving policies and of the experimental study of more complex policies support the above observations. Equivalently, the results indicate that if: 1) the workload speeds up linearly, 2) the multiprocessor system has low utilization, 3) the number of processors is small, 4) the job arrival is constant, and 5) the jobs are homogeneous, then allocating processors as quickly as possible is better.

Future work includes a more complete investigation of the impact of hyperexponential and hypoexponential distributions of the workload execution time and the workload arrival process. The effects of batch arrivals will be studied. A wider investigation of multiclass effects is planned. The usefulness of PS policies in the presence of processor failures will also be explored. The PSA policy will be compared with other processor saving policies presented in the literature. Experimental studies on various multiprocessor platforms are needed.

Acknowledgements

The authors would like to thank the Mathematical Sciences Research Section at Oak Ridge National Laboratory for providing access to the Intel Paragon xps/35 system.

References

[1] S.-H. Chiang, R.K. Mansharamani, M.K. Vernon, "Use of application characteristics and limited preemption for run-to-completion parallel processor scheduling policies," *Proc. ACM SIGMETRICS*, 1994, pp. 33-44.

[2] K. Dussa, B.M. Carlson, L.W. Dowdy, K.-H. Park, "Dynamic partitioning in a transputer environment," *Proc. ACM SIGMETRICS*, 1990, pp. 203-213.

[3] D. Ghosal, G. Serazzi, S.K. Tripathi, "Processor working set and its use in scheduling multiprocessor systems," *IEEE Transactions on Software Engineering*, Vol 17(5), May 1991, pp. 443-453.

[4] Intel Corporation, **Paragon OSF/1 User's Guide**, 1993.

[5] L. Kleinrock, **Queueing Systems**, Vol 1, Wiley Interscience, 1975.

[6] S.T. Leutenegger, M.K. Vernon, "The performance of multiprogrammed multiprocessor scheduling policies," *Proc. ACM SIGMETRICS*, 1990, pp. 226-236.

[7] S. Majumdar, D.L. Eager, R.B. Bunt, "Scheduling in multiprogrammed parallel systems," *Proc. ACM SIGMETRICS*, 1988, pp. 104-113.

[8] S. Majumdar, D.L. Eager, R.B. Bunt, "Characterization of programs for scheduling in multiprogrammed parallel systems," *Performance Evaluation*, Vol 13(2), 1991, pp. 109-130.

[9] C. McCann, R. Vaswani, and J. Zahorjan, "A dynamic processor allocation policy for multiprogrammed shared memory multiprocessors," *ACM Transactions on Computer Systems*, Vol 11(2), February 1993, pp. 146-178.

[10] C. McCann, J. Zahorjan, "Processor allocation policies for message-passing parallel computers," *Proc. ACM SIGMETRICS*, 1994, pp. 19-32.

[11] J. Ousterhout, "Scheduling techniques for concurrent systems," *Proc. 3rd International Conference on Distributed Computing Systems*, 1982, pp. 22-30.

[12] K.-H. Park, L.W. Dowdy, "Dynamic partitioning of multiprocessor systems," *International Journal of Parallel Programming*, Vol 18(2), 1989, pp. 91-120.

[13] E. Rosti, E. Smirni, L.W. Dowdy, G. Serazzi, B.M. Carlson, "Robust partitioning policies for multiprocessor systems," *Performance Evaluation*, Vol 19(2-3), March 1994, pp. 141-165.

[14] S.K. Setia, M.S. Squillante, S.K. Tripathi, "Processor scheduling in multiprogrammed, distributed memory parallel computers," *Proc. ACM SIGMETRICS*, 1993, pp. 158-170.

[15] K.C. Sevcik, "Characterization of parallelism in applications and their use in scheduling," *Proc. ACM SIGMETRICS*, 1989, pp. 171-180.

[16] K.C. Sevcik, "Application scheduling and processor allocation in multiprogrammed multiprocessors," *Performance Evaluation*, Vol 19(2-3), March 1994, pp. 107-140.

[17] E. Smirni, E. Rosti, L.W. Dowdy, G. Serazzi, "Evaluation of multiprocessor allocation policies," Tech. Report, Computer Science Dept., Vanderbilt University, Nashville, TN, (submitted for publication).

[18] A. Tucker, A. Gupta, "Process control and scheduling issues for multiprogrammed shared-memory multiprocessors," *Proc. of the 12th ACM Symposium on Operating Systems Principles*, 1989, pp. 159-166.

[19] J. Zahorjan, C. McCann, "Processor scheduling in shared memory multiprocessors," *Proc. ACM SIGMETRICS*, 1990, pp. 214-225.

[20] S. Zhou, T. Brecht, "Processor pool-based scheduling for large-scale NUMA multiprocessors," *Proc. ACM SIGMETRICS*, 1991, pp. 133-142.

PERFORMING LOG-SCALE ADDITION ON A DISTRIBUTED MEMORY MIMD MULTICOMPUTER WITH RECONFIGURABLE COMPUTING CAPABILITIES

Mike Wazlowski, Aaron Smith, Ricardo Citro, and Harvey Silverman[a]

Division of Engineering, Box D

Brown University

Providence, RI 02912

{mew,ats,rc,hfs}@lems.brown.edu

Abstract -- *Armstrong III is a 20 node multicomputer that is currently operational. In addition to a RISC processor, each node contains reconfigurable computing resources implemented with FPGAs. The in-circuit reprogrammability of static RAM-based FPGAs allows the computational capabilities of a node to be dynamically matched to the computational requirements of an application. Most reconfigurable computers in existence today rely solely on a large number of FPGAs to perform computations. In contrast, this paper demonstrates the utility of a small number of FPGAs coupled to a RISC processor with a simple interconnect. This paper describes the Armstrong III architecture and concludes with a substantive example application that performs log-scale addition.*

INTRODUCTION

Application-specific computers provide significant performance improvements over general-purpose computers by utilizing architectures that are optimized for a specific task. The former perform well on one class of application but do not usually perform well on others. Custom computing machines (CCMs) are computers that utilize field-programmable gate-arrays (FPGAs) to allow their hardware architecture to adapt to each new task. In effect, CCMs rewire themselves *in situ* for each new task. They make application-specific computers more feasible by using the same hardware for different classes of applications. CCMs are implemented with static-RAM-based FPGAs in combination with a flexible system architecture that usually includes static-RAM memory and an interconnection resource used to connect multiple FPGAs [1] [2].

The principle design goal of Armstrong III has been to demonstrate the viability of using a small number of FPGAs coupled to a RISC processor rather than just using a large body of FPGAs alone[1][2]. The fusion of a conventional RISC processor with a reconfig-

[a]This work principally funded by NSF grant MIP-9120843 and partially by MIP-9021118.

urable processing element could prove to be more cost-effective than existing CCMs that have five times as many FPGAs. The utility of this approach is demonstrated in this paper with a substantive design example.

PRISM I [3] was a first generation general-purpose computing platform with reconfigurable capabilities. An Armstrong I [4] node served as the host and an attached board with four Xilinx XC3090 FPGAs comprised the reconfigurable platform. The Armstrong III multicomputer with PRISM II processing nodes reflects the experience gained from PRISM I.

ARMSTRONG III

Armstrong III is a distributed memory MIMD multicomputer with 20 nodes that can be arranged in different connection topologies. One of the Armstrong III nodes is connected to a Sun workstation via a custom SBUS interface that provides access to hard disks and a color monitor for all nodes. Each Armstrong III node is comprised of a communications board and a PRISM II processor board. The communications board sends and receives packets to or from other nodes and is similar in function to the communications structure of Armstrong II [5]. The reconfigurable PRISM II board is the primary computation engine.

A block diagram of the communications board, is shown in Figure 1. The static-RAM is primarily for packet storage, but a small operating system kernel (13K) resides there as well. Eight serial I/O ports provide for communications with other nodes. There are two communications channels, each with four ports. Only one port from a channel may be active at a given time. A Xilinx XC3064 FPGA controls two 1K byte serial/parallel FIFOs for each channel. A hardware checksum register verifies packet integrity automatically while data is being loaded or unloaded from the FIFO. Twisted-pair cables physically connect two nodes. Two serial bit streams are simultaneously tranmitted/received at 20Mb/s for a total 40Mb/s transfer rate between nodes. The actual time to send a 256 byte packet using a simple reliable connectionless

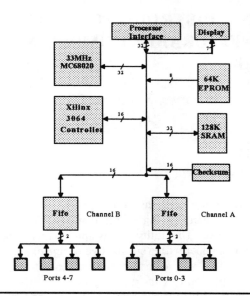

Figure 1: Communications board data path and architecture.

send/acknowledge protocol to a node that is two hops away is roughly one millisecond.

The conventional processing element of the PRISM II board is a 33MHz AMD Am29050 RISC processor as shown in Figure 2. The reconfigurable block is made up of three Xilinx XC4010 FPGAs and is show in detail in Figure 3. The 26-bit global bus connecting all three XC4010s permits sharing of control signals and internal variables. Each XC4010 is tightly-coupled to its own 128K x 32 25ns static RAM which may be accessed independently of the Am29050. The three FPGAs can function independently or, for large designs that cannot fit into a single FPGA, can be used as one large reconfigurable resource. Each XC4010 connects through buffers to both the address and data buses of the Am29050. The Am29050 coprocessor interface allows 64-bit writes with a single access using a single

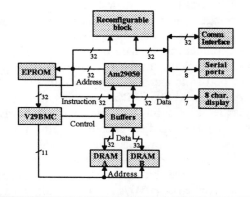

Figure 2: PRISM II board architecture.

instruction [6]. The coprocessor interface writes data on the address bus in addition to the data on the data bus. Logic can be loaded into the FPGAs at run time on a per application basis and as many times as needed while the application is running. Reconfigurability allows a greater amount of parallelism to be exploited in an application than is possible with a conventional microprocessor because the capabilities of a custom computing machine can be tailored to each application.

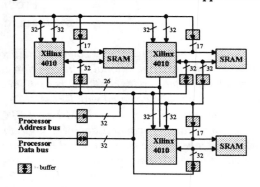

Figure 3: The reconfigurable block.

PROGRAMMING ARMSTRONG III

Creating an application for Armstrong III requires three steps:

- writing a parallel C or C++ program
- designing the custom logic for one or more FPGAs using VHDL or schematic entry
- simulating the FPGA design to ensure correctness

The output from the VHDL synthesizer or the schematic netlist is converted into Xilinx Netlist Format (XNF) and is partitioned, placed and routed by proprietary Xilinx tools [7]. A locally written utility program converts the Motorola S records generated by the Xilinx tools to a data section of an assembly language file which is then assembled. The object files produced by the C compiler and assembler are linked with an additional module that takes the programming data from the assembled data section and loads it into the FPGA at run time.

HMM TRAINING EXAMPLE

An important problem for speech-recognition research is training hidden Markov models (HMMs)[8]. A single iteration of the current LEMS HMM training program takes 26 hours to complete on a SPARCstation 10 Model 51. Run-time profiles indicate the training program spends 36% of its total execution time, by far the largest, adding probabilities on a logarithmic scale. The training program execution time can be reduced on Armstrong III by using multiple nodes and by creating an application-specific machine for logarithmic addition using the FPGAs.

A precise definition of the problem is: given $\ln(a)$ and $\ln(b)$, where a and b are probabilities, compute $\ln(a+b)$. This computation will hereafter be referred to as the *logsum*. In order to perform the computation, it is necessary to express the logsum in terms of $\ln(a)$ and $\ln(b)$. This is easily done as follows, where it is assumed that $a \geq b$,

$$\ln(a+b) = \ln\left(a(1+\tfrac{b}{a})\right)$$
$$= \ln(a) + \ln\left(1+\tfrac{b}{a}\right)$$
$$= \ln(a) + \ln(1 + e^{-(\ln(a)-\ln(b))})$$
$$= \ln(a) + f(\ln(a) - \ln(b))$$

where $f(x) = \ln(1 + e^{-x})$ may be computed with a lookup table using $\ln(a) - \ln(b)$ as the address into the table [8]. The larger of the two log probabilities is assigned to the position of $\ln(a)$ in the equation above to ensure that the table address is always positive. If the address calculated is greater than the table size, the larger operand is returned as the answer. This is valid due to the nature of the function $\ln(1 + e^{-x})$ that decreases monotonically towards zero as x increases. The table is large enough to ensure that values outside of it may be considered zero.

In the HMM training algorithm used here, probabilities on the log scale are converted to 32-bit two's-complement integers by linearly scaling them by a factor of 8192 and then rounding. This scale factor provides an acceptable number of digits of precision.

The first step towards implementing the logsum as an FPGA design is to construct a data flow graph as shown in Figure 4. The oval marked Mem represents a memory access and Max Address is a constant representing the highest address location of the table. The data flow graph exposes fine grain parallelism that may be exploited by a custom computing machine and suggests how much logic will be required. In addition, the data flow graph clarifies where pipeline registers may be added to a design.

One aspect of the HMM training algorithm requires that the logsum computation be performed on a vector of 256 probabilities. Even more fortunate is that many logsums of this length can be computed in parallel. This number of successive operations indicates that pipelining would be effective. Xilinx FPGAs are naturally suited to pipelined architectures because of the inclusion of flip-flops directly after the function

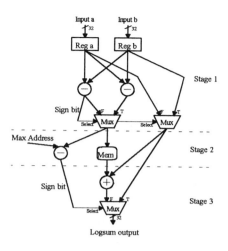

Figure 4: Data flow graph for the logsum computation.

generators in each CLB. The horizontal dashed lines in Figure 4 separate three pipeline stages labeled 1, 2, and 3. A register is added where a signal or bus crosses the dashed lines. Partitioning into stages was performed to match processor throughput with FPGA throughput. The maximum processor throughput achievable on PRISM II is 64 bits every 60ns utilizing the Am29050 coprocessor interface. Therefore, the data flow graph was partitioned into three stages less than or equal to 60ns in duration. The pipelined design has three computations in progress simultaneously giving an effective speedup of three over a non-pipelined design. However, a small penalty is incurred at the end of the computation. With a three-stage pipeline, three separate accumulated logsums will coexist. These separate logsums will have to be flushed from the pipeline and computed conventionally when the 256 probabilities have been exhausted.

To fill the pipeline, two datum are required per pipeline clock. Once the pipeline is full, just one datum is required since the accumulated logsum from the previous iteration is used as the second datum. Operations which consume just one datum per clock cycle do not exploit the maximum throughput of the Am29050, which is 64 bits per cycle. This situation is remedied by using a dual pipeline structure and two FPGAs, as shown in Figure 5. One pipeline must be filled using 64-bit writes, and then the other. Once both pipelines are full, 64-bit writes are used to keep them full, where 32 bits go to each pipeline simultaneously. A multiplexer for each logsum selects among 64 bits of input data while the pipeline is being filled and 32 bits of input data once the pipeline has been filled. This design has six accumulating logsums running concurrently.

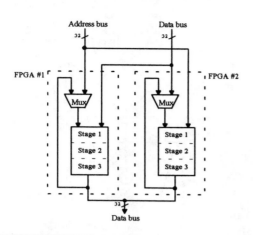

Figure 5: The dual pipeline structure for the logsum. One three stage pipeline is placed in each FPGA.

This design implementation matches maximum processor throughput with the throughput of the FPGAs. One pipeline clock pulse will be generated every time data is written to the FPGAs by the processor. The design implementation uses two of the three available FPGAs on a node and stores the required lookup tables in the local static RAMs next to each of them. The actual design was created using schematic entry.

Table 1 shows the performance of PRISM II with and without using the custom hardware, and the performance of a SPARCstation 10 Model 51 for the log-sum computation. The execution times shown are for computing the logsum of a vector of 256 probabilities. These are **measured** execution times. A SPARCstation 10 Model 51 is 3.5 times faster than a PRISM II board without utilizing the FPGAs. The logsums performed on the PRISM II board utilizing the FPGAs are 19.9 times faster than a PRISM II board without utilizing the FPGAs. A PRISM II node using the FPGAs is 5.6 times faster than the SPARC. This results in a speedup of 20% for a single node over the SPARC for the entire program.

This implementation is part of a distributed HMM training algorithm. Coarser grain parallelism is taken advantage of by distributing log-scale additions among multiple Armstrong III nodes.

Table 1: Execution times for the logsum of 256 probabilities.

	SPARC 10 Model 51	PRISM II w/o using FPGAs	PRISM II with FPGAs
Execution Times (µs)	51	179	9
Speedup	3.5	1	19.9

DISCUSSION

This paper has described Armstrong III, a multi-computer with reconfigurable computing resources on each node. The cited initial implementation of the HMM training application has shown that the limited FPGA resources of an Armstrong III node provide performance exceeding that of a state-of-the-art workstation. Recent refinements are indicating that, for the 20-node system, the time for a training iteration is reduced from 26 hours on the SPARC to roughly 3/4 of an hour on Armstrong III assuming a linear speedup.

ACKNOWLEDGMENTS

We would like to express our gratitude to Xilinx, Advanced Micro Devices, Viewlogic, Texas Instruments, Harris EDA, AMP, Augat, and the V3 Corporation for their equipment and software support.

REFERENCES

[1] D. Buell and K. Pocek (eds.), *Proceedings of the IEEE Workshop on FPGAs for Custom Computing Machines*, IEEE Computer Society Press, 1993.

[2] D. Buell and K. Pocek (eds.), *Proceedings of the IEEE Workshop on FPGAs for Custom Computing Machines*, IEEE Computer Society Press, 1994.

[3] P. Athanas and H. Silverman, "Processor reconfiguration through instruction set metamorphosis", *IEEE Computer*, vol. 26, no. 3, pp. 11-18, March 1993.

[4] J. Rayfield and H. Silverman, "System and application software for the Armstrong multiprocessor", *IEEE Computer*, vol. 21, no. 6, pp. 38-52, June 1988.

[5] P. Athanas, "Armstrong II: A loosely-coupled multiprocessor with a reconfigurable communications architecture", in *IEEE International Parallel Processing Symposium*, Anaheim, California, May 1991, pp. 723-726.

[6] Advanced Micro Devices Inc, Sunnyvale, California, *Am29050 Microprocessor User's Manual*, 1990.

[7] Xilinx Inc, San Jose, CA, *XACT Reference Guide*, 1994.

[8] H. Silverman and Y. Gotoh, *On the Implementation and Computation of Training an HMM Recognizor Having Explicit State Durations and Multiple-Feature-Set Tied-Mixture Observation Probabilities*, LEMS Monograph 1-1, Brown University, Providence, Rhode Island, 1994.

Partitioning an Arbitrary Multicomputer Architecture *

Laxmi Bhuyan, Sumon Shahed, and Yeimkuan Chang

Department of Computer Science
Texas A&M University
College Station, Texas 77843-3112
Tel. (409)845-9640
E-mail: {bhuyan, sumon, ychang}@cs.tamu.edu

Abstract— A good processor allocation strategy is the one that efficiently selects a subset of the system such that the selected subsystem is the best one in terms of processor utilization In this paper, a good way of partitioning an arbitrary multicomputer architecture is developed. The problem is formulated as a graph partitioning problem and the Kernighan-Lin (K-L) algorithm is applied. Mesh, deBruijn and hypercube structures are tested by using the proposed algorithm to show that correct subsystems are obtained.

1 Introduction

Different classes of processor allocation strategies have been proposed for multicomputer architectures. A good processor allocation/deallocation strategy is the one that efficiently selects a subset of the system such that the selected subsystem is the best one in terms of processor utilization. Many subcube allocation stratgies, such as buddy, gray code(GC) [1], tree collapsing (TC) [2], free list strategy [3] and weight allocation strategy (WAS) [4], have been developed for hypercube architectures. For mesh architecture, a two-dimensional buddy system (2DBS) was proposed as a partitioning scheme by Li and Cheng [5]. More sophisticated submesh allocation schemes in a mesh-connected multiprocessor have been developed in [6]. The main objective of all the allocation strategies is maximum utilization of the computing power of the system. In order to do that every strategy tries to minimize the *internal* and *external* fragmentations. Internal fragmentation occurs when a request gets more processors than the number of requested processors. External fragmentation occurs when there exists free processors in the system, but can not be allocated

as a possible subsystem. Although many new allocation strategies are suggested in the literature, they are quite complex to be applied in real system. On the other hand, the original Buddy strategy is simple and has proven to be quite successful through commercial implementations.

Defining subsystems for allocation to the incoming jobs is obvious for hypercube and mesh architectures. They are subcube and submesh respectively. However, the definition of a subsystem is not known for an arbitrary architecture. An arbitrary architecture can result due to an arbitrary topology or different capabilities of communication links between the processors. Also, well defined architectures become arbitrary in the presence of faults. A good technique to divide the computing system, depending on their processing power and communication cost, would be very helpful for the efficient use of the system. In this paper, a novel way of partitioning an arbitrary multicomputer system architecture is developed based on graph connectivity and application of Kernighan-Lin (KL) heuristic [7]. The technique generates a buddy tree that can be used for processor allocation with buddy strategy [1, 5]. It is shown that the proposed algorithm indeed produces subcubes and submeshes for hypercube and mesh architectures. The algorithm is also applied to a multicomputer based on the deBruijn graph [8] to generate subsystems that cannot be obtained by an intuitive partition.

The rest of this paper is organized as follows. Section 2 presents a method to convert the target multicomputer to a graphical representation. The proposed heuristic algorithm based on KL algorithm is given in Section 3. In Section 4, results and discussion of the proposed algorithm are presented. Concluding remarks are provided in the last section.

*This research is partially supported by NSF grant MIP-9301959

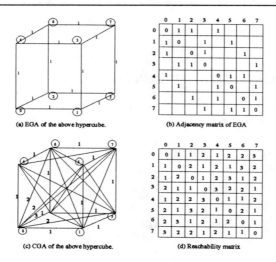

(a) EGA of the above hypercube.

(b) Adjacency matrix of EGA

(c) CGA of the above hypercube.

(d) Reachability matrix

Figure 1: Complete Graph of Architecture (CGA) for an 8-node hypercube.

2 Representing an Architecture

An arbitrary architecture can be represented graphically as $\mathbf{G(V,E)}$, where the vertices $V = \{v_1, v_2, \ldots, v_N\}$ represent the processors and the edges $E = \{e_1, e_2, \ldots, e_{|E|}\}$ represent the communication links. If costs of the edges are Considered then the graph is considered as a *weighted graph*. A weighted graph is a tuple $\mathbf{(V,E,W)}$, where $\mathbf{(V,E)}$ is a graph and W is a function from E into $\Re$, the set of all real numbers. For an edge e, $W(e)$ is called the weight of e. A weighted graph $\mathbf{G = (V,E,W)}$, with $|V| = N$ = number of processors, $|E|$ = number of links, $V = \{v_1, v_2, \ldots, v_N\}$ and $W : E \rightarrow \Re$ can be represented by an $N \times N$ matrix $\mathbf{A} = (a_{ij})$, called *adjacency matrix* for G. A is defined by,

$$a_{ij} = \begin{cases} W(v_i, v_j) & \text{if } v_i v_j \in E \\ c & \text{otherwise} \end{cases} \qquad (1)$$

where $1 \leq i \leq N, 1 \leq j \leq N$ and c is a constant whose value depends on the interpretation of weights and the problem to be solved. If a link is nonexistent, infinity or some very high number may be chosen for c because the cost of traversing a nonexistent edge is prohibitively high. Also $W(v_i, v_j) = 0$ if $v_i = v_j$. The graphical representation of an arbitrary architecture considering the link capacity will be called *equivalent graph of architecture* (EGA).

In this paper, we try to get one general representation of the arbitrary multicomputer architecture. Since the arrangement and number of communication links varies from one architecture to another, a technique needs to be developed to convert any arbitrary architecture to a general form. The graphical representation will be transformed into a fully connected

or complete graph with different weights on the links. We will call this graph the *complete graph of architecture* (CGA). The weight of each link in the CGA will specify the shortest distance by the nodes connected between the link. If the links have different weights, these are properly considered in computing the shortest distance. The CGA representation of an arbitrary architecture is obtained as follows,

- Create EGA from the definition of the arbitrary multicomputer architecture.

- Obtain an adjacent matrix that shows the connectivity of the architecture.

- Compute the shortest path from one node to another that are not connected directly in the EGA. The shortest path is computed using an extension of *Warshall's Algorithm* [9]. This shortest path represents the minimum number of links a message has to traverse to reach the destination node from the source node. Obtain the CGA.

- Obtain the adjacency matrix of the CGA, and call it as *reachability matrix*.

As an example, consider an 8-node hypercube. The EGA is shown in Figure 1(a). Then the 8×8 adjacency matrix is shown Figure 1 (b), where the row number and column number correspond to the appropriate processor number in the hypercube. This matrix has some cells filled with 1's, 0's and blanks, as obtained from EGA. A cell at row x and column y represents a one-to-one relationship between processors x and y. Thus, a 1 in cell at row x and column y means processors x and y are directly connected through a communication link. All the cells with same row and column number are filled with 0's because every processor can communicate with itself without a communication link within itself. The blank cells in the matrix means that there is no direct communication link between the corresponding processors. Next, we compute the shortest distance between the nodes by using an extension of Warshall's Algorithm [9] assuming each link has the same capacity. In Figure 1(c), the graphical representation of CGA is shown. The weight on each link of the CGA represents the minimum number of links that has to traverse from one processor to the other processor, connected through the link. If the links have diferent capabilities, the numbers would be calculated accordingly. The corresponding reachability matrix is shown in Figure 1(d). The number in a cell at row x and column y represents the minimum number of links that has to be traversed to reach from processor x to processor y.

3 Partitioning an Architecture

When a program is running on the system, it distributes the computational load to some processors and the processors communicate with each other to achieve a common goal. Generally, we may assume that the higher the communication capability of the system, the faster is the execution of the program and better is the system utilization. In a multicomputer, various programs run concurrently on different subsystems. In such an environment, the overall system utilization will be maximized if we can divide the original system such that sum of communication capabilities of all the divisions will be maximum.

Thus, we deal with the following combinatorial problem. Given a graph G with costs on its edges, partitioning the nodes into m subsets of equal size n is desired, so as to maximize the total cost of the edges cut, where the number of nodes $N = m \times n$, m and n being integers. A strictly exhaustive procedure for finding the maximal cost partition is out of question for large number of nodes. There are $\binom{N}{n}$ ways of choosing the first subset, $\binom{N-n}{n}$ ways for the second, and so on. Since, the ordering of the subsets is immaterial, the number of cases is,

$$\frac{1}{m!}\binom{N}{n}\binom{N-n}{n}\cdots\binom{2n}{n}\binom{n}{n}$$

This shows that the time complexity of the exact solution will be in $O(N!)$. For example, for $N = 32$ and $n = 8$ ($m = 4$), the number of cases is greater than 10^{15}.

Then in order to get a solution to our partitioning problem, we have to use a heuristic algorithm. We propose the Kernighan-Lin (K-L) partitioning heuristic [7] for this purpose. It is a powerful and widely used algorithm for partitioning problems and gives better result than other existing algorithms. The KL heuristic finds minimum partition cost. On the other hand, our objective is to find maximum partition cost. Thus we will slightly change the original KL heuristic algorithm. The simplest partitioning problem which still contains all the significant features of larger problems is that of finding a maximal-cost partition of a given graph of $2n$ vertices into two equal size subsets of n vertices each.

We define the *external cost of a node*, which is a member of a set of nodes in a graph, as the sum of costs of all the edges that are connecting that node in that set to any other node outside of that set. Similarly, the *internal cost of a node*, which is a member of a set of nodes in a graph, as the sum of costs of all the edges that are connecting that node in that set to any other node inside of that set. Let S be a set

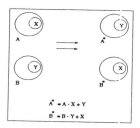

Figure 2: Interchange of sets X and Y.

of $2n$ points, with an associated cost matrix, $C = (c_{ij})$, $i, j = 1, \cdots, 2n$ and that $c_{ii} = 0, \forall i$. There is no assumption about non-negativity of the c_{ij}'s. We wish to partition S into two sets A and B, each with n points, such that the external cost, $T = \sum_{A \times B} c_{ab}$ is maximized.

In essence, the method starts with any arbitrary partition A, B of S. We try to increase the initial external cost T by a series of interchanges of subsets of A and B; the subsets are chosen by an algorithm to be described. When no further improvement is possible, the resulting partition A', B' is said to be locally maximum with respect to the algorithm. We shall indicate that the resulting partition has a fairly high probability of being a globally maximum partition. This process can then be repeated with the generation of another arbitrary starting partition A, B, and so on, to obtain as many locally maximum partitions as we desire. Given S and (c_{ij}), let us suppose A^*, B^* is a maximum cost 2-way partition. Let, A, B be any arbitrary 2-way partition. Then clearly there are subsets $X \subset A, Y \subset B$ with $|X| = |Y| \le \frac{n}{2}$ such that interchanging X and Y produces A^* and B^* as shown in Figure 2.

The problem is to identify X and Y from A and B, without considering all possible choices. The process described here is found approximately by sequentially identifying their elements. Let us define for each $a \in A$, an external cost E_a by $E_a = \sum_{y \in B} c_{ay}$ and an internal cost I_a by $I_a = \sum_{x \in A} c_{ax}$

Similarly, we define E_b and I_b for each $b \in B$. Let, $D_i = E_i - I_i$ for all $i \in S$; D_i is the difference between external and internal costs of an element in A.

Let us consider any $a \in A, b \in B$. Let z be the total cost due to all connections between A and B that do not involve a or b. Then

$$T = z + E_a + E_b - c_{ab}$$

We exchange a and b and let T' be the new cost. We thus obtain $T' = z + I_a + I_b + c_{ab}$

Therefore, the increment in cost or

$$\text{gain = new cost - old cost} = T' - T =$$
$$-(D_a + D_b) + 2c_{ab}$$

3.1 The Proposed Heuristic Algorithm

In this subsection we present the algorithm for 2-way partitioning of an Arbitrary architecture. Extension to 4-way partitioning can be found in [10].

1. Compute the D values for all elements of S.

2. Choose $a_i \in A, b_j \in B$ such that,

$$g_1 = -(D_{a_i} + D_{b_j}) + 2c_{a_i b_j}$$

is maximum; a_i and b_j correspond to the largest possible gain from a single interchange. For now, let us set a_i and b_j aside and call them a'_1 and b'_1, respectively.

3. Recalculate the D values for the elements of $A - a_i$ and for $B - b_j$, by

$$D'_x = D_x + 2c_{xa_i} - 2c_{xb_j}, \qquad x \in A - a_i$$
$$D'_y = D_y + 2c_{yb_j} - 2c_{ya_i}, \qquad y \in B - b_j$$

The correctness of these expressions is easily verified: the edge (x, a) is counted as internal in D_x, and external in D'_x, so c_{xa_i} must be added twice to make this correct. Similarly $c_{...}$ must be subtracted twice to convert (x, b_j) from external to internal.

4. Repeat the second step, choosing a pair a'_2, b'_2 from $A - \{a'_1\}$ and $B - \{b'_1\}$ such that $g_2 = -(D_{a'_2} + D_{b'_2}) + 2c_{a'_2 b'_2}$ is maximum (a'_1 and b'_1 are not considered in this choice). Thus g_2 is the additional gain when the points a'_2 and b'_2 are exchanged as well as a'_1 and b'_1; this additional gain is maximum, given the previous choices. Let us set a'_2 and b'_2 aside too.

Continue until all nodes have been exhausted, identifying $(a'_3, b'_3), \cdots, (a'_n, b'_n)$, and the corresponding maximum gains $g_1, \cdots, g_n$. As each (a', b') pair is identified, it is removed from contention for further choices so the size of the sets being considered decreases by 1 each time an (a', b') is selected.

If $X = a'_1, a'_2, \cdots, a'_k$, $Y = b'_1, b'_2, \cdots, b'_k$, then the increase in cost when the sets X and Y are interchanged is precisely $g_1 + g_2 + \cdots + g_k$. Obviously $\sum_{i=1}^{n} g_i = 0$. It is important to note that some of the g_i's are negative unless all are zero. We choose

k to maximize the partial sum $\sum_{i=1}^{k} g_i = G$. Now if $G > 0$, an increment in cost of value G can be made by interchanging X and Y. After this is done, the resulting partition is treated as the initial partition and the procedure is repeated from the first step.

If $G = 0$, we have arrived at a locally optimum partition, which we shall call a *phase 1 optimal partition*. We now have the choice of repeating with another starting partition, or of trying to improve the phase 1 application of the algorithm x times on x 2-way random starting partitions and keeping the best result, denoted as $KL - x$. The quality of a result is judged based on some additional measures, detailed in [10] that indicates where the iteration of the KL algorithm should be stopped to get the final result.

4 Results and Discussions

In this section, we present the experimental results. All the tests were performed in the same machine to have a somewhat reasonable comparison for getting various running times for the test. We used a Sparc workstation for this purpose and the machine benchmarks at 17441 dhrystones/second. We used a system size upto 16 to compare the results for exhaustive search and heuristic search. Beyond that, our exhaustive search will not be able to give a result within a reasonable amount of time. We tested hypercube, deBruijn, and other architectures. In every case we got the same partitions from the heuristic scheme as from exhaustive serach.

An nth order binary deBruijn graph [8] with 2^n nodes is defined as follows. Each node is labeled with an n-bit string. The node with label $x_0 x_1 ... x_{n-1}$ is connected with four nodes, $\alpha x_0 x_1 ... x_{n-2}$ and $x_1 ... x_{n-1} \alpha$, where $\alpha \in \{0, 1\}$. A 16-node binary deBruijn graph is shown in Figure 3. The left graph is an usual representation of a 16-node deBruijn graph with an intuitive partition. The right graph is showing the exhaustive and KL search result which is counter intuitive. Please note from the node numbers in the right half of Figure 3 that we have redrawn the deBruijn graph such that all nodes in the same group will be close to each other in the new graph. The heuristic result is obtained by applying KL algorithm only one time on a random starting partition and we use KL-1 to symbolize this. As shown in Table 1, the running time for KL is much less than that of the exhaustive search. Hence, we can conclude that we are justified in using the proposed heuristic algorithm in our problem.

Next, we consider a 256-node binary hypercube architecture as our system for verification and generate

Table 1: Partitioning data for 16-node deBruijn graph.

		m = 2	
		Time (sec)	
l	Size	Exhaustive	KL-1
0	16	13.640487	0.003240
1	8	0.060864	0.001550
2	4	0.007691	0.001961
3	2	0.006549	0.001209
4	1	0.000000	0.000000

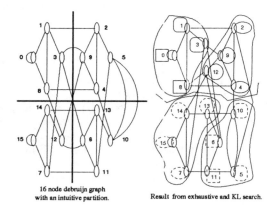

16 node debruijn graph
with an intuitive partition.

Result from exhaustive and KL search.

Figure 3: Result from exhaustive search for 16-node deBruijn.

a 2-buddy tree. The data for m=2 and KL-1 shows that all the buddies in this tree form complete subcubes. From experience, we know that this is a correct solution as implemented in various machines (iPSC, nCUBE). The total time taken to create the whole tree is 2.930951 seconds. Secondly, we consider a 256-node square mesh architecture as our system. We see that the 2 partitioning produces pure submeshes giving a good solution in case of mesh as it did for binary hypercube. The total time taken to create this tree is 1.770562 seconds. Thus, we can conlcude that our approach for partitioning an arbitrary architecture is correct.

5   Conclusion

The issue of efficient utilization of the resources in a parallel machine has been addressed by many researchers in recent years. This research focused on a new approach to solve the processor allocation problem in a parallel computer based on an arbitrary interconnection network. We have included some popular and known architectures in this paper in order to verify the potential of our new approach in solving the problem. We have tried to explain the elegance and

simplicity of using a 2-buddy tree, but more details on m-way partitioning can be found in [10].

References

[1] M. S. Chen and K. G. Shin, "Processor Allocation in an N-Cube Multiprocessor Using Gray Codes," *IEEE Transactions on Computers*, vol. 36, no. 12, pp. 1396–1407, December 1987.

[2] P. J. Chuang and N. F. Tzeng, "A Fast Recognition-Complete Processor Allocation Strategy for Hypercube Computers," *IEEE Transactions on Computers*, vol. 41, no. 4, pp. 467–479, April 1992.

[3] J. Kim, C. R. Das, and W. Lin, "A Top-Down Processor Allocation Scheme for Hypercube Computers," *IEEE Transactions on Parallel and Distributed Systems*, vol. 2, no. 1, pp. 20–30, January 1991.

[4] Y. Chang and L.N. Bhuyan, "Parallel Algorithms for Hypercube Allocation," *Proc. International Parallel Processing Symposium (IPPS)*, April 1993, pp. 105-112.

[5] K. Li and K. H. Cheng, "A Two-Dimensional Buddy System for Dynamic Resource Allocation in a Partionable Mesh connected System," *Journal of Parallel and Distributed Computing*, vol. 12, no. 5, pp. 79–83, December 1991.

[6] J. Ding and L. N. Bhuyan, "An Adaptive Submesh Allocation Strategy for Two-Dimensional Mesh Connected Systems," In *Proc. International Conference on Parallel Processing*, pp. III–193–200, 1993.

[7] B. W. Kernighan and S. Lin, "An Efficient Heuristic Procedure for Partitioning Graphs," *AT&T Bell Labs Technical Journal*, pp. 291–307, 1970.

[8] M. R. Samatham and D. K. Pradhan, "The DeBruijn Multiprocessor Network: A versatile Parallel Processing and Sorting Network for VLSI," *IEEE Transactions on Computers*, vol. 38, no. 4, pp. 323–333, April 1989.

[9] S. Baase, *Computer Algorithms: Introduction to Design and Analysis*, Addison-Wesley, 2nd edition, 1988.

[10] L. Bhuyan, S. Shahed, and Y. Chang, *Partitioning An Arbitrary Architecture*, TR, CS Dept., Texas A&M University, April 1994.

CONDITIONS OF BLOCKED BLAS-LIKE ALGORITHMS FOR DATA ALIGNMENT AND COMMUNICATION MINIMIZATION*

Hyuk-Jae Lee and José A.B. Fortes ({hyuk,fortes}@ecn.purdue.edu)
School of Electrical Engineering, Purdue University, W. Lafayette, IN 47907

Abstract——*This paper considers how automatic data alignment and algorithm partition techniques can be combined and applied to BLAS-like algorithms. It is shown that existing data alignment techniques for individual data array entries can also be applied to data blocks in partitioned algorithms if the null space of the data array indexing matrix is a boundary of computation blocks or the intersection of some of the computation block boundaries. These conditions can be used to generate different algorithm partitions and time-space transformations from which the optimal ones can be chosen for a given target architecture.*

1. INTRODUCTION

Current data alignment techniques focus on individual data entries, and cannot be directly used when many data entries in a block should be aligned collectively. However, if a block of data can be treated as a single data entry (just as a computation block can be treated as a single computation), the existing data alignment techniques can be directly used for partitioned algorithms. Previous work (e.g. [1] and references therein) in partitioning focused on uniform dependence algorithms whose data array domains are the same (within a constant offset) as computation domains. For algorithms with non-uniform dependencies, including BLAS-like algorithms, it is not trivial to partition data arrays so that each block can be treated as a single data entry. To leverage existing data alignment techniques for partitioned algorithms, this paper investigates the conditions of algorithm partitions that allow for data array blocks to be treated as a single data entry. Many partitionings and time-space mappings can be derived based on these conditions and the optimal one can be chosen for a machine with a given number of processors, memory per processor and communication costs.

Section 2 introduces terminology and definitions. Section 3 provides relations between data partitions and algorithm partitions. Conditions of partitions, data alignments, and time-space mappings for communication minimization (i.e., best data alignment) are presented in Section 4. Section 5 discusses the efficiency of various partitionings and time-space mappings for a fixed-size processor array.

2. BACKGROUND

Let $\mathcal{Z}^\nu$ denote the set of vectors with ν integer entries, and $\mathcal{Z}^{\nu\times\nu}$ be the Cartesian product of $\mathcal{Z}^\nu$. The null space of a function $\mathbf{f}$ is $\text{null}(\mathbf{f})$. Let $\mathcal{U}$ be a subset

of the domain of $\mathbf{f}$. The set of the images of entries in $\mathcal{U}$ under $\mathbf{f}$ is denoted $\mathbf{f}(\mathcal{U})$.

The algorithms of interest in this paper can be described by a nested loop containing a recurrence equation of the form:

$$V(\mathbf{f}^v(\vec{j})) = \mathbf{g}_1(V(\mathbf{f}^v(\vec{j}))) * \mathbf{g}_2(U_1(\mathbf{f}^1(\vec{j})), \cdots, U_\eta(\mathbf{f}^\eta(\vec{j})))$$

where

1. $\vec{j} \in \mathcal{J} \subset \mathcal{Z}^\nu$ is a computation point (a vector) in the computation domain $\mathcal{J}$;

2. $\mathbf{g}_1$ and $\mathbf{g}_2$ are single-valued functions and $*$ is a binary commutative operation;

3. V and U_i are data arrays; $\mathbf{f}^i(\vec{j})$ is an Euclidean indexing function, i.e., it has the general form of $F^i\vec{j}$ where $F^i \in \mathcal{Z}^{(dim(\mathcal{Y}^i)\times\nu)}$ is called the indexing matrix ($\mathcal{Y}^i$ is the domain of the ith input data array U_i) and any row of F^i is distinct with one entry valued unity and all others valued zero.

Most of algorithms that underlie the Basic Linear Algebra Subprograms (BLAS) [2] consist of one or more nested loops that satisfy the above conditions and hence the terminology "BLAS-like algorithms" is used in this paper.

A partition of a computation domain $\mathcal{J}$ is a set of nonempty disjoint subsets (called blocks) of $\mathcal{J}$ whose union equals $\mathcal{J}$. The partitions of interest in this paper are described by an equivalence relation such that $\vec{j}_1$ and $\vec{j}_2$ are in the same block if and only if $\mathbf{q}(\vec{j}_1) = \mathbf{q}(\vec{j}_2)$ for any $\vec{j}_1, \vec{j}_2 \in \mathcal{J}$, where $\mathbf{q}$ is of the form $\mathbf{q}(\vec{j}) = \lfloor (D\vec{j})/\vec{\beta} \rfloor$, $D \in \mathcal{Z}^{\nu\times\nu}$, $\vec{\beta} \in \mathcal{Z}^\nu$. The image of $\mathcal{J}$ under $\mathbf{q}$ is called *an algorithm partition* and is denoted by $\hat{\mathcal{J}}$. A block whose elements are mapped to $\hat{j} \in \hat{\mathcal{J}}$ is denoted $\mathcal{Q}^{-1}(\hat{j})$. Note that $\mathcal{Q}^{-1}(\hat{j})$ is not an inverse of $\mathbf{q}$ because $\mathbf{q}$ maps $\mathcal{J}$ onto $\hat{\mathcal{J}}$ while $\mathcal{Q}^{-1}$ maps $\hat{\mathcal{J}}$ onto $\mathcal{P}(\mathcal{J})$, the power set of $\mathcal{J}$, that is the set of all subsets of $\mathcal{J}$.

Example 1 Consider the following algorithm to compute $C = A \times B$ where A, B, and C are (8×12), (12×16), and (8×16) matrices, respectively.

```
DO i = 0, 7
  DO j = 0, 15
    DO k = 0, 11
      c(i, j) = c(i, j) + a(i, k) × b(k, j)
CONTINUE
```

Suppose that the algorithm partition is described by $\mathbf{q}(\vec{j}) = \lfloor \vec{j}/(2, 4, 3)^T \rfloor$ (i.e., D is the identity matrix). Then, the size of a block is $2 \times 4 \times 3$ while the size of partition, $\hat{\mathcal{J}}$, is $4 \times 4 \times 4$. $\mathcal{Q}^{-1}((0,0,0)^T)$ denotes the block $\{\vec{j}|(0,0,0)^T \leq \vec{j} < (2,4,3)^T\}$ while $\mathcal{Q}^{-1}((1,0,0)^T)$ denotes the block $\{\vec{j}|(2,0,0)^T \leq \vec{j} < (4,4,3)^T\}$. $\square$

* This research was partially funded by the National Science Foundation under grants MIP-9500673 and CDA-9015696.

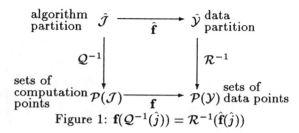

Figure 1: $\mathbf{f}(\mathcal{Q}^{-1}(\hat{j})) = \mathcal{R}^{-1}(\hat{\mathbf{f}}(\hat{j}))$

A data partition $\hat{\mathcal{Y}}$ of a data domain $\mathcal{Y}$ is defined by an equivalence relation such that $\vec{y}_1$ and $\vec{y}_2$ are in the same block if and only if $\mathbf{r}(\vec{y}_1) = \mathbf{r}(\vec{y}_2)$ for any $\vec{y}_1, \vec{y}_2 \in \mathcal{Y}$ where $\mathbf{r}$ is of the form $\mathbf{r}(\vec{y}) = \lfloor (R\vec{y})/\vec{\beta}' \rfloor$, $R \in \mathcal{Z}^{dim(\mathcal{Y}) \times dim(\mathcal{Y})}$, $\vec{\beta}' \in \mathcal{Z}^{dim(\mathcal{Y})}$. A block whose elements are mapped to $\hat{y} \in \hat{\mathcal{Y}}$ is denoted by $\mathcal{R}^{-1}(\hat{y})$.

Example 1 (continued) Suppose that the matrix A partition is described by $\mathbf{r}^a(\vec{y}) = \lfloor \vec{y}/(2,3)^T \rfloor$. Then, the size of a block is 2×3 while the size of the partition is 4×4. $\mathcal{R}^{-1}((0,0)^T)$ denotes the block $\{(0,0)^T \leq \vec{j} < (2,3)^T\}$ while $\mathcal{R}^{-1}((1,0)^T)$ denotes the block $\{(2,0)^T \leq \vec{j} < (4,3)^T\}$. □

3. RELATION BETWEEN COMPUTATION AND DATA PARTITIONS

Definition 1 A function $\hat{\mathbf{f}} : \hat{\mathcal{J}} \to \hat{\mathcal{Y}}$ is an *indexing function* of a data partition accessed in an algorithm partition if $\mathbf{f}(\mathcal{Q}^{-1}(\hat{j})) = \mathcal{R}^{-1}(\hat{\mathbf{f}}(\hat{j}))$, for any $\hat{j} \in \hat{\mathcal{J}}$. □

The condition of Definition 1 can be stated as follows: the data entries accessed by the computations of block $\mathcal{Q}^{-1}(\hat{j})$ should be the same as those in the data block indexed by the image of $\hat{j}$ under the indexing function $\hat{\mathbf{f}}$. In other words, the diagram of Fig. 1 commutes.

Example 1 (continued) Let the indexing function of a partitioned matrix A be $\hat{\mathbf{f}}^a(\hat{j}) = \left(\begin{smallmatrix} 1 & 0 & 0 \\ 0 & 0 & 1 \end{smallmatrix} \right)$. Consider block $\mathcal{Q}^{-1}(0,0,0) = \{\vec{j}|(0,0,0) \leq \vec{j} < (2,4,3)^T\}$. Since the original indexing function is $\mathbf{f}^a(\vec{j}) = \left(\begin{smallmatrix} 1 & 0 & 0 \\ 0 & 0 & 1 \end{smallmatrix} \right)$, it follows that $\mathbf{f}^a(\mathcal{Q}^{-1}((0,0,0)^T)) = \{\vec{y}|(0,0)^T \leq \vec{y} < (2,3)^T\}$. On the other hand, $\hat{\mathbf{f}}^a((0,0,0)^T) = (0,0)^T$ and $\mathcal{R}^{-1}(\hat{\mathbf{f}}^a((0,0,0)^T)) = \{\vec{y}|(0,0)^T \leq \vec{y} < (2,3)^T\}$. Therefore, $\mathbf{f}^a(\mathcal{Q}^{-1}((0,0,0)^T)) = \mathcal{R}^{-1}(\hat{\mathbf{f}}^a((0,0,0)^T))$ Hence, it satisfies Definition 1. □

Next example illustrates that it may not be possible to find an indexing function that satisfies Definition 1.

Example 2 Consider the rank-one update algorithm:
```
DO i = 0, 8
    DO j = 0, 8
        c(i, j) = c(i, j) + a(i) × b(j)
    CONTINUE
```
Suppose that the algorithm partition is given by $\mathbf{q}(\vec{j}) = \lfloor \left(\begin{smallmatrix} 1 & -1 \\ 0 & 1 \end{smallmatrix} \right) \vec{j}/(3,3)^T \rfloor$. Consider the entries of array $\vec{a}$ that are accessed by block references. For block $(0,0)$, $a(0), a(1), a(2), a(3)$, and $a(4)$ are accessed. For block $(1,0)$, $a(3), a(4), a(5), a(6)$ and $a(7)$

are accessed. Note that $a(3), a(4)$ are shared by both blocks and should be allocated to the processors that use these blocks. Hence, this data "partition" is not valid because the blocks are not disjoint. □

Existing data alignment techniques are based on relations among indexing functions, time-space mappings, and data alignments. To reuse the existing data alignment techniques in partitioned algorithms, one must find the correct indexing functions of blocked data arrays. Example 2 shows that some algorithm partitions do not allow the existence of a valid data partition. Therefore, it is necessary to find conditions of algorithm partitions that guarantee the existence of valid data partitions and derive techniques to identify them. Here, the particular desirable case when the indexing functions in the partitioned algorithms are the same as in the original algorithms (i.e., non-partitioned) is considered. For this case, Proposition 1 provides conditions on algorithm partitions that guarantee the existence of a valid data partition.

Let D be a $\nu \times \nu$ matrix and I be a set of integers such that $I \subset \{1, 2, \cdots, \nu\}$. Then D_I denotes the matrix consisting of the ith columns of matrix D for all $i \in I$. Let $\vec{a}$ be a ν-dimensional vector. $\vec{a}_I$ denotes a vector consisting of the ith entries of $\vec{a}$ for all $i \in I$.

Proposition 1 Given an algorithm partition defined by $\mathbf{q}(\vec{j}) = \lfloor (D\vec{j})/\vec{\beta} \rfloor$, there exists a data partition that makes the indexing function of data blocks referenced in the partitioned algorithm be the same as the indexing function used for array references in the original algorithm (i.e., $\hat{\mathbf{f}} = \mathbf{f}$) if there exists a set $I \subset \{1, 2, \cdots, \nu\}$ such that the columns of D_I^{-1} form a basis of null$(\mathbf{f})$. In this case, the data partition is defined by $\mathbf{r}(\vec{y}) = \lfloor ((FD_c^{-1})^{-1}\vec{y})/\vec{\beta}_c \rfloor$ where $C = \{1, 2, \cdots, \nu\} - I$. □

A boundary is generated by $\nu - 1$ columns of D^{-1}. Hence, there exist $\left(\begin{smallmatrix} \nu - dim(null(\mathbf{f})) \\ \nu - dim(null(\mathbf{f})) - 1 \end{smallmatrix} \right) = \nu - null(\mathbf{f})$ boundaries containing null$(\mathbf{f})$. Therefore, the intersection of these boundaries forms null$(\mathbf{f})$. In other words, null$(\mathbf{f})$ should be parallel to the intersection of these boundaries.

Example 1 (continued) Null$(\mathbf{f}^a)$ is generated by the second column of $D_2^{-1}(= (0,1,0)^T)$. Thus, $\hat{\mathbf{f}}^a(\hat{j}) = \mathbf{f}^a(\hat{j})$. Since $C = \{1,3\}$, $[F(D_c^{-1})]^{-1} = \left(\begin{smallmatrix} 1 & 0 \\ 0 & 1 \end{smallmatrix} \right)$ and $\vec{\beta}_c = (2,3)^T$. Hence, the data partition is $\mathbf{r}(\vec{y}) = \lfloor \vec{y}/(2,3) \rfloor$. □

Example 2 (continued) Null$(\mathbf{f}^a)$ is generated by $(0,1)^T$ an $D^{-1} = \left(\begin{smallmatrix} 1 & 1 \\ 0 & 1 \end{smallmatrix} \right)$. Thus, neither D_1^{-1} nor D_2^{-1} generates null$(\mathbf{f}^a)$, i.e., null$(\mathbf{f}^a)$ is not parallel to any boundary. Hence, Proposition 1 is not satisfied. □

4. DATA ALIGNMENTS OF PARTITIONED ALGORITHMS

Existing data alignment techniques focus on finding conditions of time-space mapping and data alignment function to minimize communication. If an algorithm partition and a data array partition satisfy the conditions in Proposition 1, each computation block can be treated as a single computation unit and each data block as a single data entry. Hence, a time-space mapping of an algorithm partition can be defined as $\hat{\mathbf{T}} : \hat{\mathcal{J}} \rightarrow (t, \mathcal{X})$ where t and $\mathcal{X}$ are time and processor index, respectively. In addition, data alignment of blocked arrays are defined on the data partition. The focus of this paper is modular mappings and K−expanded modular data alignments (EMDAs) which are described by $\hat{\mathbf{T}}_{\hat{m}}(\hat{j}) = (\hat{T}\hat{j})_{mod\ \hat{m}}$ and $\{\hat{\mathbf{p}}_{m,k}(\hat{y}, t) = (\hat{P}\hat{y} + \hat{p} + \hat{v}t + mk\hat{v})_{(mod\ b_X)} | k = 0, 1, \cdots, (K-1), m \in \mathcal{Z}\}$, respectively. $\hat{T}$ and $\hat{m}$ are called the transformation matrix and the modulus vector, respectively [3]. $\hat{P}$, $\hat{p}$, and $\hat{v}$ are called pattern alignment, offset alignment, and mobility, respectively [4]. Note that "^" appears in notation used for algorithm partitions, data partitions, and functions defined on them.

Proposition 2 gives conditions of partitioned algorithm mappings ($\hat{\mathbf{T}}$) and blocked data alignments that guarantee optimal data alignment (they are similar to conditions for optimal data alignment in non-partitioned algorithms [4]). Also as a consequence of Proposition 1, other alignment techniques proposed elsewhere for non-partitioned algorithms can be used for partitioned algorithms.

Let $T_{(*,1)}$ denote the first column of matrix T. $T_{(*,(2:\nu))}$ and $T_{((2:\nu),*)}$ denote the matrices consisting from the 2nd to the νth columns of T, and from the 2nd to the νth rows of T, respectively.

Proposition 2 Let $\hat{\mathbf{T}}_{\hat{m}} : \hat{\mathcal{J}} \rightarrow \mathcal{X}$ be a modular time-space mapping of the algorithm partition $\hat{\mathcal{J}}$ and $\{\hat{\mathbf{p}}_{\hat{m}_1,k} | k = 0, 1, \cdots, K-1\}$ be a K-EMDA of the data partition $\hat{y}$ where $\hat{\mathbf{p}}_{\hat{m}_1,k}(\hat{y}, t) = (\hat{P}\hat{y} + \hat{p} + \hat{v}t + \hat{v}\hat{m}_1k)_{(mod\ b_X)}$. For any $\vec{j} \in \mathcal{J}$, there exists $\hat{\mathbf{p}}_{\hat{m}_1,k}(\hat{y}, t)$ such that $(\hat{T}_{((2:\nu),*)}\hat{j})_{mod\ b_X} = \hat{\mathbf{p}}_{\hat{m}_1,k}(\hat{\mathbf{f}}(\hat{j}), t)$, for $\hat{j} = \mathbf{q}(\vec{j})$ and $\hat{\mathbf{f}}(\hat{j}) = \mathbf{r}(\mathbf{f}(\hat{j}))$ if (1) $\hat{P}^{-1} = \hat{F}\hat{T}^{-1}_{*,(2:\nu)}$, (2) $\hat{v} = -\hat{P}\hat{F}\hat{T}^{-1}_{*,1}$, (3) $\hat{p} = \vec{0}$, (4) $K = \text{lcm}_{\{i|\hat{v}_i \neq 0\}}(\text{lcm}(\hat{m}_1\hat{v}_i, (b_X)_i)/|\hat{m}_1\hat{v}_i|)$. □

Example 1 (continued) Let the time-space mapping be $\hat{\mathbf{T}}_{(4,4,4)}(\hat{j}) = \left(\begin{pmatrix} -1 & -1 & 1 \\ 1 & 0 & 0 \\ 0 & 1 & 0 \end{pmatrix} (\hat{j}) \right)_{(mod\ (4,4,4))}$. From Proposition 2, data alignments of matrices A and B must be $\hat{\mathbf{p}}^a(\hat{y}, t) = \left(\begin{pmatrix} 1 & 0 \\ -1 & 1 \end{pmatrix} \hat{y} + (0, -1)^T t \right)_{(mod\ (4,4))}$, and $\hat{\mathbf{p}}^b(\hat{y}, t) = \left(\begin{pmatrix} 1 & -1 \\ 0 & 1 \end{pmatrix} \hat{y} + (-1, 0)^T t \right)_{(mod\ (4,4))}$, respectively. Hence, these matrices are initially stored as shown in Fig. 2 (a). All the entries stored in processors are computed and then matrix A shifts left and matrix B shifts up for the next computation. This computation step and communication step are repeated until matrices A and B are restored as initially. This algorithm for matrix multiplication corresponds to the partitioned (blocked) version of Cannon's algorithm. □

Example 3 Consider the matrix multiplication al-

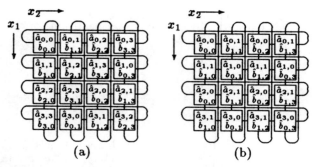

Figure 2: Initial data alignments of the blocked matrix multiplication algorithms: $\hat{a}_{i,j}$ and $\hat{b}_{i,j}$ denote $(i, j)^{th}$ blocks of domains of matrices A and B, respectively

gorithm in Example 1 again. Suppose that computation partition is defined by $\mathbf{q}(\vec{j}) = \lfloor \vec{j}/(2, 3, 6)^T \rfloor$ and the data partitions of matrices A and B are defined by $\mathbf{r}^a(\vec{y}) = \lfloor \vec{j}/(2, 6)^T \rfloor$ and $\mathbf{r}^b(\vec{y}) = \lfloor \vec{j}/(6, 3)^T \rfloor$, respectively. Suppose that the time-space mapping is $\hat{\mathbf{T}}_{(2,4,4)}(\hat{j}) = \left(\begin{pmatrix} -1 & -1 & 1 \\ 1 & 0 & 0 \\ 0 & 1 & 0 \end{pmatrix} (\hat{j}) \right)_{(mod\ (2,4,4))}$. From Proposition 2, data alignments of matrices A and B must be 2-EMDAs such that $\{\hat{\mathbf{p}}^a_{2,k}(\hat{y}, t) = \left(\begin{pmatrix} 1 & 0 \\ -1 & 1 \end{pmatrix} \hat{y} + 2k(0, -1)^T t \right)_{(mod\ (4,4))} | k = 0, 1\}$ and $\{\hat{\mathbf{p}}^b_{2,k}(\hat{y}, t) = \left(\begin{pmatrix} 1 & -1 \\ 0 & 1 \end{pmatrix} \hat{y} + 2k(-1, 0)^T t \right)_{(mod\ (4,4))} | k = 0, 1\}$, respectively. The initial data storage patterns are shown in Fig. 2 (b). □

5. COMMUNICATION OVERHEADS

Given that the partitions of interest are determined by $\mathbf{q}(\vec{j}) = \lfloor (D\vec{j})/\vec{\beta} \rfloor$, different partitions are possible by choosing D and $\vec{\beta}$. Matrix D is often determined by other constraints such as those in Proposition 1 or in [6, 1]. Therefore, this section assumes that D is given, and investigates $\vec{\beta}$ that minimizes communication overheads. Some components of $\vec{\beta}$ have their values chosen to make the size of a partition match the size of an available processor array. The values of other components of $\vec{\beta}$ are chosen to satisfy other constraints such as available memory space.

Consider the problem of which components of $\vec{\beta}$ are chosen to make the algorithm partition fit an available processor array. A consequence of Condition (2) of Proposition 2 is that the mobility $\hat{v}$ is $\vec{0}$ if and only if $T_1^{-1} \in \text{null}(\mathbf{f})$. Therefore, it is desirable to choose T_1^{-1} (the projection vector [7], that is the direction along which computations are mapped into processors) as the basis of null($\mathbf{f}$). Therefore, the choice of the projection vector is limited by the data arrays used in an algorithm. For most of BLAS algorithms, the number of optimal projection vectors is relatively small enough to examine all possibilities in a reasonable amount of time. For example, in matrix multiplication, there are six possible choices of the projection vector (Proposition 3 in [5]). Depending on the projection vector, the number of data entries in a single communication can vary. For the matrix multiplication in Exam-

projection	$(\pm 1,0,0)^T$	$(0,\pm 1,0)^T$	$(0,0,\pm 1)^T$
no. data entries	14	20	18

Table 1: The size of a single communication

$\vec{\beta}$	(2,4,3)	(2,4,4)	(2,4,6)	(2,4,12)
no. data entries	26	32	44	80

Table 2: The number of data entries in a single block

ple 1, Table 1 shows the size of a single communication when the projection vector is $(\pm 1,0,0)^T$, $(0,\pm 1,0)^T$, or $(0,0,\pm 1)^T$, respectively.

The larger the blocks stored in a single processor, the less number of communications are required. However, block size is limited by available memory space. Moreover, there exist special cases for which the necessary memory space increases even though the size of blocks decreases. This is because alignment of a partitioned data array must satisfy conditions in Proposition 2 and therefore, cases exist when more than one blocks must be stored in a single processor (See [4]).

Example 4 Consider the matrix-matrix multiplication algorithm in Example 1 and a processor array of size (4×4). Suppose that the algorithm blocks are mapped to processors along the direction, $(0,0,1)^T$. Then, $\beta_1 = 2$ and $\beta_2 = 4$ (in order to accommodate an (8×16) projection in (4×4) processors).

Consider how β_3 affects communication overheads. If $\beta_3 = 6$, the size of the partition is $(4 \times 4 \times 2)$. Suppose that $\hat{T}_{\hat{m}}(\hat{j}) = \left(\begin{pmatrix} -1 & -1 & 1 \\ 1 & 0 & 0 \\ 0 & 1 & 0 \end{pmatrix} \hat{j} \right)_{(mod\,(2,4,4))}$ is used for the time-space transformation. Note that the only difference from the modular mapping in Example 1 is the first entry of the modulus vector, $\hat{m}_1$. This implies that this algorithm requires two time units[†] to complete computations. Communications are not necessary inside a block but only between blocks. Hence, only one communication (i.e., one message with several data entries)[‡] is necessary. Note that the partitioned version of Cannon's algorithm in Example 1 needs four time units and therefore three communications are required. The trade-off of this example is the size of a block which is twice as large as that in Example 1. Hence, approximately twice as much memory as that of Example 1 is necessary in this example.

Suppose that $\beta_3 = 12$. Then, the size of the data partition is $(4 \times 4 \times 1)$. In this case, this algorithm needs just one time unit and therefore no communication is necessary. However, it requires approximately four times as much storage as that of Example 1.

Now, consider the number of data entries in a single block. For matrix 'A', the number of data entries in a block with size $(\beta_1 \times \beta_2 \times \beta_3)$ is $\beta_1 \times \beta_3$. The numbers of data entries of matrices 'B' and 'C' are $\beta_2 \times \beta_3$ and $\beta_1 \times \beta_2$, respectively. If $\vec{\beta} = (2,4,3)^T$, the data entries in a single block of matrices A, B, and C are six, twelve, and eight, respectively. Hence, in total, twenty six data items are contained in a single block. Similarly, the numbers of data entries in a single block can be computed for other partitions. Table 2 shows the number of data entries in a single block.

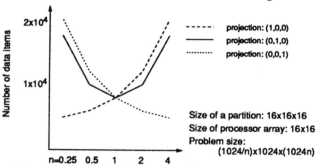

Figure 3: Communication overheads vs. partitions

Suppose that the size of an available memory space is thirty two. Then, the partition with $\vec{\beta} = (2,4,6)^T$ is not valid while all data entries can be stored if $\vec{\beta} = (2,4,4)^T$. However, in this case, since $\hat{m}_1(= 3)$ is relative prime with $\hat{m}_2$ or $\hat{m}_3$, Proposition 2 shows that more than one blocks have to be stored in a single processor [4]. Therefore, more than thirty two data entries must be stored in a single processor. ☐

Fig. 3 shows how communication overheads depend on the choice of coordinates to be mapped into a space domain. The matrix multiplication algorithms of sizes $((1024/n) \times 1024 \times (1024n))$, $n = 0.25, 0.5, 1, 2, 4$ are considered. This figure shows that the communication overheads depend on the choice of coordinates when the computation domain is not cubic. As the difference among the sizes along three coordinates increases, the difference of communication overheads for possible projections also increases.

REFERENCES

[1] P. Boulet, et. al., "(Pen)-ultimate tiling ?", *Integration, the VLSI Journal*, vol. 17, pp. 33-51, Aug. 1994.

[2] J.J. Dongarra, et. al., "A set of level 3 basic linear algebra subprograms," *ACM Trans. Math. Software*, vol. 16, no. 1, pp. 1-17, Mar. 1990.

[3] H.-J. Lee and J.A.B. Fortes, "On the injectivity of modular mappings," *Proc. ASAP94*, pp. 236-247, Aug. 1994.

[4] H.-J. Lee, and J.A.B. Fortes, "Data alignments for modular time-space mappings of BLAS-like algorithms," *ASAP95*, July 1995.

[5] H.-J. Lee and J.A.B. Fortes, "Toward data distribution independent parallel matrix multiplication," *IPPS95*, April 1995.

[6] R. Schreiber and J.J. Dongarra, *Automatic blocking of nested loops*, Tech. Rep. 90.38, RIACS, Aug. 1990.

[7] A. Darte, "Regular partitioning for synthesizing fixed-size systolic arrays," *Integration, the VLSI Journal*, vol. 12, pp. 293-304, Dec. 1991.

[†]The time unit is the time required to finish all computations in a single block. It increases as the block size increases.

[‡]Each message contains all data generated and/or used by a block that is needed by another block in the destination processor. Message size increases with block size.

TABLE OF CONTENTS- FULL PROCEEDINGS

(R): Regular Papers
(C): Concise Papers